690
Ra
$12.95
Ramsey, Dan
The complete foundation and floor framing book
DATE DUE
EAU CLAIRE DISTRICT LIBRARY
DEMCO

"Then the Lord answered Job out of the storm. 'Where were you when I laid the earth's foundation? Tell me, if you understand. Who marked off its dimensions? Surely you know! Who stretched a measuring line across it? On what were its footing set, or who laid its cornerstone—while the morning stars sang together and the angels shouted for joy?' "

(Job 38:1, 4-7, NIV)

The COMPLETE FOUNDATION and FLOOR FRAMING BOOK

Dan Ramsey

FIRST EDITION
THIRD PRINTING

Library of Congress Cataloging in Publication Data

Ramsey, Dan, 1945-
The complete foundation and floor framing book.

Includes index.
1. Foundations. 2. Floors. 3. Framing (Building)
I. Title.
TH5201.R3 1987 690'.11 87-14716
ISBN 0-8306-0378-6
ISBN 0-8306-2878-9 (pbk.)

TAB BOOKS Inc. offers software for sale. For information and a catalog, please contact TAB Software Department, Blue Ridge Summit, PA 17294-0850.

Questions regarding the content of this book should be addressed to:

Reader Inquiry Branch
TAB BOOKS Inc.
Blue Ridge Summit, PA 17294-0214

Contents

Acknowledgments

The author acknowledges the extensive assistance of these resources: American Plywood Association; Cooperative Extension Service, Iowa State University; U. S. Department of Agriculture; National Forest Products Association; and the U.S. Navy Training Center. Thanks so much.

Introduction

The foundation is the most important construction element of a home, barn, or other structure. No matter how well the structure is built, if the foundation is weak, the structure soon will begin to fall apart and the invested time and money will be lost.

The Complete Foundation and Floor Framing Book clearly illustrates how to plan and install a good foundation for your residence, garage, and commercial or outdoor structure. It covers every aspect, from planning and laying out your foundation to selecting tools and materials, excavating, installing concrete and masonry foundation walls and piers, pouring concrete slab, and more. There's even a chapter on new technologies for installing permanent wood foundations at reduced cost.

The topic of floor framing is literally interlocked with structural foundations. *The Complete Foundation and Floor Framing Book* extensively covers the planning and installation of structural floors, including beams, joists, bracing, and subflooring.

Included in this comprehensive book are numerous charts and tables that will help you plan, estimate, and install foundations and floor framing. Information is presented on estimating concrete requirements, footing widths, forms materials, pier sizes, rebar requirements, concrete blocks, mortar, framing members, board feet, subflooring requirements, and much more. Finally, there's a comprehensive glossary of foundation and floor framing terms that will clearly explain and illustrate the terms you will use in construction.

The Complete Foundation and Floor Framing Book is truly a complete book for the first-time homebuilder, as well as the professional builder who needs to solve problems quickly.

Chapter 1

Foundation Basics

All structures require a foundation that is secure enough to keep the walls and roof permanently in place. The foundation can be made of poured concrete, concrete or cinder blocks, other masonry materials, or pressure-treated wood.

The foundation of a house generally extends several feet below ground level. If there is a basement, entire walls are used for the foundation. In houses built on piers, however, individual foundations are poured or built under each pier. In a slab house, a foundation is used along the perimeter.

The base of the foundation is called a *footing*. This footing, usually made of poured concrete, should be wider than the foundation thickness. The footing is laid below the frost line, and the foundation is built above the footing.

It is essential that the foundation be constructed soundly so that the rest of the structure remains secure and stable. A shifting foundation or one with bad cracks can cause the upper structure to weaken. Therefore, follow local building code requirements related to the thickness and depth of the foundation before you build any structure like a garage or an additional wing onto your home.

TYPES OF FOUNDATIONS

A foundation serves several purposes. It supports the weight of the house and other vertical loads, such as snow. It stabilizes the house against horizontal forces, such as wind. It is a retaining wall that supports the earth fill around the house. Often, it is a basement or cellar wall that might be a barrier to moisture, heat loss, or sound transmission.

The most common foundation is the *continuous wall* (Fig. 1-1). It can be built of stone, clay tile, block, brick, concrete, treated wood, metal, or other material.

Continuous walls are used to support heavy loads or to enclose a crawl space or basement. If enclosure of space is the main objective, then the wall can be built of lighter, more porous insulating materials that will reduce heat loss and sound transmission.

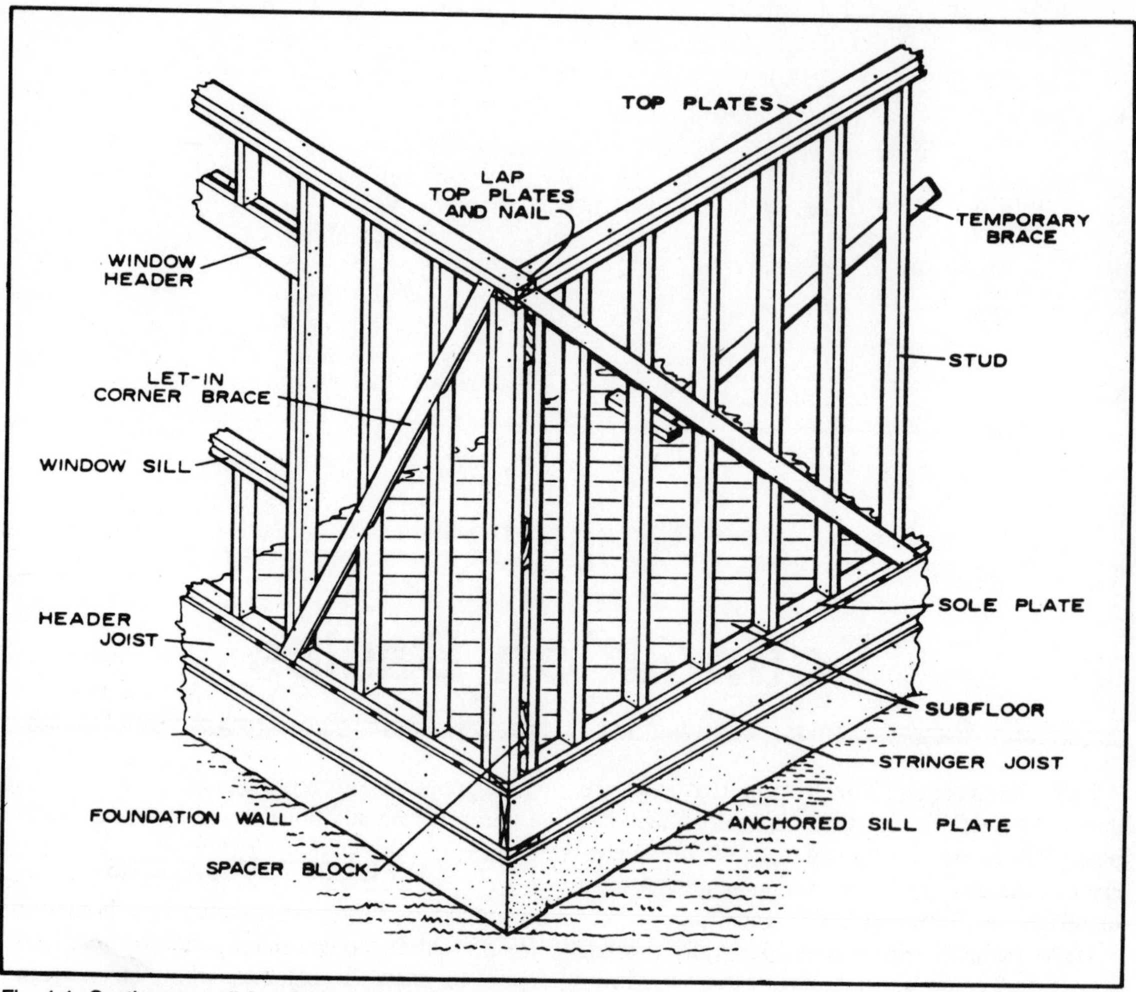

Fig. 1-1. Continuous wall foundation with platform construction.

A *step foundation* is a continuous wall of variable height. It is used on steep grades or for houses with partial basements.

The *pier foundation* (Fig. 1-2) is a series of piers that support the house. It is generally masonry, but sometimes it is made of other materials. The *pole*, or *post foundation*, is a special kind of pier built of pressure-treated wood. It is often used on steep terrain where there is considerable variation in the height of the piers and where a regular masonry pier might bend or break.

Beams placed between the piers of a pier foundation support the house (Fig. 1-3). The size of such a beam depends on the load it must carry and the distance between piers. The space between piers is generally enclosed with curtain walls that carry no load and whose main purpose is to enclose the space and act as a barrier to wind, heat, moisture, and sometimes, animals.

A *grade beam foundation* (Fig. 1-4) is a pressure-treated wood or reinforced concrete beam that is submerged to a depth of about 8 inches below grade. It can be supported on a stone fill or on underground piers that extend into the ground below the frost line. The grade beam is especially useful in dry climates or in well-drained soils where the house can be built close to the ground.

The *slab foundation* (Fig. 1-5) is a special foun-

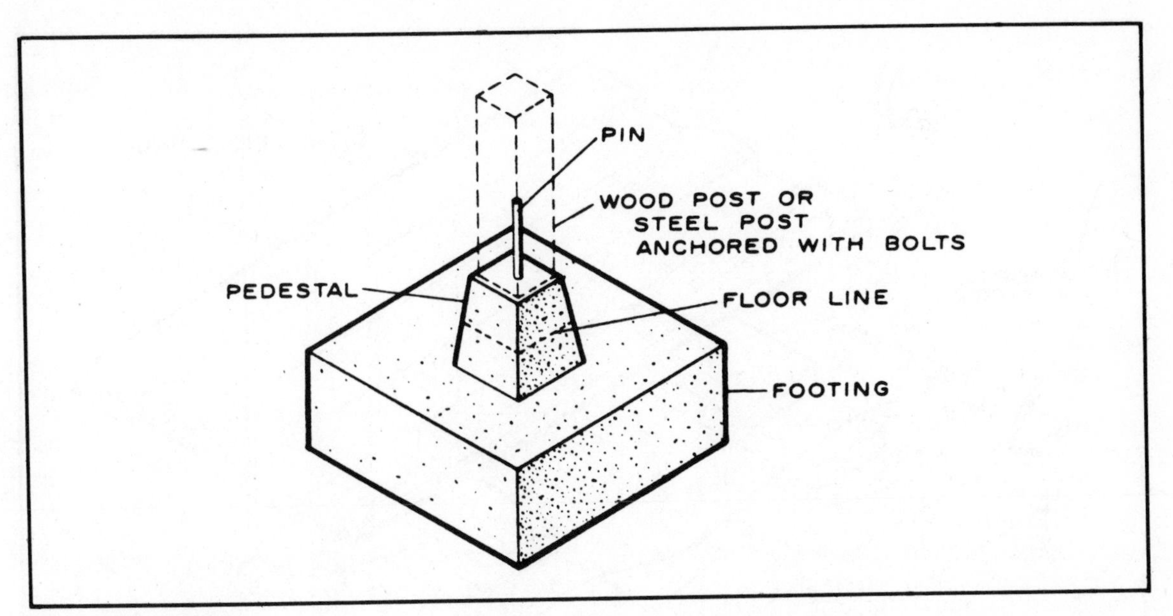

Fig. 1-2. Pier foundation.

dation that floats on top of the soil and also serves as the floor of the house. The slab is thickened under all the walls to support their heavier loads. All slab floors are not slab foundations; many are simply concrete floors. A separate foundation supports the wall loads.

Every foundation must support the weight of the house and its contents. This load can vary considerably, depending on the type of construction, the kind of furniture, and the special uses to which the house is subjected. In colder climates, the foundation must carry the ice and snow that can accumulate on the roof. If the foundation loads are heavy, reinforced concrete will provide the strongest wall. Wider masonry walls will carry heavier loads than narrow ones.

Many houses are so heavy that the foundation must be widened at the bottom to keep them from sinking into the soil. The widened bottom on the foundation is called a *footing* (Fig. 1-6). Its size depends on the kind of soil under it. Soil strengths vary from 1,000 to 12,000 pounds per square foot. Footings usually are designed for 1,000 pounds per square foot, but if you know your soil type, you might design smaller footings.

The thickness of the footing depends on how far it protrudes beyond the foundation wall. A com-

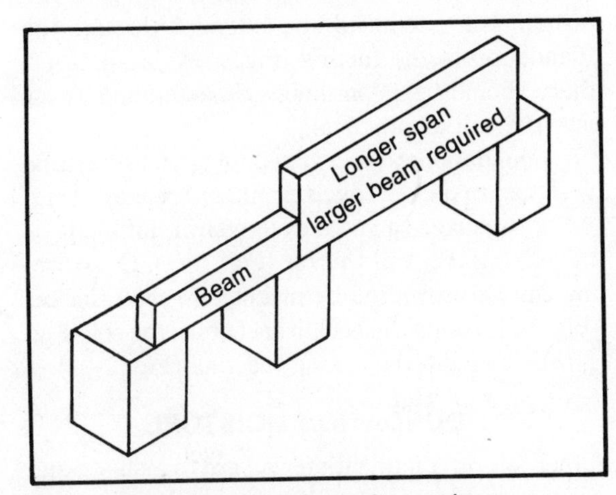

Fig. 1-3. Beam size depends on pier spacing.

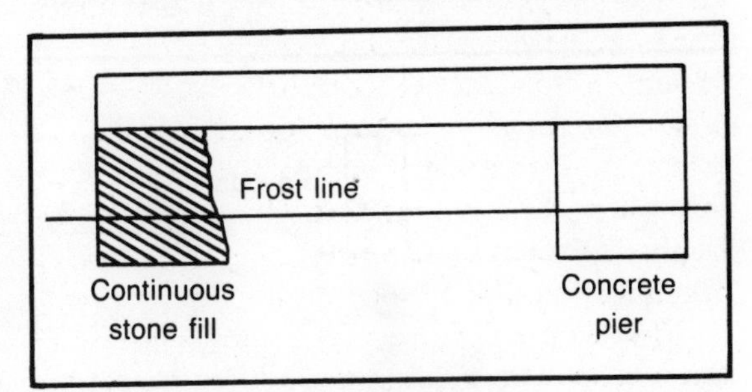

Fig. 1-4. Grade beam foundation.

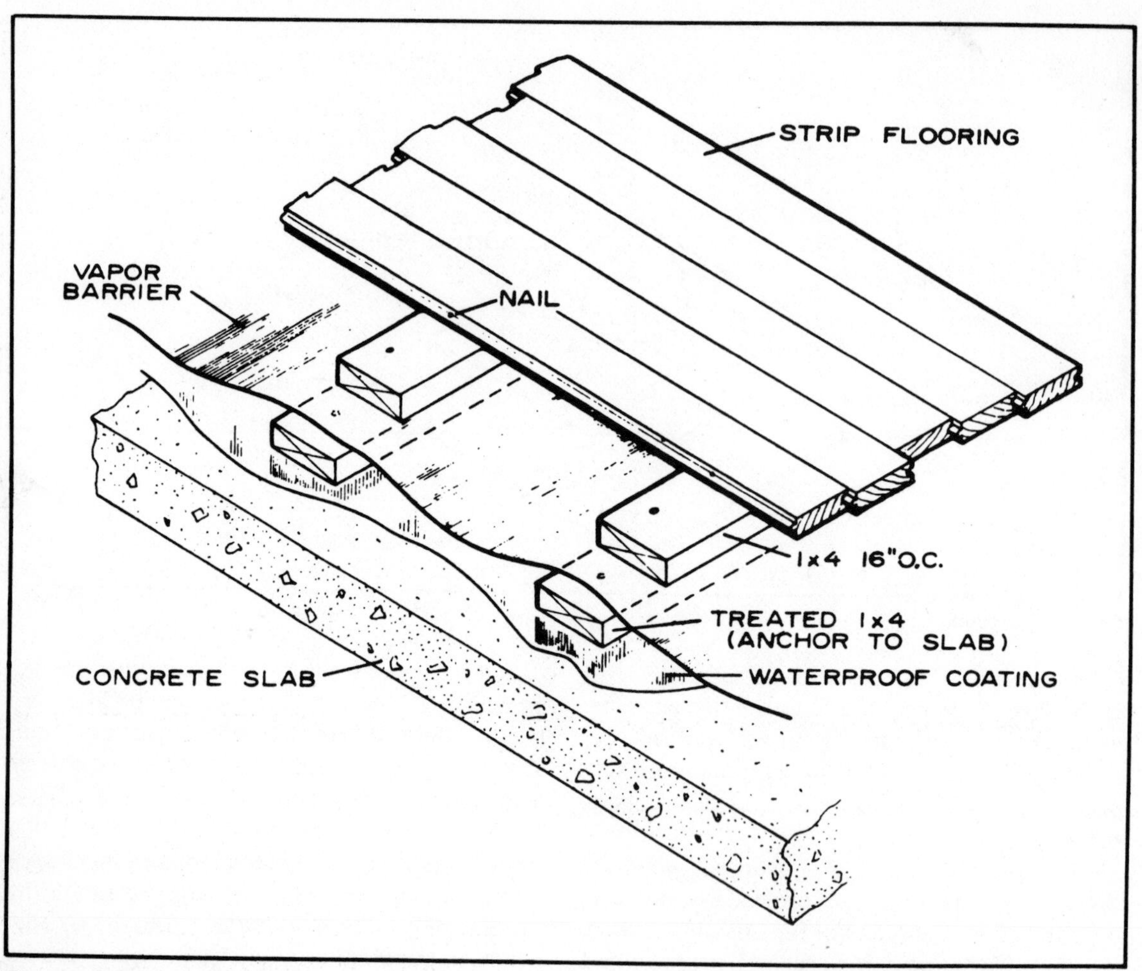

Fig. 1-5. Slab foundation.

mon rule is that the thickness should be twice as great as the largest projection (Fig. 1-7)

Because wind can lift or slide houses off their foundations (Fig. 1-8), houses must be securely fastened to the foundation. For masonry walls, the fastening device should be extended through the foundation to the footing (Fig. 1-9). In all cases, there should be a continuous tie extending as far into the soil as practical.

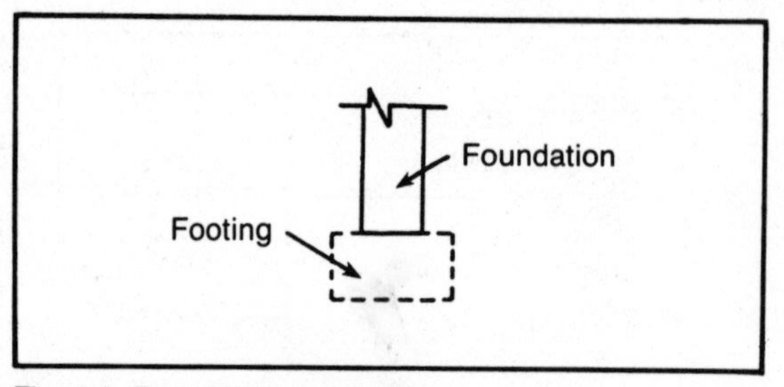

Fig. 1-6. Foundation and footing.

Foundations acting as retaining walls must be designed to prevent overturning or breakage (Fig. 1-10). Breakage can be prevented by reinforcing or by making the wall thicker (Fig. 1-11). Overturning can be prevented by making the wall thicker (Fig. 1-12), tying the wall to anchors in the soil (Fig. 1-13), or counterbalancing the wall (Fig. 1-14).

FOUNDATION MOISTURE

Other jobs of a foundation wall are to insulate and to moistureproof. Basements traditionally have

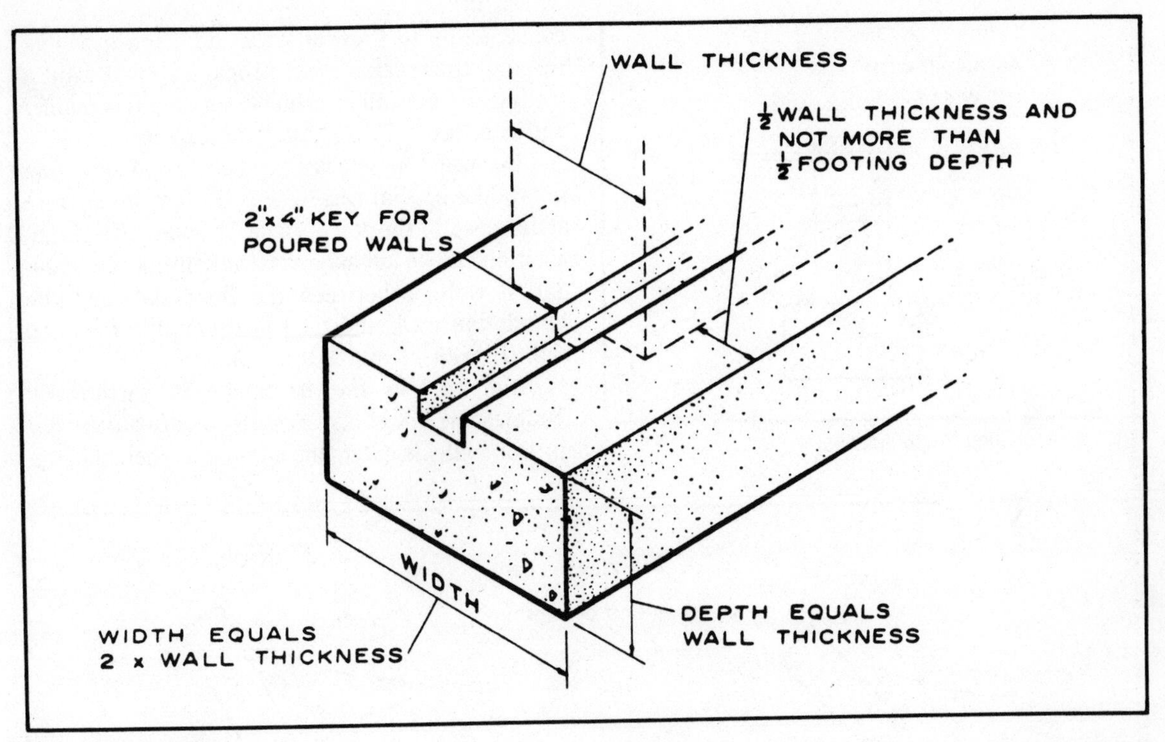

Fig. 1-7. Wall footing.

been used as cellars to store foods and supplies, especially in the colder climates. In recent years, however, and with less need for long-term food storage, houses are being built without basements or with basements to be used as family living areas or game rooms. If you are going to use your basement as a living area, you will want a dry basement that can be kept at a desired temperature.

The temperature generally can be maintained by installing a proper size of heating system. If you are going to heat economically, however, the foundation or basement wall must be of reasonably tight

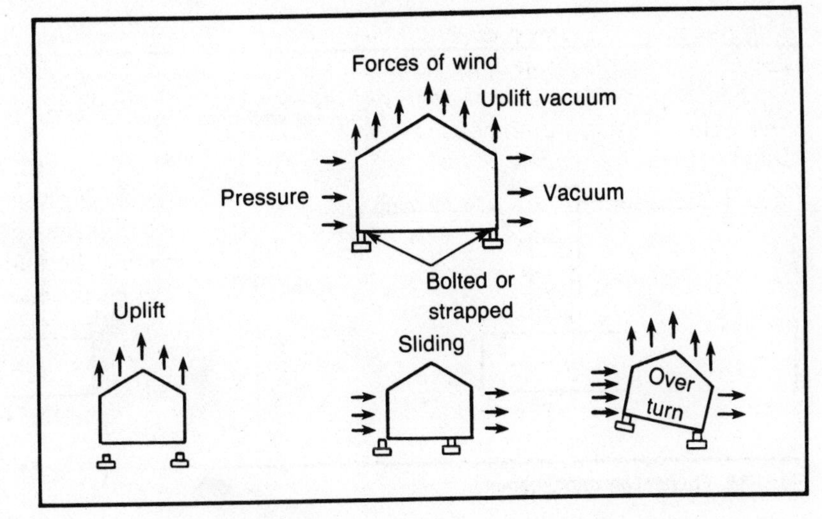

Fig. 1-8. Wind and foundation.

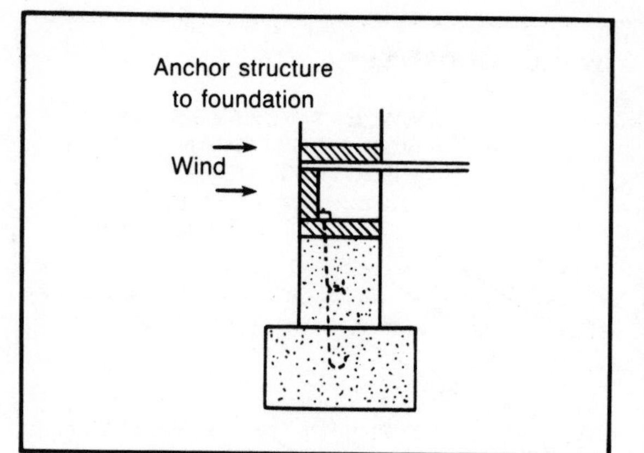

Fig. 1-9. Anchoring the foundation.

construction to prevent warm air from escaping through the cracks. Most people also will want a well-insulated wall to achieve maximum economy and comfort.

Because concrete is a good heat conductor, you must take special care when you install concrete slabs close to the soil surface. A sheet of rigid insulation can be installed vertically inside the foundation wall or between the floor slab and the foundation wall, and then horizontally under the slab (Figs. 1-15 and 1-16).

Moisture in the basement is particularly troublesome in wet climates. Hot and relatively dry air from outside enters the basement where its tem-

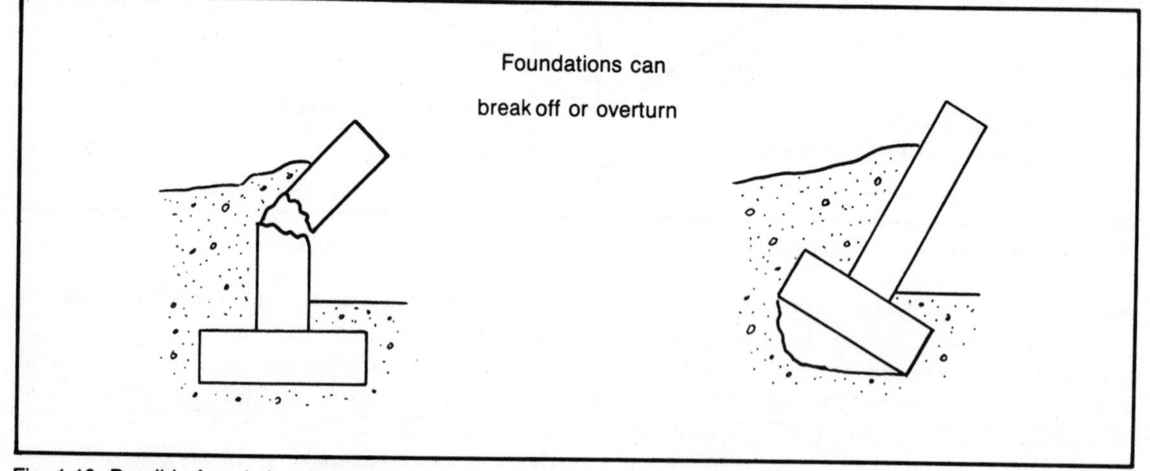

Fig. 1-10. Possible foundation damage.

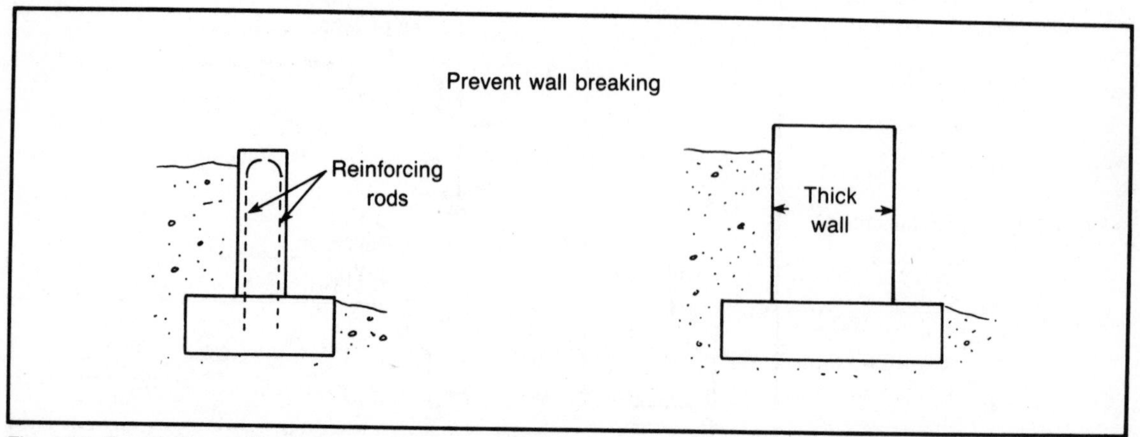

Fig. 1-11. Foundation reinforcement.

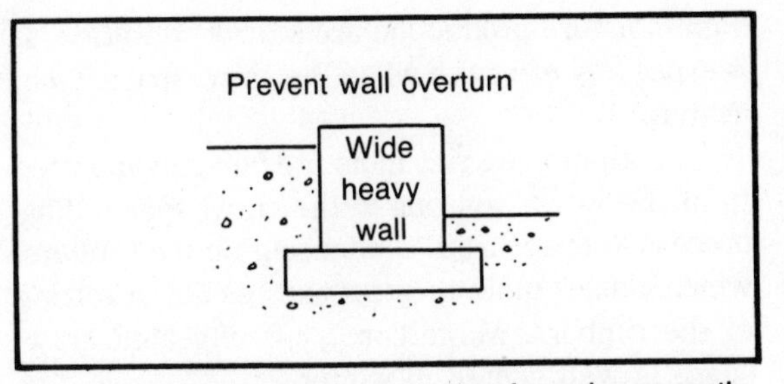

Fig. 1-12. Wider foundations are both safer and more costly.

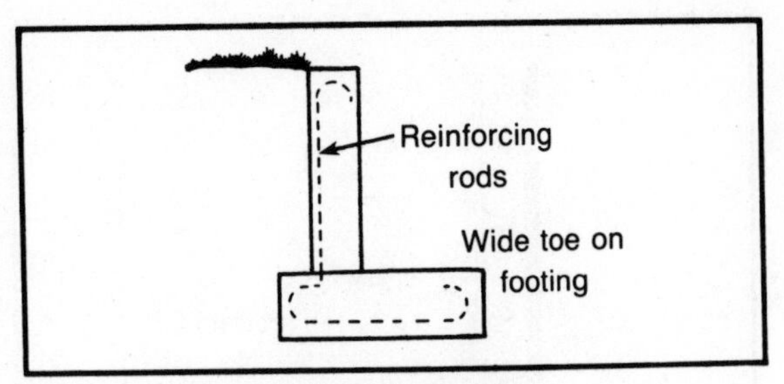

Fig. 1-14. Counterbalance foundation.

perature is reduced by several degrees. As air cools, it can't hold as much moisture; thus, cooling the air causes it to become quite humid. If such humid air comes in contact with an even cooler surface, such as the wall or the floor of the basement, it becomes so humid that it deposits some of its moisture upon the cold surface. This is called *sweating*.

Sweating creates an atmosphere conducive to the growth of mold and fungi, which can be quite objectionable. This problem generally goes away in the winter months when the outside air is colder than the inside air. There are exceptions, however. One example is moisture being produced in the house by extensive boiling of water or by frequent use of hot showers. Another is moisture from the soil soaking through the basement wall and evaporating into the air.

You can deal with high humidity. Dehumidifiers can remove several quarts of moisture per day from the air. For dehumidification to be economical and effective, however, the atmosphere must be closed. Doors and windows should be kept shut. Any outside air getting into the basement will bring more moisture in and give the dehumidifier that much more work to do.

Another way to deal with moisture is to eliminate cold surfaces in the basement or house. You can use insulation to accomplish this purpose. You can cover floors with felt paper and tile; you can insulate and finish walls.

If you insulate the wall or floor, you must provide a vapor barrier to prevent moisture in the air from flowing through the insulation to the cold surfaces where it will condense under the insulation. A plastic film, some paints, and several other materials can be used as vapor barriers.

One common example of moisture passing through insulation is the condensation of moisture under a carpet on a basement floor. Many people

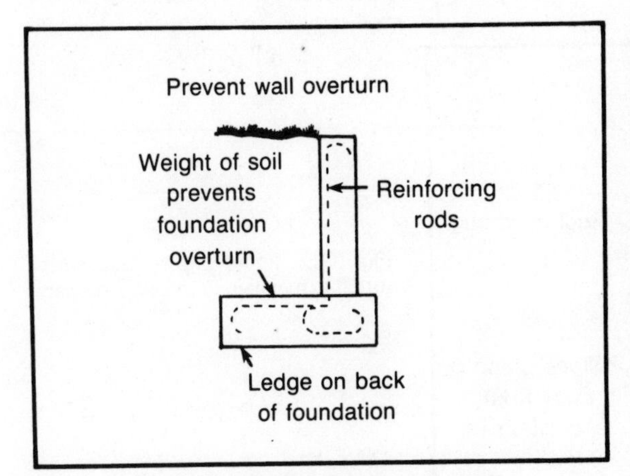

Fig. 1-13. Anchoring the foundation to the soil.

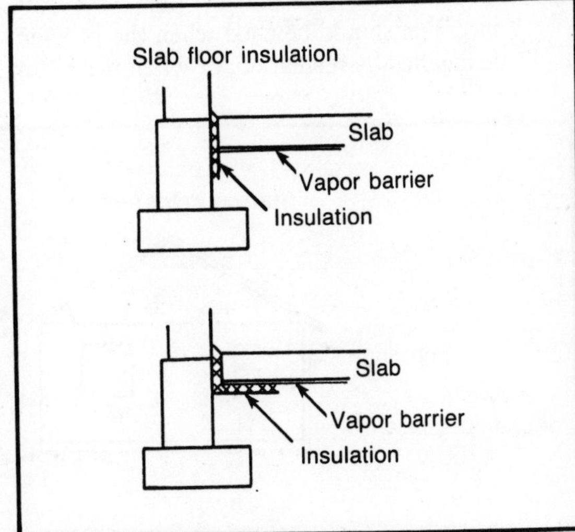

Fig. 1-15. Slab floor insulation.

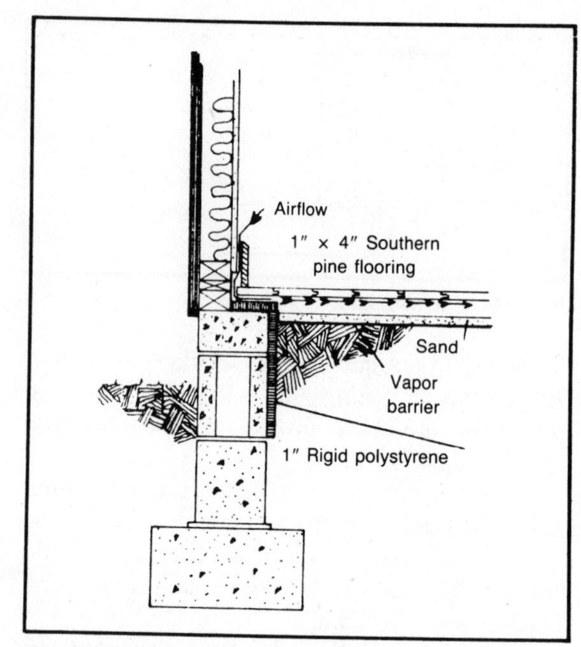

Fig. 1-16. Detail of slab floor insulation.

must take basement carpets up during the summer months to avoid such condensation.

In winter, ventilation can remove excess moisture from the basement or crawl space. Ventilation becomes extremely important if moisture is being produced there. Ventilation brings in cold air and exhausts warm air, however, so excessive ventilation will carry off a good deal of heat.

Ventilation should be used when the problem can't be handled by insulation, or when periods of high moisture production are of short duration. It is especially effective when the crawl space is not heated.

Ventilation carries moisture being evaporated from the warm soil out of the crawl space. This process keeps it from condensing on the timbers, which causes mold to grow and results in rotting of the timbers. Make sure your unheated crawl space is well vented in winter.

Heating the crawl space and insulating the foundation wall can eliminate the need for vents and probably reduce total heat loss from the house. In this way, air from the living area can be used to ventilate the crawl space, and moisture carried out of the soil can humidify the dry air in the living area.

Another problem is moisture that flows out of the soil into the basement in liquid form, either through the wall or the floor. The smart builder will give the foundation every possible advantage (Fig. 1-17) against ground or surface moisture by:

- ☐ Providing the house with gutter and downspouts to carry roof moisture away from the foundation wall
- ☐ Sloping the grade away from the house on all sides
- ☐ Using swales or open drainage to carry off surface water
- ☐ Backfilling behind the foundation wall with porous fill and providing drain tile at the base of the footing below the basement floor level to drain moisture away from the house

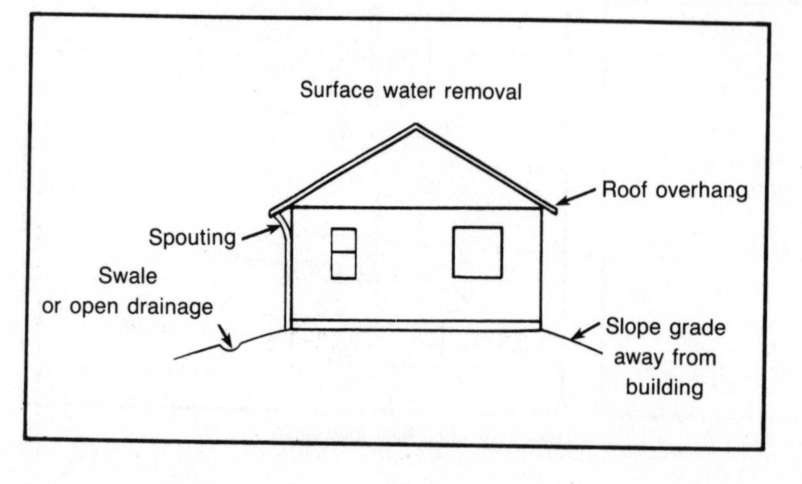

Fig. 1-17. Improving surface water runoff to maintain foundation integrity.

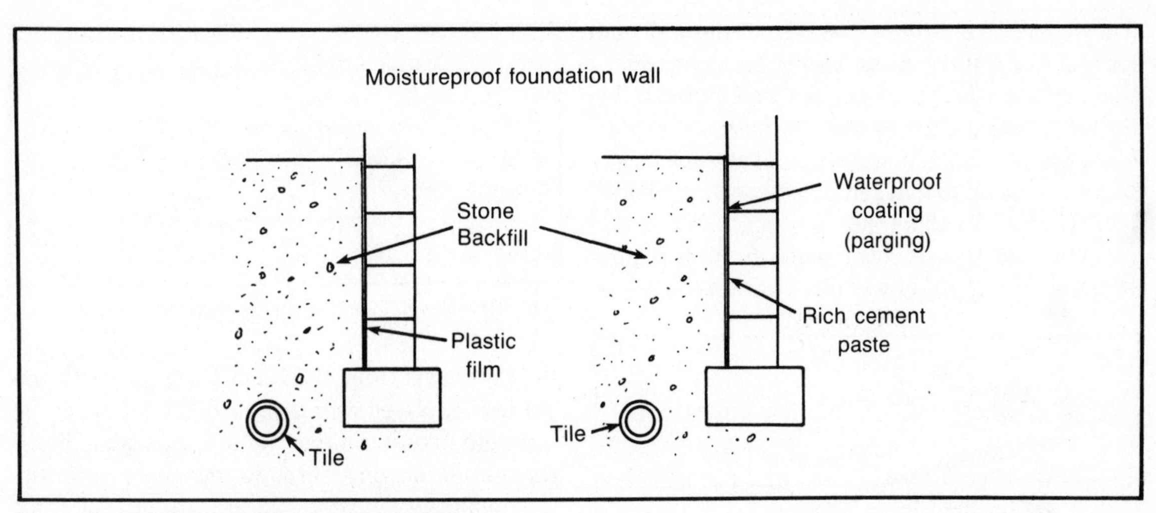

Fig. 1-18. Parging or plastering your foundation.

Parging, or plastering, the outside of the foundation wall with a rich cement paste will prevent moisture that comes in contact with the wall from soaking into or through it (Fig. 1-18). Waterproof paints and coating also should be used over the parging to increase moisture resistance.

If you have surface moisture leaking through your basement wall, the most effective solution is to dig down on the outside of the wall and install a drain field below the basement floor. Then parge and paint the wall and backfill with a coarse aggregate, such as cinders, gravel, sand, or stone. If the drain field can't be emptied by gravity, install a sump pump to carry the moisture away from the bottom of the foundation wall.

Because this solution is expensive, most homeowners look for an alternate method. Soil moisture flowing down along the outside of the basement wall and under the house can be removed by breaking a hole about 2 feet square and 1 to 2 feet deep in the basement floor and using a gravity drain or a sump pump to carry the moisture away.

Several patching and plastering materials have been developed for sealing cracks on the inside of basement walls. Most are ineffective. Those that are good will stop the flow in one place, but the moisture will simply back up and find a new flow path.

If you can be satisfied with a wet wall, some systems have been developed to keep the floor dry.

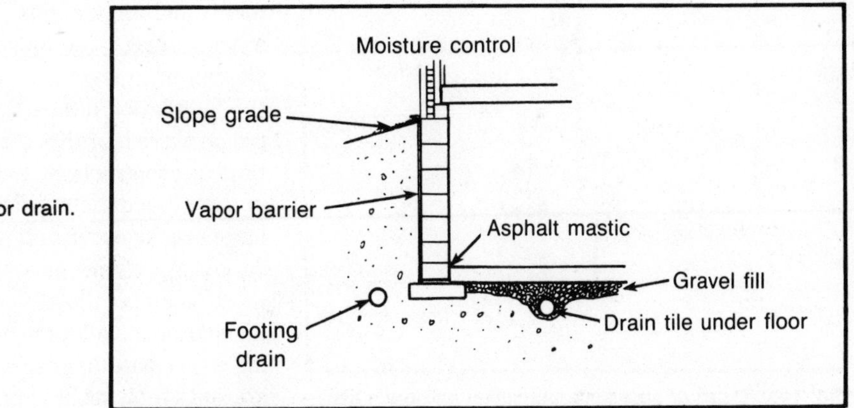

Fig. 1-19. Basement floor drain.

They collect the water at the base of the wall and carry it to a drain or sump. Using this same collection system, you can reduce the wall moisture by drilling a series of holes near the base of the wall just above the collection system. These holes enable the moisture to flow through the wall at a lower level (Figs. 1-19 and 1-20).

You can protect floor slabs from moisture through the use of a gravel fill and drain tile below the floor level (Figs. 1-21 and 1-22). A plastic vapor barrier placed over the gravel fill and under the concrete floor is another aid in keeping soil moisture from flowing through the floor into the basement.

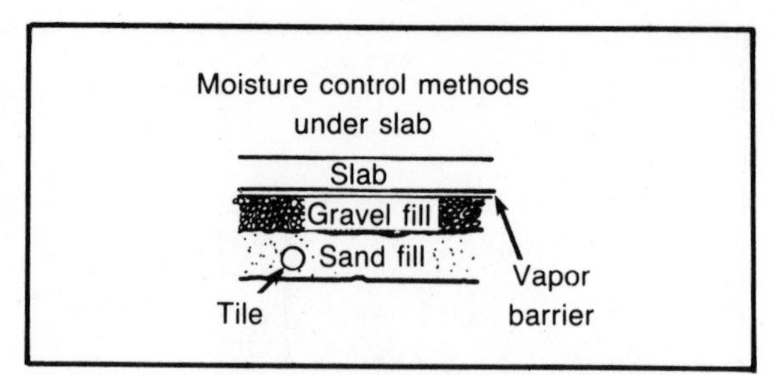

Fig. 1-21. Gravel fill and drain tile below the slab.

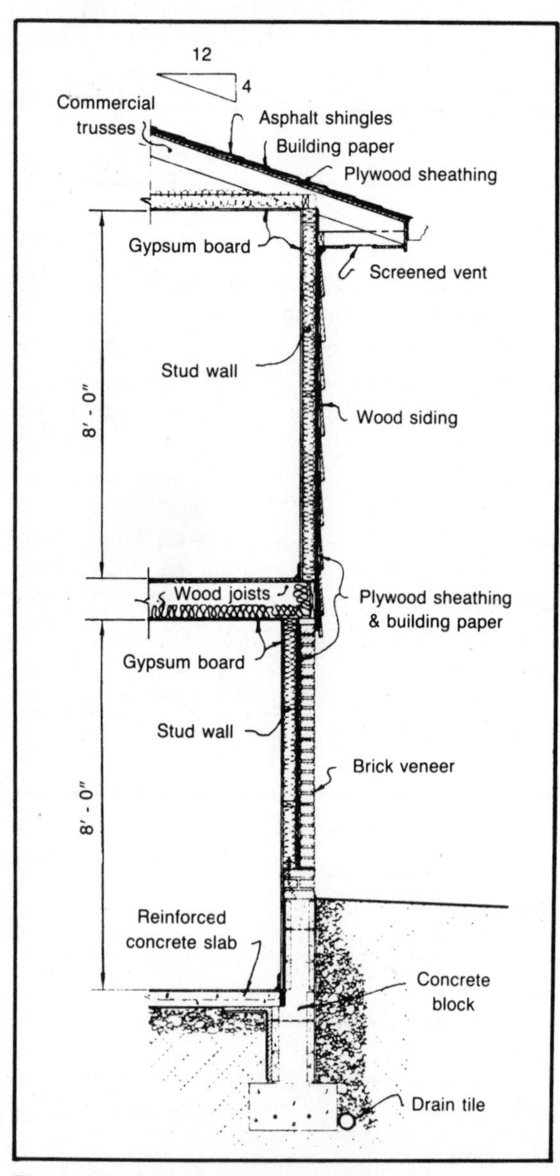

Fig. 1-20. Detail of basement foundation and drain tile.

You can insulate and finish basement walls in several different ways. If you decide to use rigid insulation, you can attach 1/2 inch to 2 inches of foam plastic or other materials to the wall with a recommended adhesive. Preferably, the rigid insulation should be a vapor barrier or should have a vapor barrier attached to it. Some rigid-foam panels have a plastic-treated paper surface that stops water vapor.

After you have installed insulation, you can attach panel board or other wall finishes to the insulation with a good grade of panel adhesive. You can likewise use adhesive to attach baseboard and moldings. You can surface-mount electrical outlets or countersink the wires into the insulation.

If you are building or buying a house, take a little extra care and save yourself the enormous problems that are common to poorly installed basements and foundations. When you are purchasing a house, check the foundation carefully. Look for sloping or unlevel floors, which can indicate that the foundation has settled, that timbers have rotted or moved, or that the floor was not level at the time of construction. You might find some cracks in most foundation walls that you can tolerate, but large cracks occur only if there is excessive foundation movement or settling.

Remember that if you follow the recommended foundation installation procedures, chances of having a wet basement are remote. Find out if there are soil problems in your area that can cause foun-

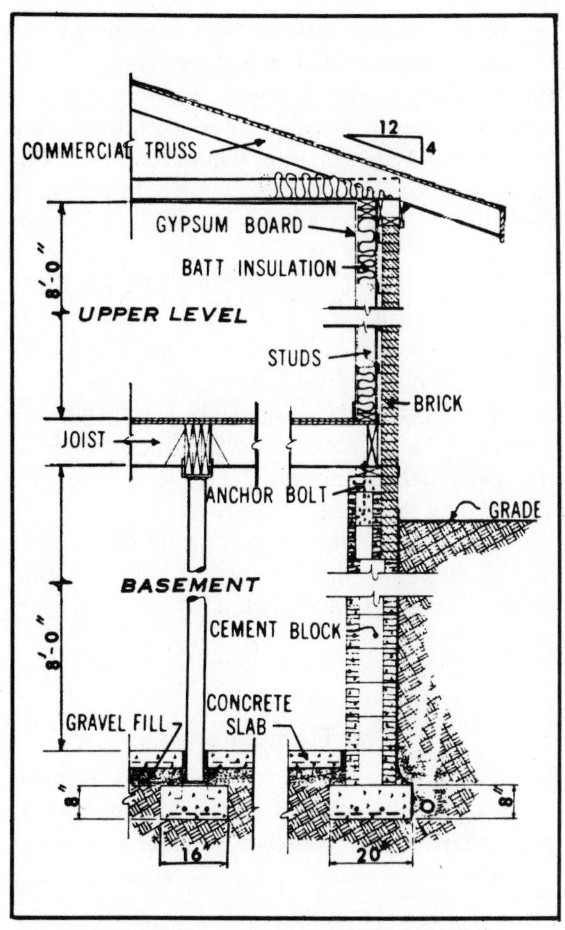

Fig. 1-22. Detail of foundation wall with gravel fill.

dation headaches and seek advice on handling such problems before you install your foundation.

SELECTING YOUR FOUNDATION

With this overview of the various types of foundations and a discussion of moisture prevention, you're now ready to select the foundation that best fits your building needs. Let's take a more detailed look at the available types of foundations and their construction.

Continuous Walls

Continuous wall foundations are used for buildings that are subjected to heavy loads, for buildings that must be warmed, and for buildings that must be rodentproof. They are built out of either concrete or unit masonry, or of pressure-treated wood.

Generally, foundation walls that are 8 inches thick and are built of concrete, concrete block, brick, or clay tile are ample for wooden buildings that are no heavier than a house. Rubble stone walls, for the same weight of building, must be 12 to 16 inches thick.

Light frame buildings are anchored to continuous foundations with 5/8-inch bolts spaced 8 feet apart. They extend 12 inches into concrete walls or 24 inches into unit masonry walls.

The cores of hollow masonry units (concrete block and clay tile) are filled around the anchor bolts with concrete. Two rows of hollow units are filled under girders or other concentrated loads. A building hint: crumpled paper stuffed in the hollow cores prevents the concrete from dropping lower than needed.

The foundation wall must be reinforced at all corners and at the junctions of the outer wall and inner walls to prevent cracking. When hollow units are used for foundations, special joint reinforcement below grade level can be used to counteract the pressure of the soil on the foundation.

A concrete rodent shield is shown in Fig. 1-23. The shield is about 1 foot beneath the ground and extends about 1 foot horizontally from the building.

The crawl space (beneath the floor of the building and enclosed by the foundation) should be well ventilated to reduce humidity. The vent area should be at least 1 square foot per 15 lineal feet of foundation wall. Ventilators that can be closed during the winter are best.

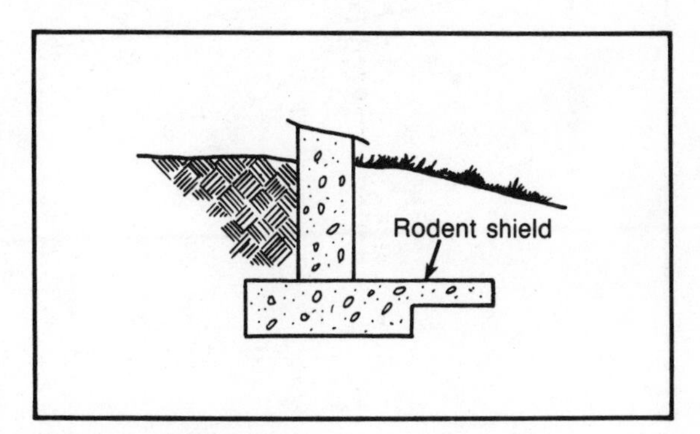

Fig. 1-23. Rodent shield.

Basement Walls

Basement walls usually serve both as bearing and as retaining walls. As *bearing walls,* they support the building; as *retaining walls,* they resist the lateral pressure of the outside earth. Basement wall thicknesses should be determined by a structural engineer.

A height of 6 3/4 feet from basement floor to the first floor joists is considered a minimum convenience height. Greater height might be necessary to accommodate a furnace, warm-air ducts, and plumbing, and still leave adequate head room.

Unless areaways are provided, the top of the foundation wall should be at least 2 feet above grade. This height permits windows in the basement for light and ventilation.

Step Foundations

A stepped foundation (Fig. 1-24) is a variation of a continuous wall foundation. It is used where the ground slopes or where there is a basement under only part of the building. The foundation is stepped down gradually to keep the footing on solid ground and to avoid undermining the higher part of the foundation with excavations at the time of construction.

Where the foundation is built on sloping rock, level steps are cut in the rock. These steps prevent the foundation from slipping. Slight slopes are sometimes merely heavily clipped; sometimes they are doweled. Where outcroppings of rock have been exposed to weathering for some time, the surfaces are likely to be decomposed or loose. This loose surface must be cut away.

For average soil, a vertical step (**V** in Fig. 1-24) of not more than 2 feet in a horizontal distance (**H** in Fig. 1-24) of 4 feet is generally satisfactory. In any case, the horizontal distance should be over 2 feet and the vertical step should be less than three-fourths of the horizontal distance. For example, if the horizontal distance is 6 feet, the vertical step should be less than 4 1/2 feet.

There should be a projection on the vertical part of the step. It should be as wide as the footing on the horizontal part of the step and at least 6 inches thick.

Pier Foundations

Pier foundations are more economical than continuous wall foundations. Standard piers are built of brick, concrete, rubble stone, or hollow masonry

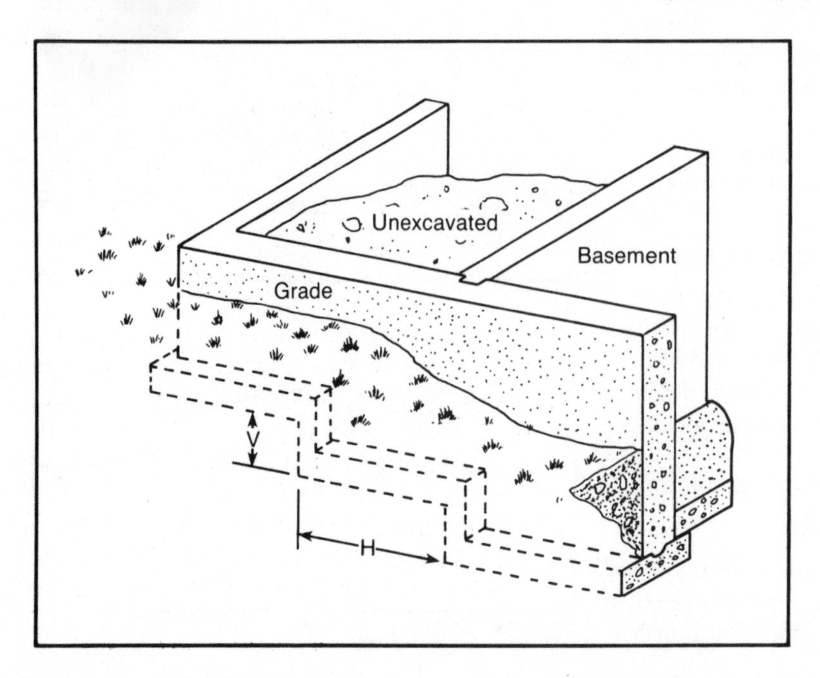

Fig. 1-24. Stepped foundation.

filled with concrete. More slender piers can be made of reinforced concrete or masonry. If you require a warm building floor, you can close the space between the outside piers with curtain walls.

Pier spacing is largely a matter of economy and building weight. Because beam strength varies inversely to the square of the beam span, a beam spanning 6 feet is four times as resistant to bending as the same beam would be if it spanned 12 feet. Closer spacing of piers is usually more economical than using heavier beams.

Concrete and concrete block piers that are 3 to 4 feet high must be reinforced at each corner with vertical steel rods 3/8 inch in diameter. The rods should be at least 2 feet long and hooked on the ends. One end is buried 5 inches in the footing (Fig. 1-25) to bond the pier to the footing; the rest of the rod is extended into the pier.

Concrete and unit masonry piers 4 to 6 feet high need additional reinforcing, as shown in Fig. 1-26. The advice of a structural engineer is required for reinforcing larger piers.

It is easier to reinforce 4- to 6-foot piers with two-piece reinforcing rods than with one-piece rods that run the full height of the pier. Short dowels of reinforcing rod are set into the footing so that they extend 18 inches above the footing. The dowels are set so that they will align with the corresponding pier-corner reinforcing rod. The dowels are wired to the pier-corner rod.

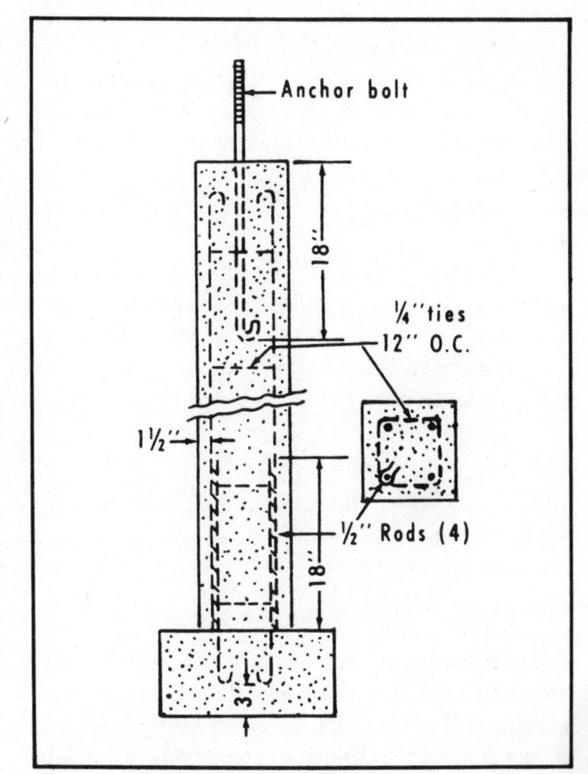

Fig. 1-26. Reinforcing larger piers using dowels and baling wire.

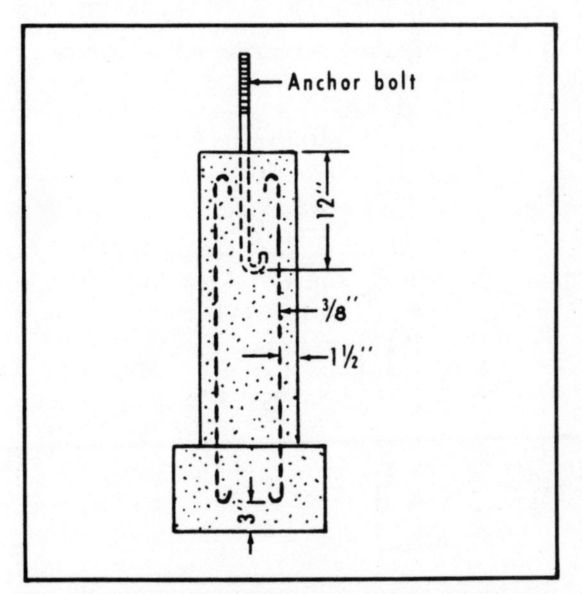

Fig. 1-25. Method of reinforcing and anchoring short piers.

Bolts, 5/8 inch in diameter, anchor the building to the piers. The bolts, with a hook bent in the lower end, are set 12 to 18 inches into piers of concrete and 3 feet into piers of unit masonry. They extend above the pier far enough to fasten down the building's sills or girders. Figure 1-17 shows a method of anchoring wood posts to piers.

Buildings with pier foundations are more frequently victims of wind uplift than are buildings with other types of foundations. Usually this is because the piers are not heavy enough and are not sufficiently anchored in the ground. A pier footing should move at least 1 cubic yard of soil if it is pulled from the ground. It should be at least 15 inches square and should be at least 2 feet deep. The footing, pier, and building sill or post must be fastened together securely.

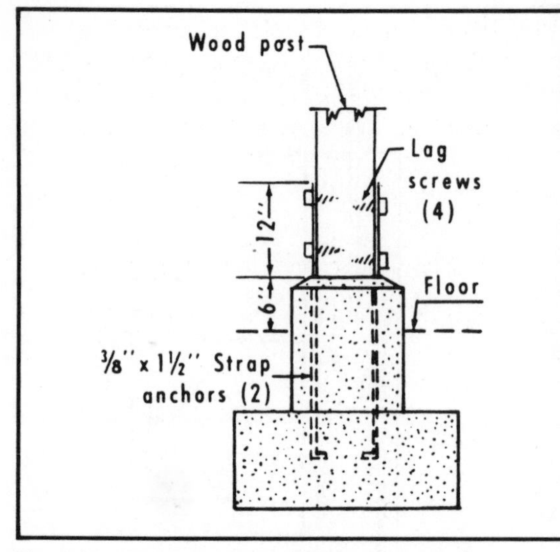

Fig. 1-27. Anchoring wood posts to concrete piers.

Pole Foundations

Pole foundations are economical and sturdy. Pressure-treated wood poles—7 to 10 inches in diameter and extending from concrete footing pads to the roof of a building—anchor the building and resist overturning and racking wind forces. Pole foundations are commonly used in farm and storage structures.

The depth to which the poles are set, the type of backfill around the poles, and the type of anchorage on the poles determine the building's resistance to wind forces. Soil type, soil moisture, and quality of backfill tamping determine the foundation design load that can be carried by the poles.

Usually a hole 20 inches in diameter affords adequate room for the pole and concrete footing pad, gives some room for movement of the pole to obtain proper alignment, and does not cramp the tamping tool used for packing the backfill. In most areas, poles should be set 5 feet deep.

A backfill of well-tamped, crushed stone is better than a backfill of earth. A pole set in crushed stone will move about one-half as much as a pole set in earth. A concrete backfill is required when movement of the pole at the ground line cannot be tolerated. Poles set 5 feet in any type of backfill that is reasonably well tamped will hold against normal wind uplift.

Grade Beam

A *grade beam* is a rectangular, reinforced-concrete beam that serves as a continuous foundation. It does not extend into the soil more than a few inches, but is supported on reinforced concrete piers.

Where soft soil underlays a more compact soil, a good foundation often can be secured with a grade beam. Holes are drilled through the soft soil, and the holes are filled with reinforced concrete piers. These piers support the reinforced beam in the soft soil.

The grade beam itself extends about 8 inches above grade and about the same distance below grade. It is placed over a fill of loose gravel, cinders, or similar porous material that will drain water from beneath the beam.

The concrete piers are 10 inches in diameter, and spaced 8 feet on center for one-story buildings and 6 feet on center for two story buildings. The piers are reinforced with 5/8-inch reinforcing rod that extends through the grade beam (Figs. 1-28 through 1-31).

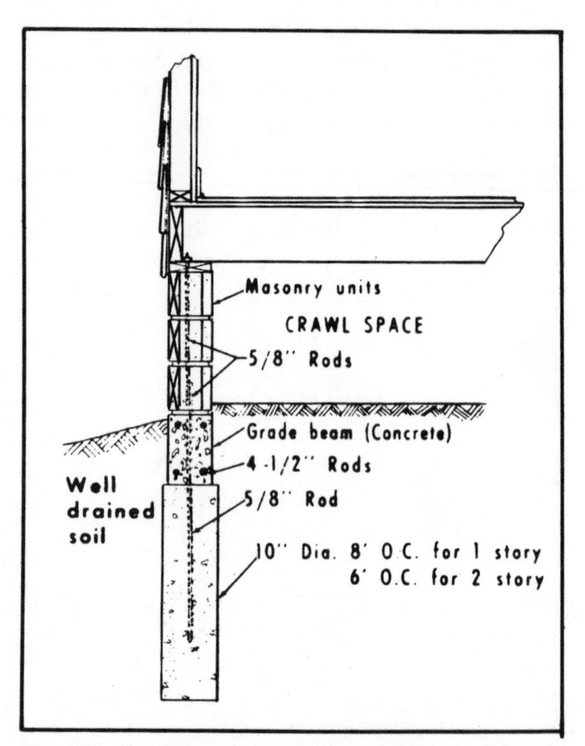

Fig. 1-28. Reinforcement of masonry units.

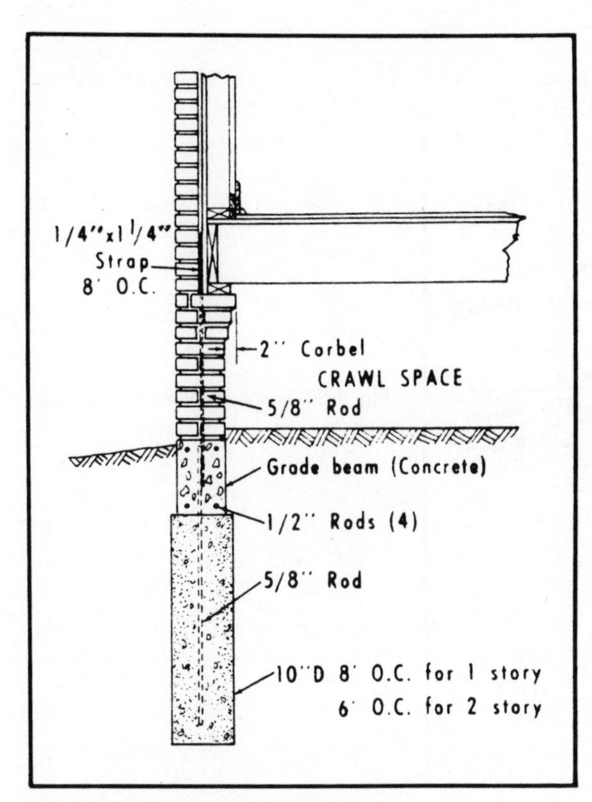

Fig. 1-29. Reinforcement of brick.

The cross-sectional size of the grade beam for the average one-story residence should be about 8 × 16 inches. The size for building other than dwellings depends on the weight of the building and the length of the beam. It should be determined by an engineer or builder who is familiar with reinforced concrete design.

Slab Foundations

Slab floors and slab foundations are different. *Slab floors* are cast independently of the foundation and are usually isolated from the foundation by 2 inches of rigid insulation.

Slab foundations are both foundation and floor, and are cast as one reinforced unit. The building is anchored to the slab with 5/8-inch bolts that are set in the slab when the slab is cast. The weight of the slab anchors the building to the ground.

The soil under the slab must be compacted to prevent the slab from settling and cracking. If earth fill is required under the slab, or under a part of it, the fill must thoroughly settle and be as firm as undisturbed soil before the slab is poured. The soil or fill under the slab should be covered with 4 to 5 inches of gravel, topped with sand, to give a well-drained subgrade to slab.

FOUNDATION MATERIALS

Solid concrete walls and brick walls have a high resistance to crushing. They are best for buildings with heavy loads.

Hollow units—clay tile or concrete block (except cinder block)—are strong enough for foundation walls for ordinary buildings. Cinder block is too porous to be suitable below grade and in direct contact with damp earth.

Alkaline soil deteriorates concrete. Local builders and county extension agents can offer information on how to work with alkaline soil. Brick and clay tile are unaffected by alkaline or acid soil.

Freezing will damage concrete if the concrete freezes before it has set and cured. Unit masonry also must be protected from freezing until the mortar is cured.

Concrete

Concrete is the best material for foundations. Directions for making concrete footings and walls and methods for building forms are explained in Chapter 3.

In general, a 1:3:5 concrete mixture (1 part cement, 3 parts sand, and 5 parts gravel) is used for foundation walls and footings. A 1:2 1/2:3 1/2 mix is used where watertightness is essential, as in basements, or where the concrete will be reinforced with steel.

If some unmuddied water is standing in the trenches, use a stiff mixture of concrete. Stiff concrete will displace the excess water. When it is necessary to pump water out of the foundation trench to keep the trench from flooding, pour the foundation with stiff, high-early-strength concrete, especially that part of the wall below water level.

The sleeves for pipes, blocking for windows, anchor bolts, and other inserts must be placed in

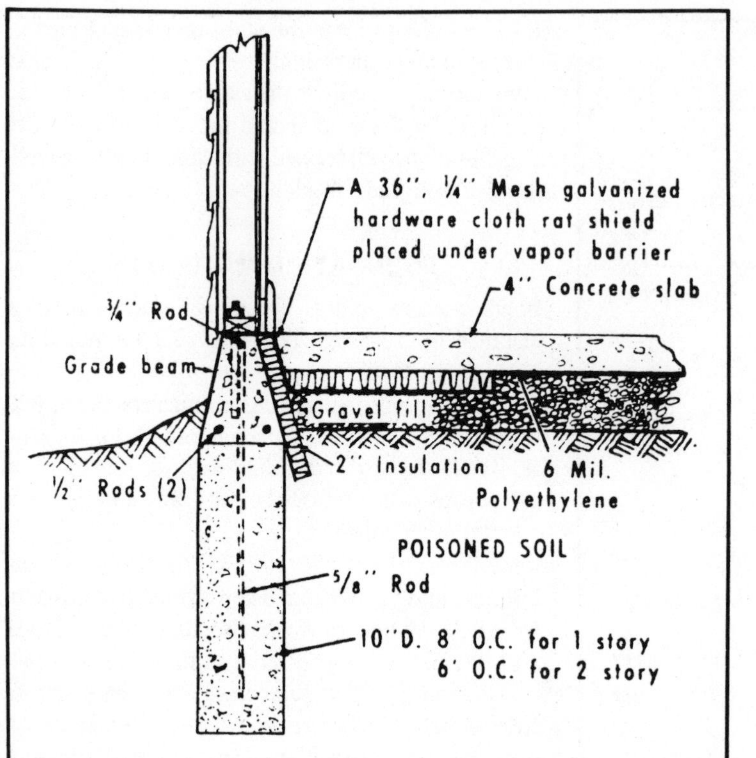

Fig. 1-30. Reinforcement of concrete slab.

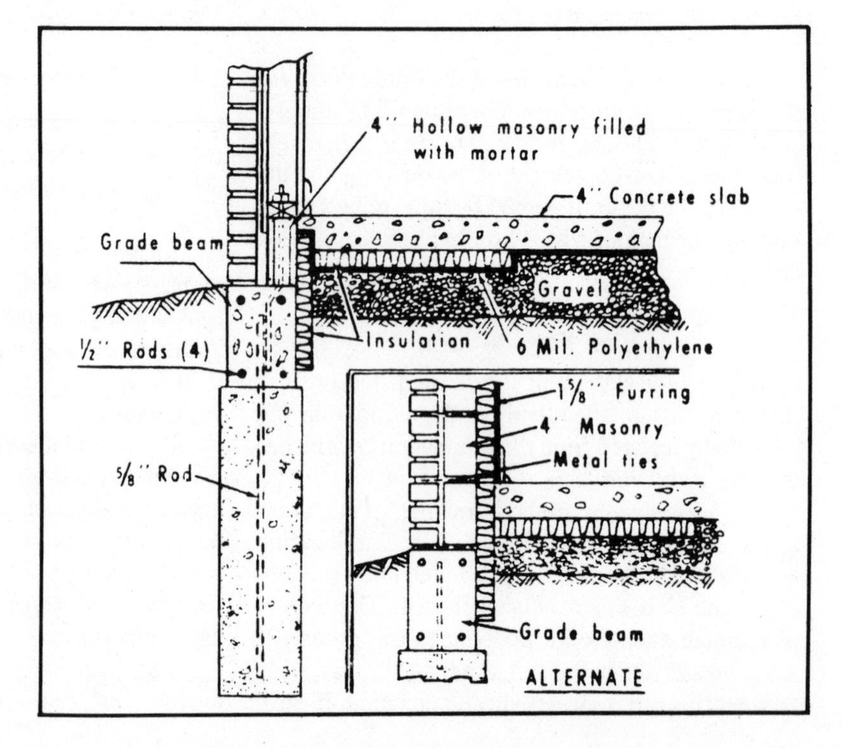

Fig. 1-31. Reinforcement of slab with added insulation.

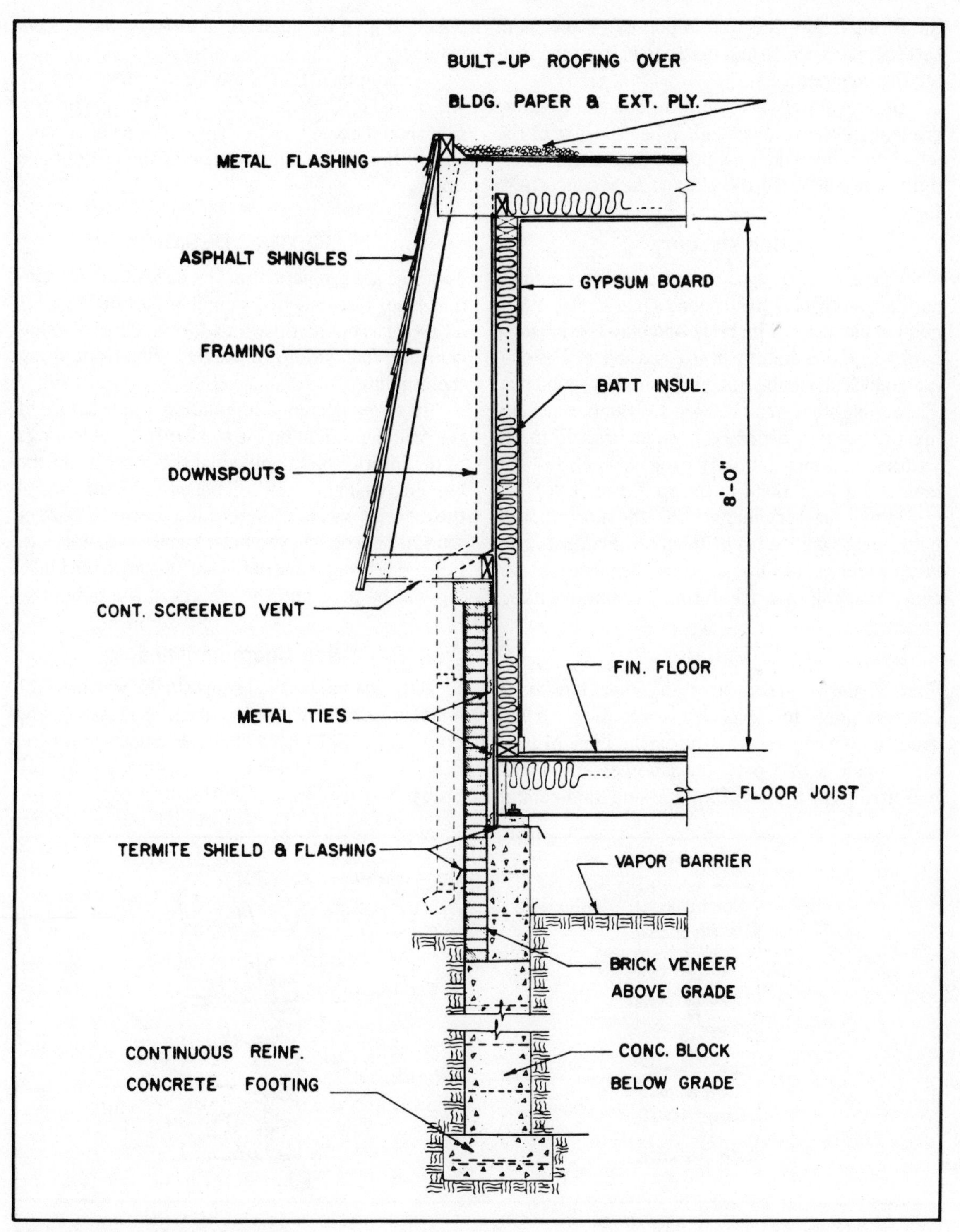

Fig. 1-32. Concrete block foundation.

the forms before the wall is poured. Place them carefully; it is nearly impossible to change them after the concrete sets.

Basement walls are poured as one unit to above grade, especially in wet soil. When the size of the job or equipment does not permit this, make watertight joints between the old and new concrete.

Unit Masonry

There are several types of unit masonry for foundations; brick, clay tile, concrete block (Fig. 1-32), and rubble stone. The brick and clay tile must be hardburned and uniform in size and shape. The concrete block should be suitable for underground use. The rubble stone must be hard and nonporous, and must not readily decompose when exposed to the weather. It is difficult to lay irregular rock and still maintain a good bond in the wall (Fig. 1-33).

In unit masonry foundations, the mortar joints between the units must be filled completely to prevent water penetration. Water penetration can cause the wall to crack during freezing weather.

Mortar

Type M mortar—a high-strength cement mortar—is recommended for masonry foundations below grade and in contact with the earth. Type M mortar is made with 4 parts portland cement, 1 part hydrated (type S) lime, 15 parts sand, and enough water to give the mixture a smooth, plastic consistency.

A common defect of mortar is oversanding, often caused by careless measuring. The sand proportion should never be less than 2 1/2 times, or more than 3 times, the total volume of the cement and lime.

FOOTING DESIGN

Footings are projections at the base of the foundation. They distribute the weight of the building over an area larger than the foundation. They also anchor pier foundations to prevent wind from lifting the building.

In a well-designed foundation, the pressure of the building weight on the soil beneath the footings of the interior piers will equal the pressure of the building weight on the soil beneath the footings of the exterior walls. To determine the proper size of foundation footings, you must know the characteristics and bearing value of the soil on which the building will be built and the weight of the building.

Soil Characteristics

Ideal foundation-bed soil supports the weight of the building, neither swells nor shrinks excessively, and does not heave from frost action. Such soil is rare, however. Dry, well-compacted, sandy clay soil probably comes closest to the ideal soil.

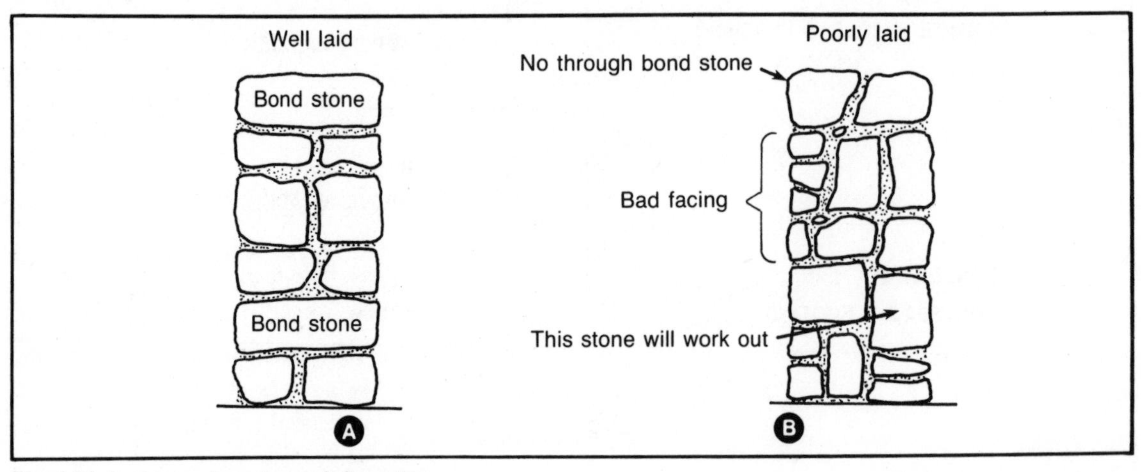

Fig. 1-33. Laying an irregular rock formation.

Clay soils become plastic when wet. Under moderate pressure, wet clay squeezes from beneath the foundation. The bearing value of clay soil can be improved, however, if it is drained and compacted by ramming in a layer of gravel or cinders.

Before you build on dry clay, make certain it will not swell excessively when it is wet. Some clays swell enough to lift buildings. Frost heaving can be reduced if the foundation bed is below the frost line (freezing depth).

Sand shrinks and settles when it becomes dry, and swells or flows when it becomes wet. Either change can make a foundation insecure.

If you build on sand, make sure the moisture content of the sand will remain constant. Otherwise, the sand under the foundation will move.

Avoid filled ground unless it is well settled. Filled ground generally causes trouble—the variation in fill depth and the different soils and waste materials used for fill cause the filled ground to settle unevenly. The different materials used for fill also have different bearing values.

Spongy or peaty soils must have buildings specially designed for them. If you have spongy or peaty soil, hire a structural engineer to design both your foundation and your building. It's cheap insurance.

Occasionally a relatively thin layer of rock overlays soft clay or loose sand. Such a bed is unsafe for heavy buildings and for concentrated pier loads. Make sure that rock at your site is not merely a large boulder which might loosen under the weight of the building.

You will need to investigate the soil to determine its characteristics. The best time to investigate is during a wet period, either in the early spring or late fall. Make preliminary probes with a 1/2- or 5/8-inch rod about 12 feet long. By driving the rod through the soil, you can determine the type, depth, and direction of slope of the underlying strata, and discover any buried boulders or rock layers.

If the preliminary probing indicates questionable conditions, investigate further. Dig to the depth of the proposed footing and then, using a soil auger, drill a hole 3 feet below that point. Note the various strata through the excavation and the hole.

Determine the safe load-bearing capacity of your soil as follows:

	Pounds/sq. ft.
Hard rock	30,000
Soft rock	16,000
Gravel or coarse sand, well consolidated	12,000
Dry, hard clay or coarse, firm sand	8,000
Moderately dry clay or coarse sand and clay	4,000-6,000
Ordinary clay and sand	3,000-4,000
Soft clay, sandy loam, or silt	1,000-2,000

Building Weight

There are two types of building weight on a foundation: dead loads and live loads. The *dead load* is the weight of the building itself: roof, floors, walls, foundation. The *live load* is the weight of things housed in the building and the forces acting upon the building, such as persons, furniture, animals, grain, wind, and snow. Dead loads and live loads for estimating weight are shown in Table 1-1.

Two examples are given for estimating building weight. One example is for the one-story house shown in Fig. 1-34; the other is for the barn shown in Fig. 1-35.

The weight on piers and walls must be calculated independently so that in neither case will the footing be too small. In general, the weight carried by a pier is the weight of the adjacent half of the beams (or girders) that are supported by the pier. This weight is represented by the shaded area in Fig. 1-34.

Note that only one-half of the total live load is used in designing the foundation of a house; it would be exceptional for all of the floor area of the house to be fully loaded at the same time. The total live load, however, is used to determine the size of girders and beams because the full load might be concentrated on any one of them. One-half the total live load is also used for the roofs and walls of storage buildings. The total live load is used for stored items and floors of storage buildings.

Table 1-1. Dead Loads.

Dead Loads (Approximate)

Roofs	Pounds per square foot
Gable, sheathed with ¾-inch boards, supported 2 feet on center, 15-pound felt, 210-pound asphalt shingles	7
Gable, trussed, 5 feet 4 inches on center, 2- by 4-inch purlins, 28-gage corrugated steel	4
Gable, added weight:	
Asbestos shingles	3
Built-up roof	5
Slate	7

Walls	Pounds per square foot
Stud framing, plates and sills, 2 by 4's 16 inches on center	2
Stud framing, 2 by 4's, 3 feet on center, 28-gage corrugated steel	2
Stud wall, plastered both sides	18
Stud wall, plastered one side	10
Stud wall, sheathed and sided with wood	7
Stud wall, ⅜-inch gypsum board both sides	6
Brick veneer	40
Brick, 9 inches thick	84
Clay tile, 4 inches thick	18
Clay tile, 6 inches thick	28
Clay tile, 8 inches thick	34
Concrete block, light aggregate, 4 inches thick	20
Concrete block, light aggregate, 8 inches thick	38
Concrete block, heavy aggregate, 4 inches thick	30
Concrete block, heavy aggregate, 8 inches thick	85

Floors	Pounds per square foot
Double-on 2 by 10's, 16 inches on center	7

Miscellaneous Materials	
Concrete or rubble stone, per inch of thickness	12
Gypsum board, ½ inch thick	2

Live Loads (Approximate)	
First floor in dwellings	40
Second floor in dwellings	30
Attic floor in dwellings (habitable for storage only)	20
Roofs, in general	20 to 40
Assembly halls (where crowds collect)	100

Weight of Produce	Pounds per cubic foot
Apples, carrots, potatoes	40
Beans, wheat, shelled corn	48
Ear corn, husked	28
Oats	26
Bran	16
Loose hay	4 to 5
Chopped hay or ordinary baled hay	10 to 13
Baled straw	8
Lime, fertilizer	55 to 60

Footing Area

A common fault in construction is to make the footings beneath the piers too small in relation to the footing beneath the exterior wall of the building. To find out how large a footing should be, you must know two things: the bearing value of your soil and the weight the footing must support. In the house example in Fig. 1-34, the wall load was 1,077 pounds per lineal foot. The pier load was 4,664 pounds per pier.

Table 1-1 shows approximate dead loads. In Table 1-2 you will find the width of footing required to bear the wall load; in Table 1-3 you will find the size of footing required to bear the pier load. For example, if the soil bearing value is 1,000 pounds per square foot, the width of the wall footing must be 14 inches. The pier footing must be 27 inches square.

If the soil bearing value is 3,000 pounds per square foot, the wall footing can be less than 8 inches; however, a 12-inch footing will make construction easier. Because the 12-inch wall footing

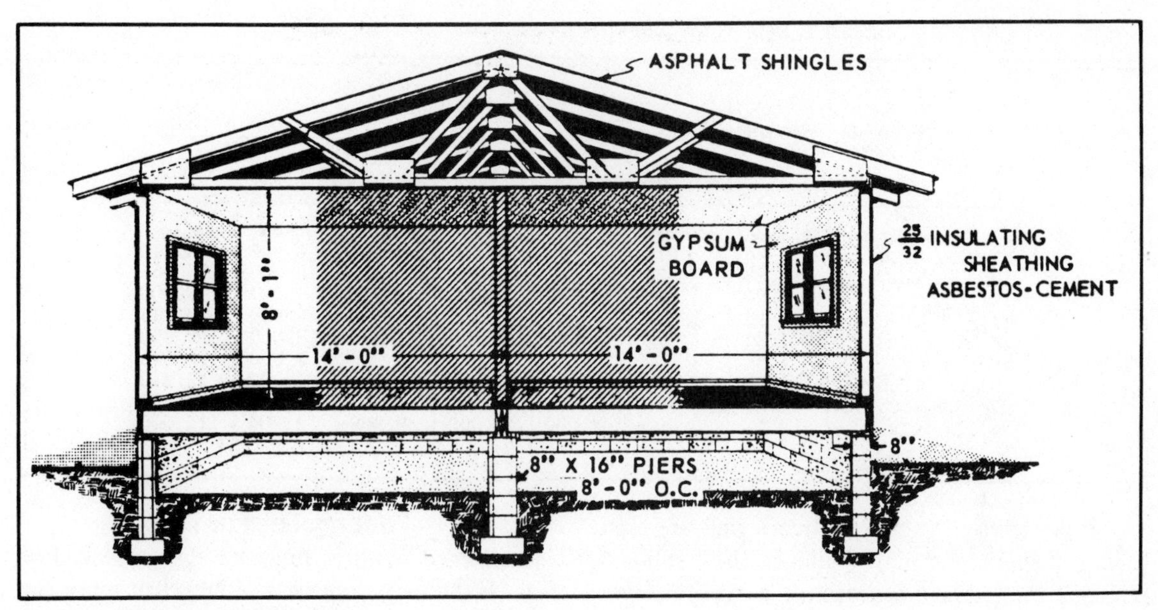

Fig. 1-34. Estimating building weight.

is larger than necessary, the pier footing also must be larger than necessary to equalize the bearing. Therefore, because a 12-inch wall footing will support approximately 2.8 times the wall load (2.8 × 1,077 = 3,016 pounds per lineal foot), the pier footing also must be large enough to support 2.8 times the pier load (2.8 × 4,664 = 13,059 pounds per pier). The pier footing must be 27 inches square.

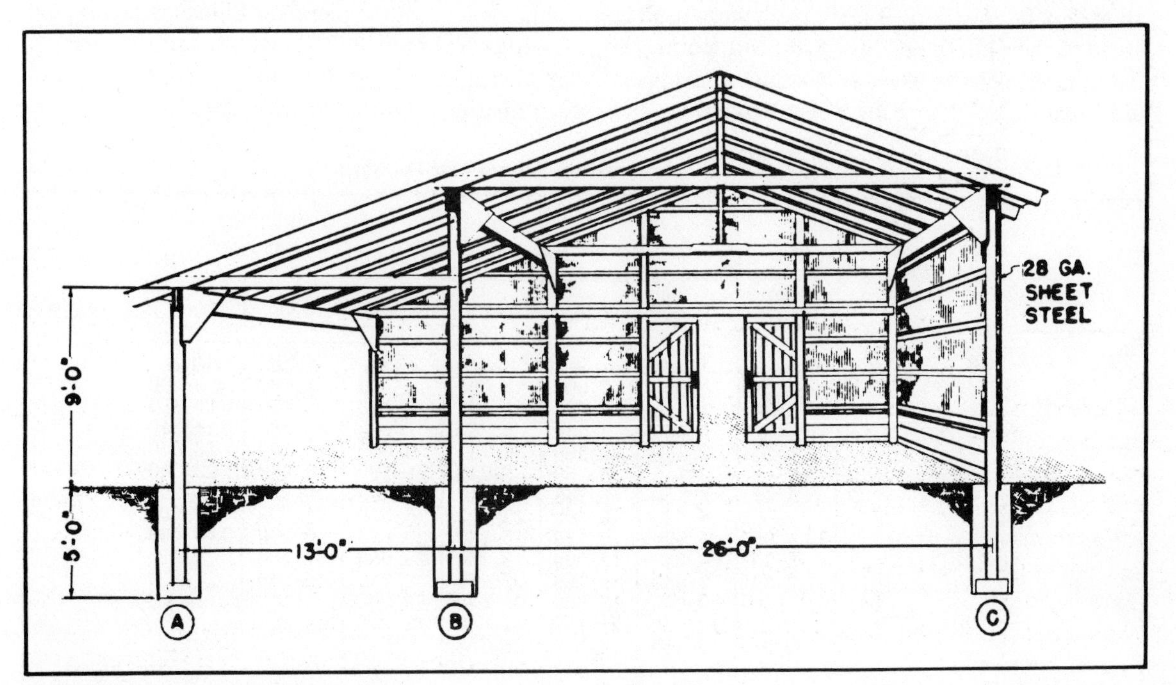

Fig. 1-35. Estimating outbuilding weight.

Table 1-2. Safe Total Load, per Linear Foot, on Wall Footings.

Width of footing (inches)	Bearing area	Soil bearing value in pounds					
		1,000	2,000	3,000	4,000	6,000	8,000
	Square feet	*Pounds*	*Pounds*	*Pounds*	*Pounds*	*Pounds*	*Pounds*
8	0. 66	670	1, 340	2, 000	2, 670	4, 000	5, 340
10	. 83	835	1, 665	2, 500	3, 335	5, 000	6, 665
12	1. 00	1, 000	2, 000	3, 000	4, 000	6, 000	8, 000
14	1. 16	1, 165	2, 335	3, 500	4, 665	7, 000	9, 335
16	1. 33	1, 330	2, 670	4, 000	5, 330	8, 000	10, 670
18	1. 5	1, 500	3, 000	4, 500	6, 000	9, 000	12, 000
20	1. 67	1, 670	3, 335	5, 000	6, 670	10, 000	13, 340
22	1. 83	1, 835	3, 665	5, 500	7, 335	11, 000	14, 665
24	2. 0	2, 000	4, 000	6, 000	8, 000	12, 000	16, 000

If piers, rather than a continuous wall, are used under the perimeter of the building, the combined areas of the footings beneath the perimeter piers must equal the footing width shown in Table 1-2 for a continuous wall multiplied by the length of the wall. For example, the necessary width of the wall footing for the house on soil with a bearing value of 1,000 pounds per square foot was 14 inches. On the 48-foot side of the house, the footing would have an area of 56 square feet. If six piers were used instead of the wall, the piers would need footings of 9.3 square feet each. The pier footings would measure about 3 feet 1 inch by 3 feet 1 inch.

Footing Thickness

Unreinforced footings must be at least 6 inches thick. If the footing projects 4 inches or more beyond the wall or pier, the footing thickness must be at least 1 1/2 times the projection. For instance, if the footing under the 8-inch foundation of the house is 14 inches wide, it will project only 3 inches beyond each side of the wall and must be 6 inches thick. The footings for the piers, however, were 27 inches square; if the piers are 12 inches square, the footings will project 7 1/2 inches and will need to be at least 11 1/2 inches thick.

Footings reinforced with steel do not have to

Table 1-3. Safe Total Load on Square Pier Footings.

Size of footing (inches)	Bearing area	Soil bearing value in pounds per square foot					
		1,000	2,000	3,000	4,000	6,000	8,000
	Square feet	*Pounds*	*Pounds*	*Pounds*	*Pounds*	*Pounds*	*Pounds*
12	1. 0	1, 000	2, 000	3, 000	4, 000	6, 000	8, 000
14	1. 36	1, 360	2, 720	4, 080	5, 440	8, 160	10, 880
16	1. 77	1, 780	3, 560	5, 340	7, 120	10, 680	14, 240
18	2. 25	2, 250	4, 500	6, 750	9, 000	13, 500	18, 000
20	2. 78	2, 780	5, 560	8, 340	11, 120	16, 680	22, 240
22	3. 37	3, 370	6, 740	10, 110	13, 480	20, 220	26, 960
24	4. 00	4, 000	8, 000	12, 000	16, 000	24, 000	32, 000
27	5. 06	5, 060	10, 120	15, 180	20, 250	30, 370	40, 500
30	6. 25	6, 250	12, 500	18, 750	25, 000	37, 500	--------
33	7. 55	7, 560	15, 120	22, 680	30, 240	45, 360	--------
36	9. 00	9, 000	18, 000	27, 000	36, 000	--------	--------
39	10. 56	10, 560	21, 120	31, 680	42, 240	--------	--------
42	12. 25	12, 250	24, 500	36, 750	--------	--------	--------

Table 1-4. Suggested Depths for Placing Bottoms of Footings.

State	Light buildings		Farmhouse[1]		Heavy permanent barns and storage		Local considerations
	A	B	A	B	A	B	
	Inches	*Inches*	*Inches*	*Inches*	*Inches*	*Inches*	
Alabama	12	12	18	18	18	18	Reinforce footings and floor, and use piles in Blackbelt area.
Alaska	48 to 60	60 to 72	48 to 60	60 to 72	48 to 60	60 to 72	In nonpermafrost areas place polystyrene on outside of the foundation walls.
Arizona	12	20	18	36	18	36	Closeness of irrigation a factor.
Arkansas	12	12	16	16	18 to 24	18 to 24	Continuous foundations preferred.
California	6	12 to 18	8 to 12	18 to 24	12	24 to 30	
Colorado	12	18	18	24	18	24	Protect from roof water.
Connecticut	--------	(2) 24	--------	30 to 48	--------	30 to 48	
Delaware	18	24	24	30	30	30	Consult county building code.
Florida	surf	surf	surf	6 to 12	surf	6 to 12	Wide footings near surface; sandy soil.
Georgia	6	12	12 to 18	18	12 to 18	18	Conditions variable; seek local advice.
Idaho	12	18	24	36	36	48	Reinforce in wet cold locations.
Illinois	12	18	24	36	36	48	Reinforcement advised.
Indiana	18 to 24	18 to 24	24 to 36	24 to 36	36	36	
Iowa	18	20	36	42	36	42	
Kansas	24	24	60	60	48	48	Reinforce; heavy footings needed on swelling and shrinking soils.
Kentucky	18 to 24	18 to 24	18 to 24	30	30	30	
Louisiana	2 to 12	2 to 12	2 to 12	2 to 12	2 to 12	2 to 12	Wide footings on alluvial soils.
Maine[3]	48 to 60	60 to 72	48 to 60	60 to 72	48 to 60	60 to 72	
Maryland[4]	--------	--------	--------	--------	--------	--------	Conditions variable; seek local advice.
Massachusetts	24 to 48	24 to 48	24 to 48	24 to 48	24 to 48	24 to 48	Soil conditions fairly uniform.
Michigan	(4) 18	24	36	36	36	36	
Minnesota	12	18	60	60	(5) 18	(6) 36	
Mississippi	9	9	(7)	(7)	(7)	(7)	
Missouri	12	18	18	24	24	30	
Montana	18	18	44	44	30	40	Seek local advice.
Nebraska	12	18	18	24	18	24	Guard against roof water and rooting animals.
Nevada	0 to 6	18	0 to 6	18	12	24	
New Hampshire	36	48	72 to 96	72 to 96	48	72	Greater depth is for masonry.
New Jersey	6 to 8	24 to 30	16	36	(8) 16	(8) 36	
New Mexico	12	(4)	15 to 18	20 to 24	18 to 20	24 to 30	
North Carolina	12	12	12	18 to 24	12	18 to 24	
North Dakota	18	18	--------	--------	--------	--------	Reinforce.
Ohio	18	24	36	42	36	42	Do.
Oklahoma	12	18	18	24	24	24	Reinforce masonry for swelling, shrinking and heaving soil.
Oregon[11]	--------	--------	--------	--------	--------	--------	
Pennsylvania	36	--------	48 to 72	48 to 72	48	48	
South Carolina	10 to 12	12	14	18	14	18	
South Dakota	18	18	54	54	24	24	For frame buildings use continuous foundations.
Do	42	42	60	60	48	48	For masonry buildings use continuous foundations.
Tennessee	12	12	24	24	24	24	Guard against termites.
Texas	12	20	20	30	30	34	
Vermont	12	12	60	60	60	60	Conditions vary widely; carry to firm soil.
Virginia	24	24	24	24	24	24	
Washington	--------	--------	--------	--------	--------	--------	Conditions variable; seek local advice.
West Virginia	18 to 24	24 to 30	18 to 24	24 to 30	24	30	
Wisconsin	30	42	36	48	(9) 36	(10) 42	
Wyoming	24	30	36	42	36	42	

[1] Where depth is 48 inches and over, basements are generally used.
[2] For temporary buildings.
[3] Use buttress on outside face of wall or use a footing; less depth is required in gravelly soils.
[4] For snow-protected ground.
[5] Wooden barns.
[6] Masonry barns.
[7] Depth to uniform soil.
[8] Footings for storage structures reinforced.
[9] 48 inches if building is unheated.
[10] 54 inches if building is unheated.
[11] Conditions vary (climate, elevation, soil and soil moisture) seek local advice.

be so thick. Their design should be referred to a structural engineer, however.

Footing Depth

Footing depths considered safe are given in Table 1-4. They are considered sufficient to prevent damage by frost, but are not necessarily the total depth to which frost penetrates. These depths are given as a guide only. It is best to check with local building codes and with local engineers, builders, or other authorities before you build.

The safe depth varies with the depth to which frost penetrates and the effect of frost on the soil. Great variations in frost depth can occur within a single state.

In any case, set the footings below the topsoil and on firm ground. If they are placed too close to the surface, rodents can burrow under them, and wind and rain can erode the soil beneath them.

FOOTING CONSTRUCTION

Poured concrete is recommended for foundation footings for heavy buildings. Concrete fills up irregularities in the bed and arches over soft spots in subsoil. Also, concrete footings provide a level and smooth surface for starting unit masonry foundations.

Two 1/2-inch reinforcing rods, spaced not more than 8 inches apart and running the length of the footing, should be embedded about 3 inches into the footing when the subgrade is not uniform. Reinforcing is especially important at the corners of the footing to prevent cracking.

Large footings are more economical if they are made with reinforced concrete. The required thickness of reinforced concrete is less than the required thickness of unreinforced concrete. Reinforced footings, however, require an engineer's advice.

Concrete can be safe by *stepping* (or *corbelling*) large, thick footings (Fig. 1-36).

Sometimes in firm ground the trench is sloped out at the bottom and the footing is poured as part of the wall foundation. By using this method, separate footing construction is avoided.

If the foundation will be built of brick, concrete block, clay tile, or stone, the footing can be built of the same material. A bed of cement mortar (1 part cement, 3 parts sand), about 1 inch thick, is placed at the bottom of the trench to smooth the subgrade and provide a bed to lay the first *course,* or row, of the footing.

Brick footings also can be corbelled. Start brick footings with a double course of brick at the base; you can make steps with each successive course. For heavy buildings, the steps are made in every second row; with the lower row is a *stretcher row* (bricks running the length of the wall) and the upper row is a *header row* (bricks running across the wall). A brick step should not be more than 2 inches wide over a header course and 1 inch wide over a stretcher course.

The footing must be bonded to the foundation. Embed rods 1/2 inch in diameter and 8 to 10 inches long, spaced 16 inches apart, one-half their length in the footing. They project into the foundation wall above the footing when the foundation wall is built.

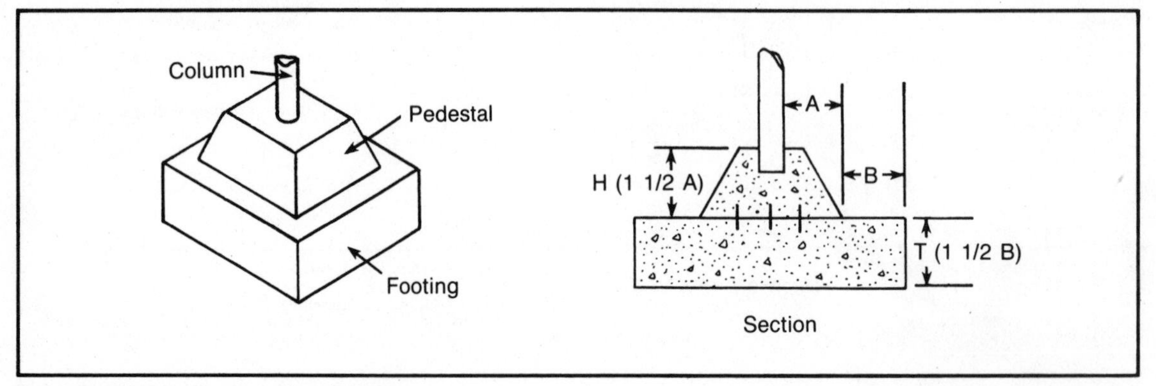

Fig. 1-36. Stepping or corbelling footings.

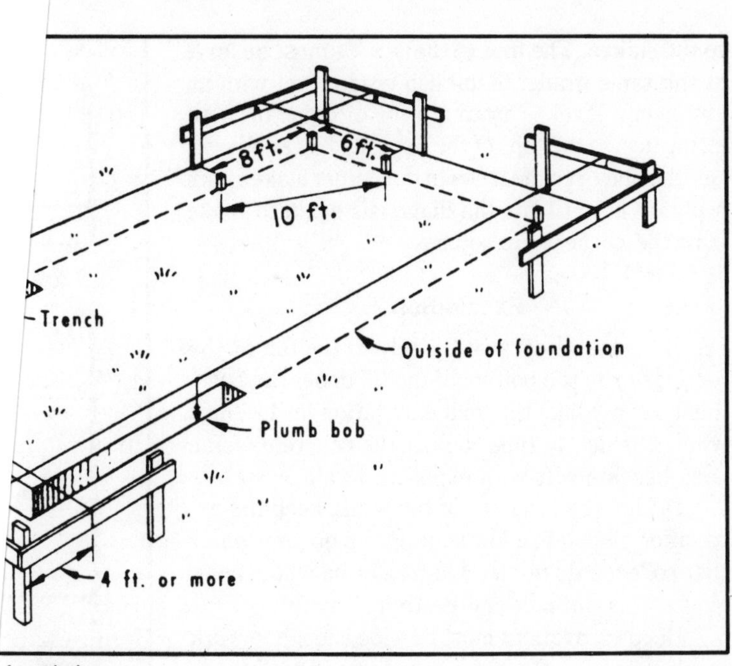

foundation.

ıgs to the foun-
tones half their
become part of
in the founda-
out 2 feet apart.

CONSTRUCTION PRACTICES

There is more to constructing a good foundation than knowing the correct footing size and selecting a suitable wall material. The building must be laid out, the excavation dug, the drainage tile installed, and the basement walls waterproofed. If all these steps are done properly, you will save both time and money.

Building Layout

The first step in laying out a building foundation is to clear the site. Remove the sod. Remove all tree stumps from the site—they harbor termites and should not be used for fill or left under the building.

The second step in the building layout is to establish grade lines to keep the foundation true and level. Figure 1-37 shows the arrangement for batter boards and lines for a rectangular building. Locate small stakes at each corner of the building. Indicate the outside line of the foundation wall with tacks driven in the stake tops. Measure the *diagonals,* the distance between stakes that are catercorner from each other. If the diagonals are the same length, the corners are square.

Square the corners of the layout. To do this, measure along one end a distance in 3-foot units, then measure along an adjacent side the same number of 4-foot units. The diagonal of these distances will equal the same number of 5-foot units when the corner between the end and the side is square. For example, start at a corner and measure along the end of the building two 3-foot units (6 feet). Mark the distance with a small stake. Then measure from the same corner along one side of the building two 4-foot units (8 feet). Mark that distance with a small stake. Now if the corner from which you began your measurements is square, the distance between the two small stakes will be two 5-foot units, or 10 feet.

After you have located and squared the corners, drive three 2- × -4-inch stakes into the ground 3 to 4 feet beyond each corner, as shown in Fig. 1-37. Nail 1- × -6-inch boards (*batter boards*) horizontally to the stakes. The tops of the 1- x -6s must be level at the same grade. (Establish grade level with an engineer's level.) Fasten a piece of twine or stout string across the tops of the opposite batter boards. Set the line over the tacks in the corner stakes with a plumb bob. Check the diagonals again to make sure the corners are square.

Excavation

Dig the general excavation only to the top of the footings or to the bottom of the fill under the basement floor. Make the final excavation for footings when it is nearly time to pour the concrete; some soils become soft with exposure to air or water.

During the progress of the work, keep the excavation sloped to a low spot. Pump out any water that collects; do not let it saturate the whole area. Wet soil is difficult and costly to handle.

Deep excavations must be wide enough to work in when you are constructing and waterproofing the foundation wall. Where soil slides easily, it is sometimes more economical to make the excavation wider than it is to install forms, cribbing, or other braces. The general method of shoring up sliding soil, however, is illustrated, in Fig. 1-38. This method can be adapted to most conditions.

Avoid handling the soil a second time. Spread as much soil as possible in its final position and reserve only an amount sufficient for backfilling.

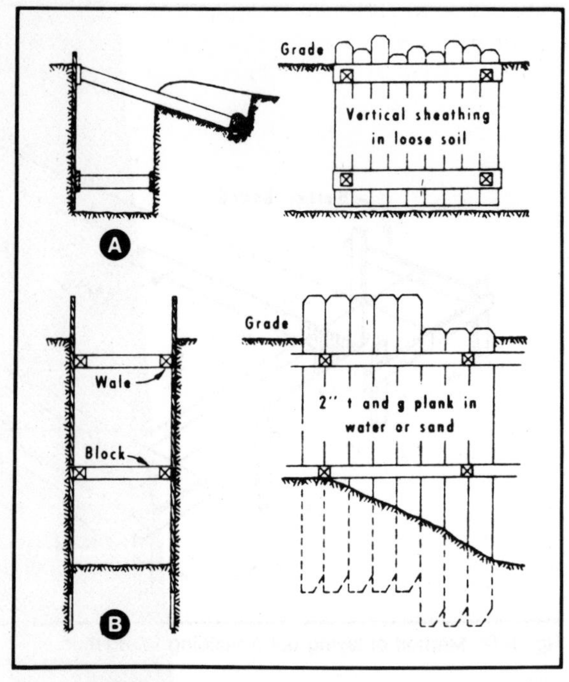

Fig. 1-38. Shoring up when constructing or waterproofing your foundation.

Drainage

Draining water away, and keeping it away, from the foundation is important. Equip the building with gutters and downspouts to carry water from the roof. Grade the finished site for surface runoff, sod or page the ground for about 10 feet around the building, and intercept and divert groundwater from higher elevations.

The footings of houses with basements should be protected with drain tile. Lay 4-inch agricultural drain just above the bottom of the footing. Slope the tile at least 1/8 inch per lineal foot and backfill over it with 18 inches of gravel. Make sure the tile outlet is unclogged and water flows freely from the end of the tile to the ground surface or a sewer drain.

Drainage tile for a grade beam foundation is laid slightly differently. Lay it just outside the line of piers and 8 to 16 inches below the bottom of the grade beam. Slope the tile at least 1 inch in 20 feet and outlet the low end on the ground surface. Install the tile after the piers and grade beam have set and cured.

Waterproofing

Waterproofing is least expensive and most effective if it is done to the exterior of the foundation wall at the time of construction. Where the soil has only a moderate amount of moisture or where the foundation drainage will not allow the water in the soil to build up pressure, it is usually enough to waterproof only the exterior of the foundation walls by applying two coats of hot coal-tar pitch.

A better waterproofing is the application of two coats of cement mortar, each 3/8 inch thick. The mortar is a mixture of 1 part portland cement (or 2/3 cement and 1/3 lime to accelerate setting) and 2 1/2 to 3 parts sand by volume. One cubic foot of mortar will cover 30 to 32 square feet of wall. Both coats of mortar must cover the foundation wall from the top of the projection of the footing to several inches above grade at the top.

Before you apply the mortar, clean the foundation wall thoroughly. Remove dirt, grease, oil, and loose particles of mortar or concrete. Then moisten the wall surface.

Scratch the first coat of mortar before it hardens. A board with nails driven through it, like a sharp rake, makes an excellent scratcher. Scratching is essential to make the second coat stick to the first coat.

Allow the first coat to harden for at least 24 hours. Then dampen it down and apply the second coat. Keep the second coat damp for at least 48 hours.

You can waterproof the foundation further by coating the cured mortar with hot coal-tar pitch. It takes 20 to 25 pounds of tar to coat 100 square feet of wall.

FOUNDATION REPAIRING AND REMODELING

Foundation repairing or remodling is frequently a hard job requiring the services of an engineer, builder, or house mover. Often it is necessary to lift the building from the existing foundation and support the building in the air until the repair is completed. If an entire foundation is replaced, the new foundation must be expertly constructed to fit the building. If only part of a foundation is replaced, the new part must not only fit the building, but be bonded to the rest of the foundation. Digging a basement under a house usually means that the workmen must work in cramped, close quarters. If the house has a chimney above the area of the new basement, the chimney will need special support to keep it from crashing down on the floor of the new basement.

Some foundation repair jobs are alike; others are one-of-a-kind situations. There are a few general pointers, however, that will make many of the jobs safer and easier.

Raising and Supporting Buildings

Before you raise a building, unload it. Remove furniture and appliances from houses, or machinery and feed from barns and storage sheds.

Next disconnect electric and telephone wires, plumbing, and masonry steps and porches. Remove the nuts from the bolts that anchor the sills to the foundations.

Make holes through the foundation walls near the original building piers. Slip temporary sills —8- × -8-inch or 12- × -12-inch timbers—through the holes. The temporary sills must extend beyond the walls of the building far enough to be supported on cribbing.

Jack up the temporary sills to transfer the weight of the building from the foundation to the timbers. Support the temporary sills on *cribbing,* or blocking, built out of 6 × 6s or 4 × 4s. Remove the jacks. Build the cribbing carefully so it will not rock or tip. Set it on firm, dry ground; 2- × -10-inch planks laid close together on the ground will serve as a footing and distribute the load of the cribbing. When only a small part of the building is raised, post supports—wood 6 × 6s, 4 × 4s, or pipes—are often sufficient and less of an obstruction to work than temporary sills and cribbing.

If you do not have a sufficient number of jacks to raise the whole building at one time, raise one side a little and set cribbing to hold the raised side in place. Then progress around the building, raising each side by stages, until the whole building is raised high enough to build the new foundation.

Rebuilding Masonry Foundation Walls

Masonry foundation walls of large buildings are safely and easily rebuilt if alternate sections 4 feet long and 8 feet apart are replaced one at a time. Remove a 4-foot section of foundation wall and excavate to the bottom of the footing. Use the same construction methods for rebuilding the section that you would use for building a new foundation.

The new footing can be stepped if the ground slopes, but each step of the footing must be level. Concrete is the best material for the footing; the foundation wall, however, is more easily built from unit masonry.

Adding a Basement

Although it is usually poor economy to add a large

basement to an existing house, a small basement 8 to 10 feet wide with sufficient space for a furnace can be added without great cost or labor. Refer to Fig. 1-39.

The farther the basement is under the house, the greater the amount of work involved, so locate it near one outside foundation wall. If the basement will be used for heating-plant space, locate it so the chimney will be accessible from the basement.

Support the building girder, or sill, that spans the length of the basement with I beams or heavy wood beams. These beams replace the piers that supported the building girder before the basement was dug. The ends of the supporting beams must rest in slots in the top of the new basement walls.

You must allow a safe distance between basement walls and foundation walls that are parallel to them. If the basement wall is too close to the foundation wall, the pressure exerted on the soil by the foundation wall will be exerted laterally on the basement wall. The safe distance depends on the type of soil and on the difference in height between the footing of the foundation wall and the footing of the basement wall.

In loose, sandy soil, do not put the new basement wall closer to a parallel foundation wall than 1 1/2 feet for each vertical foot distance between the footings of the two walls. For example, if the new footing is 4 feet lower than the old footing, the foundation wall and the basement wall must be 6 feet apart.

The safe distance per vertical foot difference in damp clay is 2 feet; in mixtures of sand, dry clay, gravel, and ordinary soil it is 1 1/2 feet; in decomposed rock, cinders, or ashes it is 1 foot. If the soil becomes very wet, greater distance is necessary and must be determined by an engineer.

When the sidewalls of the basement are built and the building girders are supported on beams, excavate the basement. Remove the earth from the excavation through an opening in the foundation wall. This opening later can become an outside door to the basement.

Build the rear wall of the basement after the excavation is completed. If the foundation of the house is made entirely of piers (with or without curtain walls on the perimeter of the house), four basement walls will be needed.

When a chimney is within the area of a new basement, it must be extended to below the basement floor. During the excavation, support the chimney on two or three steel beams. Support the

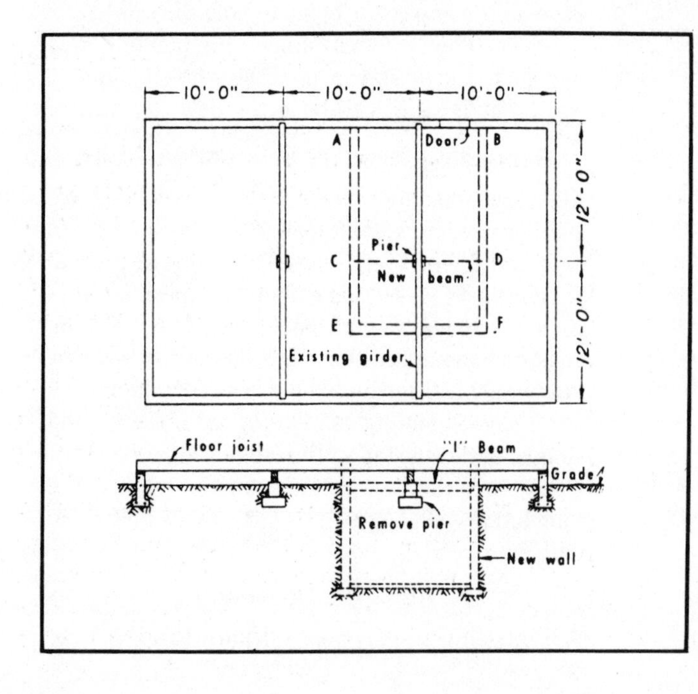

Fig. 1-39. Adding on a small basement.

ends of the beams on cribbing. The exact placement of the beams and cribbing depends on the headroom and work area under the house. Ordinarily you will cut holes in the masonry of the chimney just above the footing, insert the beams through the holes, and block up the beams. Then extend the chimney to a new footing below the basement floor. After the new chimney extension is strong enough to support the weight of the existing chimney, withdraw the beams and fill the holes.

When you extend piers to a lower level, support the building girders on blocking, tear out the old piers, and build new ones. Start with a footing below the basement floor and build up. Pipe or structural steel piers are easier to build than masonry piers.

When you extend a wall or chimney to a lower level with unit masonry, you probably will have to omit the top course of units on the extension because of the unevenness of the bottom of the old footing. You can fill this space with stiff concrete. A boxlike form will keep the concrete in place until it sets.

Figure 1-40 shows forms for poured concrete extensions. Chisel off the projection of the footing plumb with the wall in 12- to 18-inch sections. Then pour the concrete through the chiseled out sections.

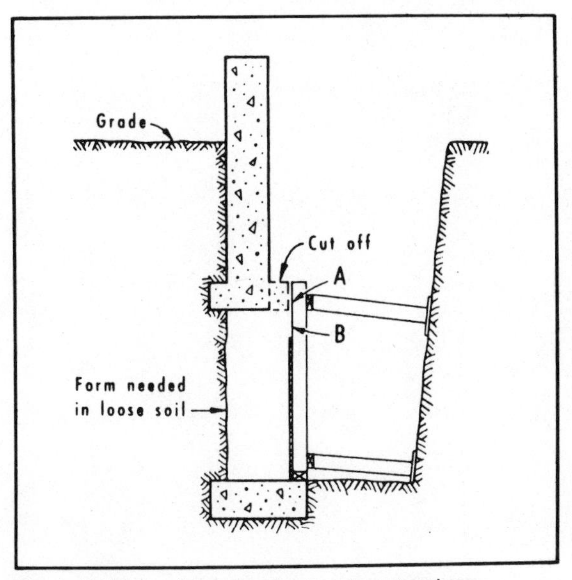

Fig. 1-40. Forms for poured concrete extensions.

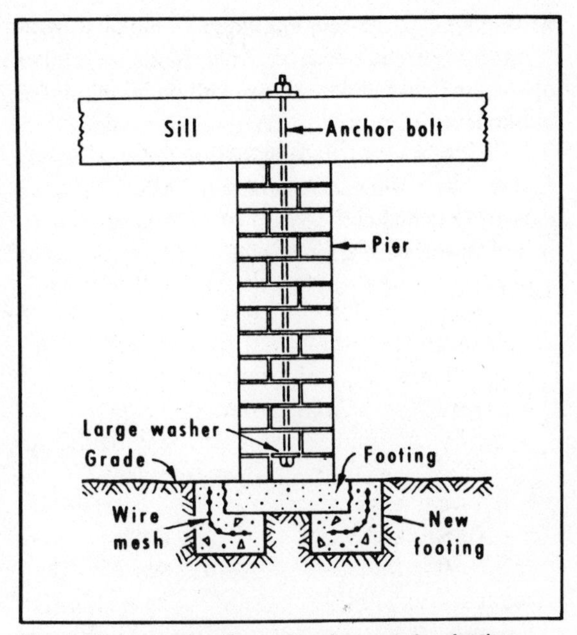

Fig. 1-41. Increasing the width of an existing footing.

Repairing Defects

Foundations usually settle excessively or unevenly when their footings are not deep enough to withstand frost heaving or erosion, or when the piers and walls have too little bearing area for the type of soil and building load. Settling also might be caused by rotted wood posts and sills or by defective masonry.

Figure 1-41 illustrates a good method for increasing footing width. It is, however, more economical to relieve an overloaded foundation by installing extra piers than it is to increase the width of the existing footings. If you add piers to your house foundation, make sure the sills and girders of the house actually rest on both the new and old piers. If the new piers settle a little when the building is lowered onto them, shim them with wedges of slate or flat pieces of hard tile. Occasionally a pier is made about 1/2 inch higher than its final grade to allow for settling.

Where the soil under the footing is eroded but the foundation wall is not damaged, ram a mixture of damp (not wet) sandy clay and 5 to 8 percent, by volume, portland cement under the footing. This mixture will provide a firm, secure bearing. Then bank soil against the foundation; bank it high enough for protection from frost. Slope the soil to divert surface and roof water, and pave or sod the banked soil to protect it from wind erosion.

Exposed wood usually deteriorates at joints first and causes metal fasteners to loosen. Remove loose spikes and driftbolts. Treat the holes and affected areas with a preservative. Plug the holes with wood dowels or tar. Replace the spikes or bolts in new locations.

Cut away sections of unsound sills and replace them with plank patches. If the whole timber has been weakened, replace it.

Untreated wood posts rot rapidly at the ground line. Replace them with treated posts or with masonry piers.

Chapter 2

Location and Excavation

Making a hole in the ground sounds easy, but it isn't. Locating and digging out or excavating the ground for a foundation requires careful planning and action. This chapter, first takes a broad look at the procedures, then describes how to implement those procedures efficiently.

SITE CONDITION

Before excavating for the new home, determine the subsoil conditions by performing test borings or by checking existing houses constructed near the site. You might encounter a rock ledge, necessitating costly removal; a high water table might require design changes from a full basement to crawl space or concrete slab construction. If the area has been filled, the footings should always extend through undisturbed soil. Any variation from standard construction procedures will increase the cost of the foundation and footings. Thus, it is good practice to examine the type of foundations used in neighboring houses. This might influence the design of the new house.

HOUSE PLACEMENT

After the site is cleared, the location of the outer walls of the house is marked out. The surveyor marks the corners of the lot after surveying the plot of land. The corners of the proposed home also should be roughly marked by the surveyor.

Before you decide on the exact location of the house, check local codes for minimum setback and side-yard requirements. The location of the house is usually determined by such codes. In some cases, setback might be established by existing houses on adjacent property. Most city building regulations require that a plot be a part of the house plans, so its location is determined beforehand.

The next step, after the corners of the house have been established, is to determine lines and grades as aids in keeping the work level and true. The batter board (Figs. 2-1 and 2-2) is one of the tools used to locate and retain the outline of the house. The height of the boards is sometimes established to conform to the height of the foundation wall.

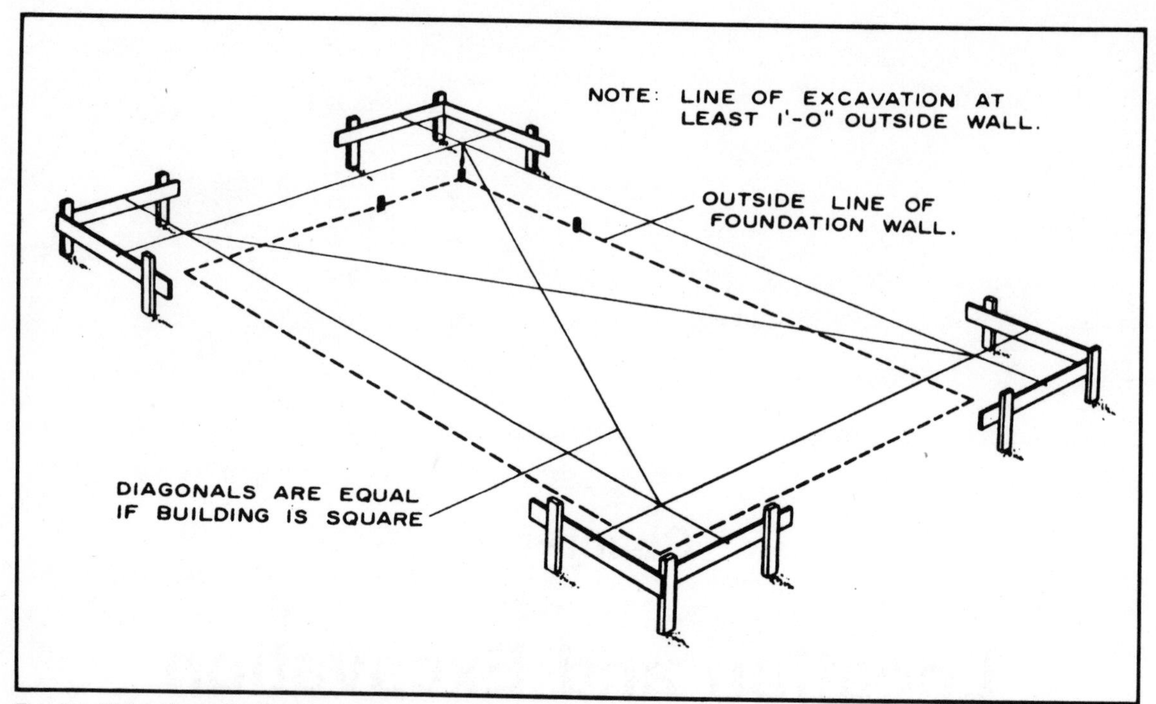

Fig. 2-1. Using the batter board to stake out the foundation.

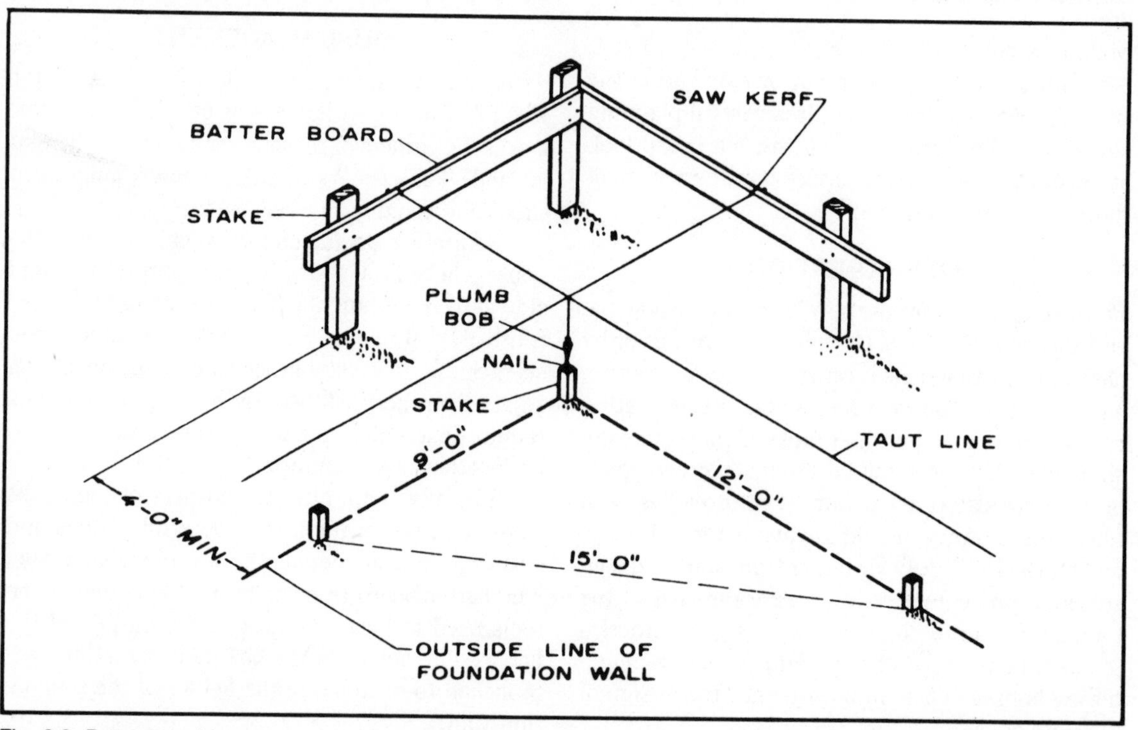

Fig. 2-2. Batter board layout.

First locate small stakes accurately at each corner of the house, and drive in their tops to indicate the outside line of the foundation walls. To ensure square corners, measure the diagonals to see if they are the same length. You can also square the corners by measuring along one side a distance in 3-foot units, such as 6, 9, and 12 feet, and along the adjoining side the same number of 4-foot units, such as 8, 12, and 16 feet. The diagonals then will measure the equal of 5-foot units, such as 10, 15, and 20 feet, when the unit is square. Thus, a 9-foot distance on one side and a 12-foot distance on the other side should result in a 15-foot diagonal measurement for a true 90-degree corner.

After you have located the corners, drive three 2-×-4-inch or larger stakes of suitable length at each location 4 feet (minimum) beyond the lines of the foundation. Then nail 1-×-6-inch or 1-×-8-inch boards horizontally so the tops are all level at the same grade. Next, hold twine or stout string (carpenter's chalk line) across the top of opposite boards at two corners, adjusting it so that it is exactly over the nails in the corner stakes at either end. A plumb bob is handy for setting the lines.

Cut saw kerfs at the outside edge where the lines touch the boards so that they can be replaced if broken or disturbed. After you have located similar cuts in all eight batter boards, the lines of the house will be established. Check the diagonals again to make sure the corners are square.

You can divide an L-shaped house plan into rectangles. Treating each rectangle separately or as an extension of one or more sides.

FOUNDATION WALL HEIGHT

It is common practice to establish the depth of the excavation, and consequently the height of the foundation, on ungraded or graded sites by using the highest elevation of the excavation's perimeter as the control point (Fig. 2-3). This method ensures good drainage if sufficient foundation height is allowed for the sloping of the final grade (Fig. 2-4). Foundation walls at least 7 1/3 feet high are desirable for full basements, but 8-foot walls are commonly used.

Foundation walls should extend above the finished grade around the outside of the house so that the wood finish and framing members are ade-

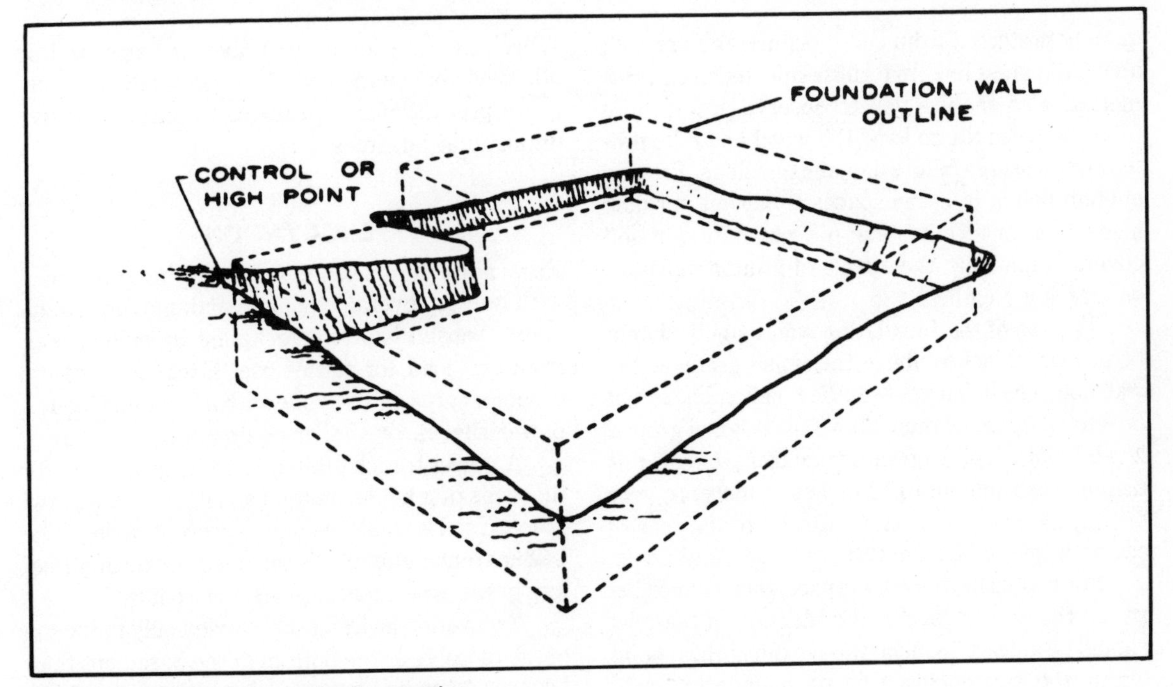

Fig. 2-3. Establishing the depth of excavation.

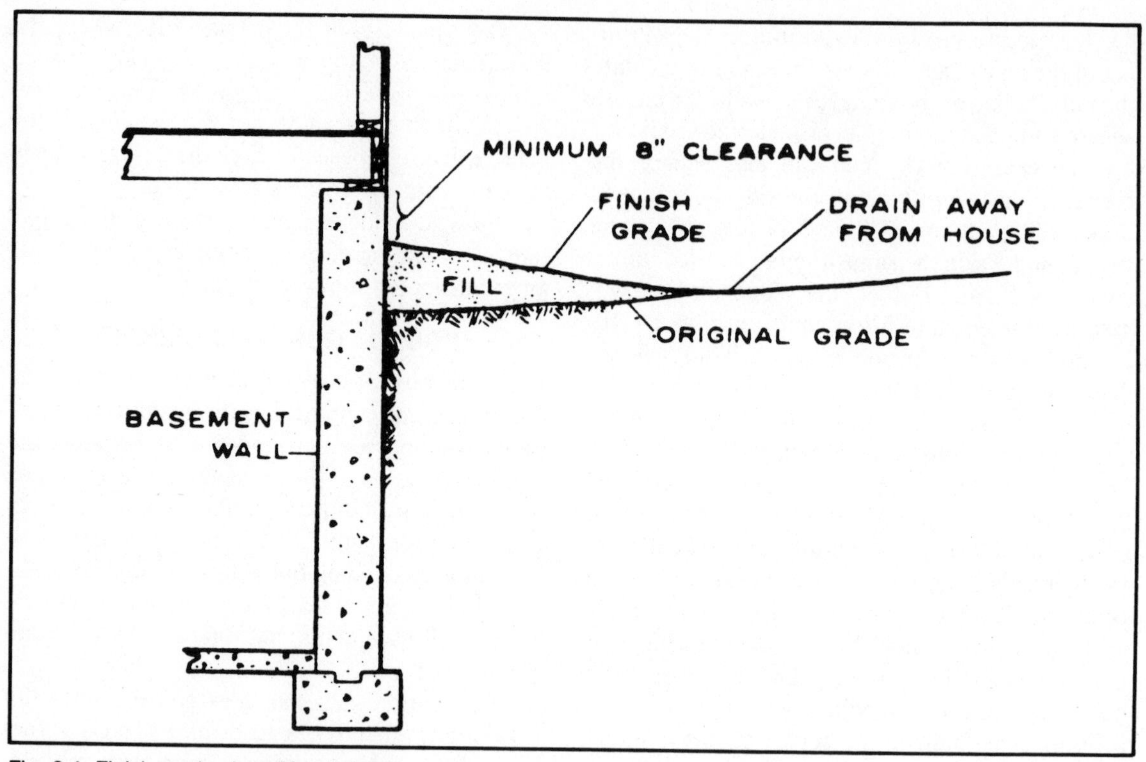

Fig. 2-4. Finish grade sloped for drainage.

quately protected from soil moisture and are well above the grass line. In termite-infested areas, this method also enables you to observe any termite tubes between the soil and the wood and take protective measures before damage develops. Provide enough height in crawl spaces to permit periodic inspection for termites and for installation of soil covers to minimize the effects of ground moisture on framing members.

The top of the foundation wall usually should be at least 8 inches above the finish grade at the wall line. The finish grade at the building line might be 4 to 12 inches or more above the original ground level. In lots sloping upward from front to rear, this distance can amount to 12 inches or more. In very steeply sloped lots, a retaining wall to the rear of the wall line is often necessary.

For houses having crawl space, the distance between the ground level and underside of the joist should be at least 18 inches above the highest point within the area enclosed by the foundation wall. Where the interior ground level is excavated or otherwise below the outside finish grade, use adequate precautionary measures to ensure positive drainage at all times.

EXCAVATION

Excavation for basements can be accomplished with one of several types of earth-moving equipment. Topsoil is often stockpiled by bulldozer or front-end load for future use. Excavation of the basement area can be done with a front-end loader, power shovel, or similar equipment.

A power trench often is used in excavating for the walls of a house built on a slab or with a crawl space, if soil is stable enough to prevent caving. The power trench eliminates the need for forming below grade when footings are not required.

Excavation is preferably carried only to the top of the footings or the bottom of the basement floor because some soil becomes soft upon exposure to air or water. Thus, it is advisable not to make the final excavation for footings until it is nearly time to pour the concrete, unless form boards are to be used.

Excavation must be wide enough to provide space to work when you are constructing and waterproofing the wall and laying drain tile in poor drainage areas (Fig. 2-5). The steepness of the back slope of the excavation is determined by the subsoil encountered. With clay or other stable soil, the back slope can be nearly vertical. With sand, an inclined slope is required to prevent caving.

Some contractors, in excavating for basements, only rough-stake the perimeter of the building for the removal of the earth. When the proper floor elevation has been reached, the footing layout is made and the earth removed. After the concrete is poured and set, the building wall outline is then established on the footings and marked for the formwork or concrete block wall.

This is an overview of the methods of locating and excavating for a foundation. In the coming pages are descriptions of how drawings and specifications are read, an explanation of how the engineer's level is used, and more detailed information on grading and excavation techniques.

PLANS AND DRAWINGS

A building project can be broadly divided into two major phases: the design phase and the construction phase. Once the architect has developed the

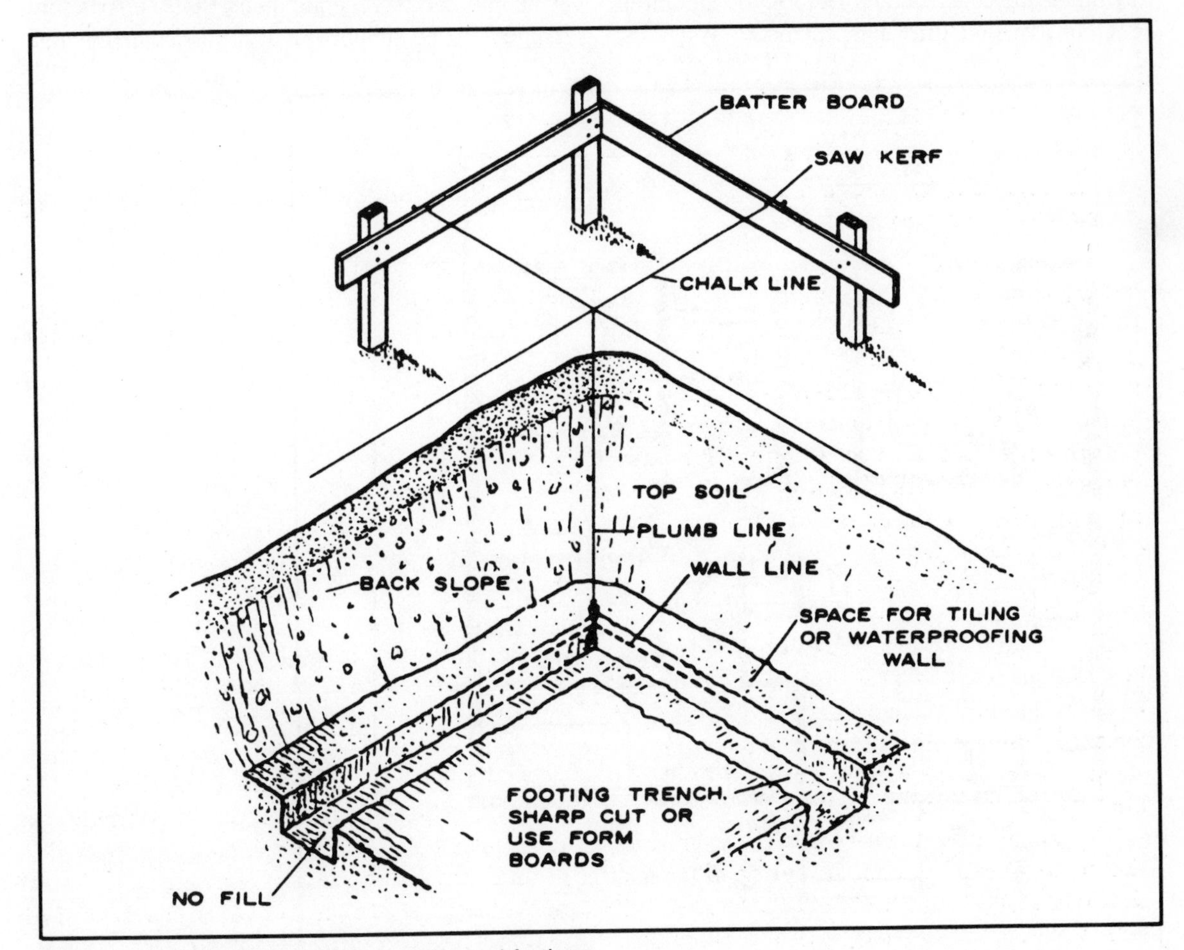

Fig. 2-5. Establishing corners for excavation and footings.

concept of the building, he puts the concepts down on a drawing called the *presentation drawing.* Presentation drawings are usually done in perspective.

Next, the architect and engineer decide upon the materials to be used in the structure and the construction methods that are to be followed. The engineer determines the loads that supporting members, and eventually the foundation, will carry and the strength qualities the members must have to bear the loads. From these determinations, the architect and engineer develop design sketches that aid the draftsmen in preparing construction drawings (Fig. 2-6).

The construction drawings, plus specifications, are the chief sources of information for those actually doing the construction. Before you can interpret construction drawings correctly, you must have some knowledge of the structure and of the terminology for common structural members.

Structure Terms

The main parts of a structure are the *load-bearing structural members,* which support and transfer the loads on the structure while remaining in equilibrium with each other. The places where members are connected to other members are called *joints.* The total of the load supported by the structural members at a particular instant is equal to the total dead load plus the total live load.

The *total dead load* is the total weight of the structure, which increases as the structure rises and remains constant once it is completed. The *total live load* is the total weight of the movable objects (such as people, furniture, equipment) that the structure happens to be supporting at a particular instant.

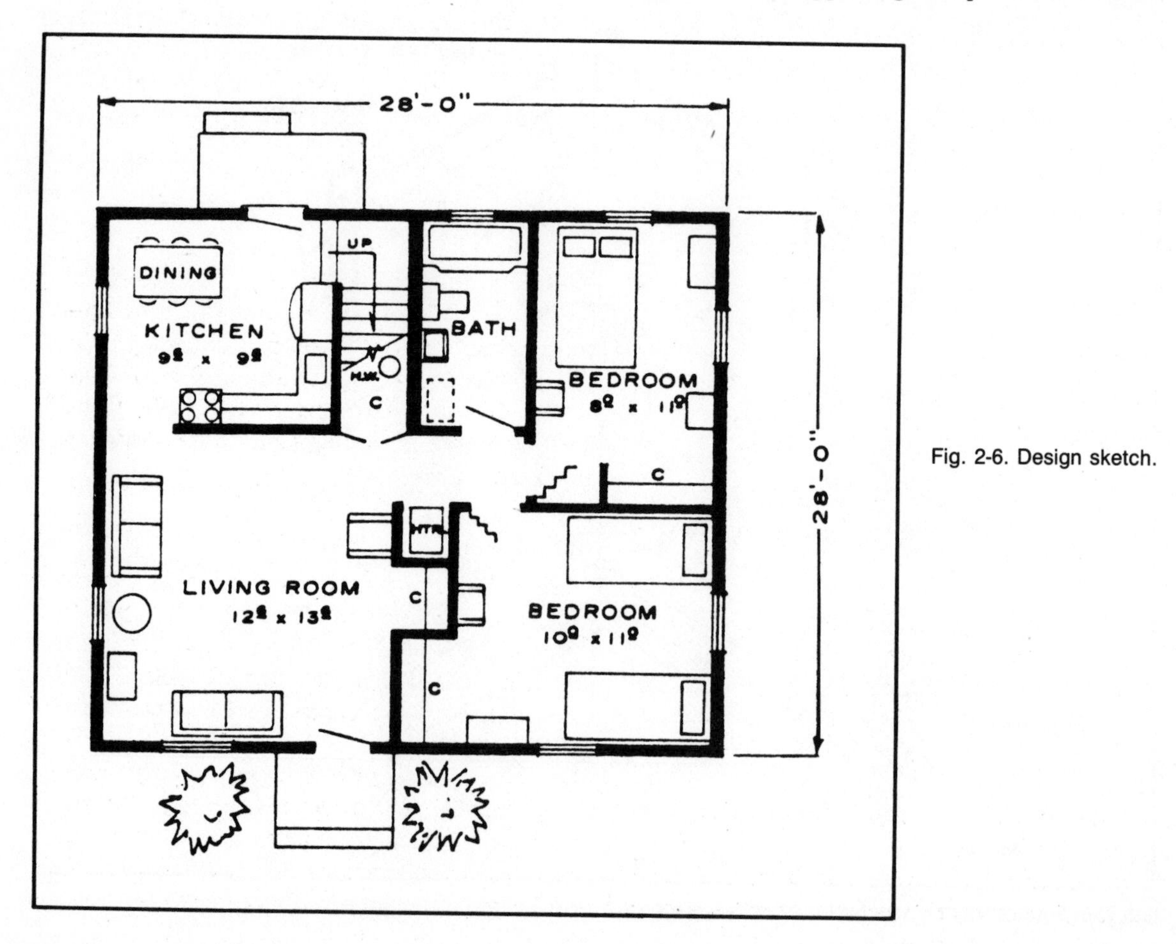

Fig. 2-6. Design sketch.

The live loads in a structure are transmitted through the various load-bearing structural members to the ultimate support of the earth as follows. Immediate or direct support for the live loads is provided by horizontal members; these in turn are supported by vertical members, which are supported by foundations and/or footings, which are, finally, supported by the earth.

The ability of the earth to support a load is called the *soil bearing capacity,* as discussed in Chapter 1. It is determined by test and measured in pounds per square foot. Soil bearing capacity varies considerably with different types of soil. Soil of given bearing capacity will bear a heavier load on a wide foundation or footing than it will on a narrow one. As you can see, the design and installation of foundations is important to the overall quality of the structure.

Structural Members

The chief vertical members in light frame construction are called *studs.* They are supported on horizontal members called *sills* or *sole plates* and are topped by horizontal members called *top plates* or *rafter plates. Corner posts* are enlarged studs located at the building corners.

In early full-frame construction, a corner post was usually a solid piece of larger timber. In most modern construction, built-up corner posts are used, consisting of various numbers of ordinary studs nailed together.

Technically, a horizontal load-bearing structural member that spans a space, and is supported at both ends is called a *beam.* Horizontal structural members that support the ends of floor beams or joists in wood frame construction are called sills, *grits,* or *girders,* depending on the type of framing being done. Horizontal members that support the wall ends of rafters are called rafter plates. Horizontal members that assume the weight of concrete or masonry walls above door and window openings are called *lintels.*

A beam of given strength, without intermediate supports below, can support a given load over only a certain maximum span. If the span is wider than this maximum, intermediate supports, such as a *column,* must be provided for the beam. Sometimes it is not feasible or possible to install intermediate supports. In this case, a *truss* can be used instead of a beam.

A beam consists of a single horizontal member. A truss, however, is a framework consisting of two horizontal members joined together by a number of vertical or inclined members. The horizontal members are called the *upper* and *lower chords.* The vertical or inclined members are called the *web members.*

The horizontal or inclined members that provide support to a roof are called *rafters.* The lengthwise (right angle to the rafters) member that supports the peak ends of the rafters on a roof is called the *ridge, ridge board, ridge piece,* or *ridge pole.* Lengthwise members other than ridges are called *purlins.* In wood-frame construction, the wall ends of the rafters are called *rafter plates.* They are in turn supported by the outside wall studs. In concrete or masonry wall construction, the wall ends of rafters can be anchored directly on the walls or in plates bolted to the walls.

CONSTRUCTION DRAWINGS

Construction drawings are drawings in which as much construction information as possible is presented graphically or by means of pictures. Most construction drawings consist of orthographic views. General drawings consist of plans and elevations drawn on a relatively small scale. Detail drawings consist of sections and details drawn on a relatively large scale.

Plans

A *plan view* is a view of an object or area as it would appear if projected onto a horizontal plane passed through or held above the object or area. The most common construction plans are plot plans (also called site plans), foundation plans, floor plans, and framing plans.

A *plot plan* shows the contours, boundaries, roads, utilities, trees, structures, and any other significant physical features pertaining to or located on the site. The locations of proposed structures are indicated by appropriate outlines or floor plans.

By locating the corners of a proposed structure at given distances from a reference line or baseline, the plot plan provides essential data for those who will lay out the building lines. By indicating the elevations of existing and proposed earth services, using contour lines, the plot plan provides essential data for the graders and excavators.

A *foundation plan* is a plan view of a structure projected on a horizontal plane passed through at the level of the top of the foundation. Figure 2-7 tells you that the main foundation of this structure will consist of a rectangular 12-inch concrete block wall 22 feet wide by 28 feet long and centered on a concrete footing 24 inches wide. In addition to the outside wall and footing, there will be two 12-inch square piers centered on 18-inch square footings and located on center 9 1/2 feet from the end wall building lines. These piers will support a ground-floor centerline girder.

A *floor plan*, or *building plan*, is shown in Fig. 2-8. Information on a floor plan includes the lengths, thicknesses, and character of the building walls at that particular floor; the widths and locations of door and window openings; the lengths and character of partitions, the number and arrangement of rooms; and the types and locations of utility installations. A *floor framing plan* is shown in Fig. 2-9.

A *wall framing plan* gives similar information about the studs, corner posts, bracing, sills, plates, and other structural members in the walls. Because it is a view on a vertical plane, a wall framing plan is not a plan in the strict sense. A *roof framing plan*

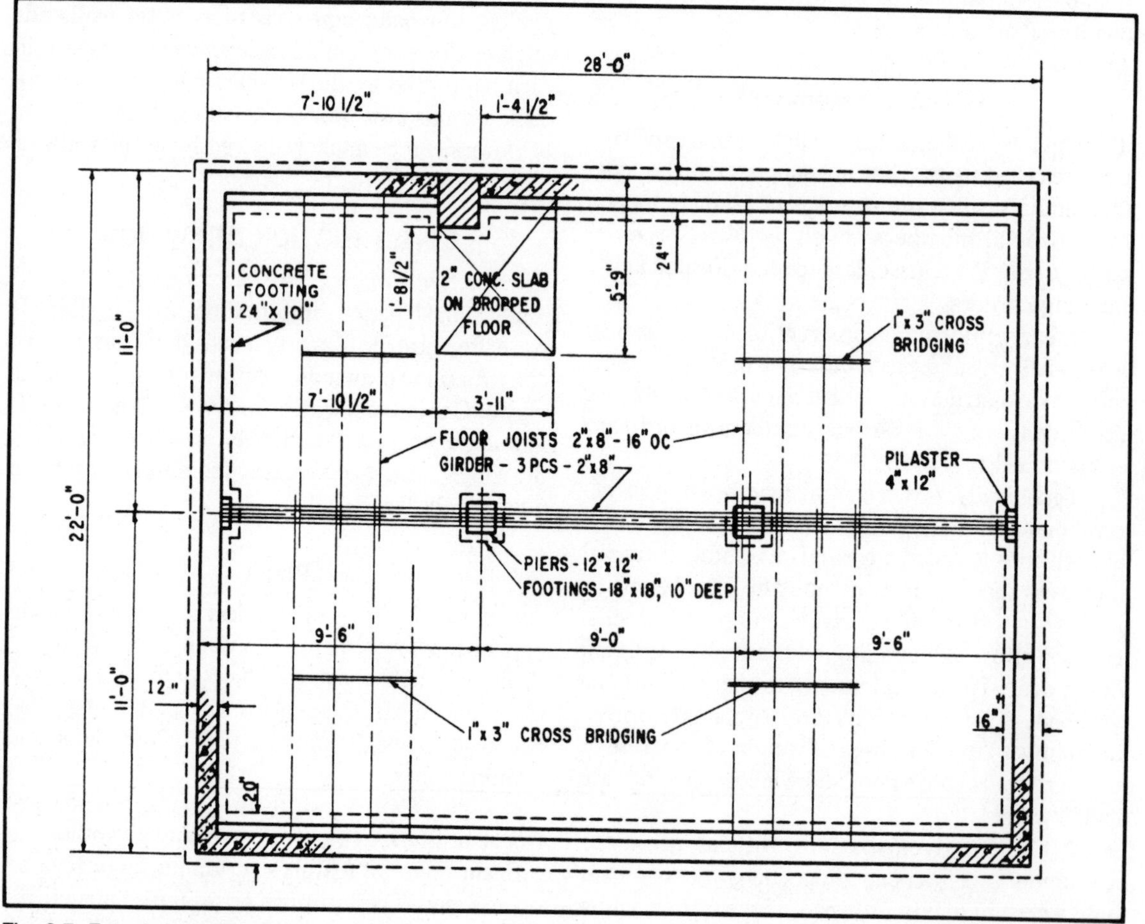

Fig. 2-7. Foundation plan.

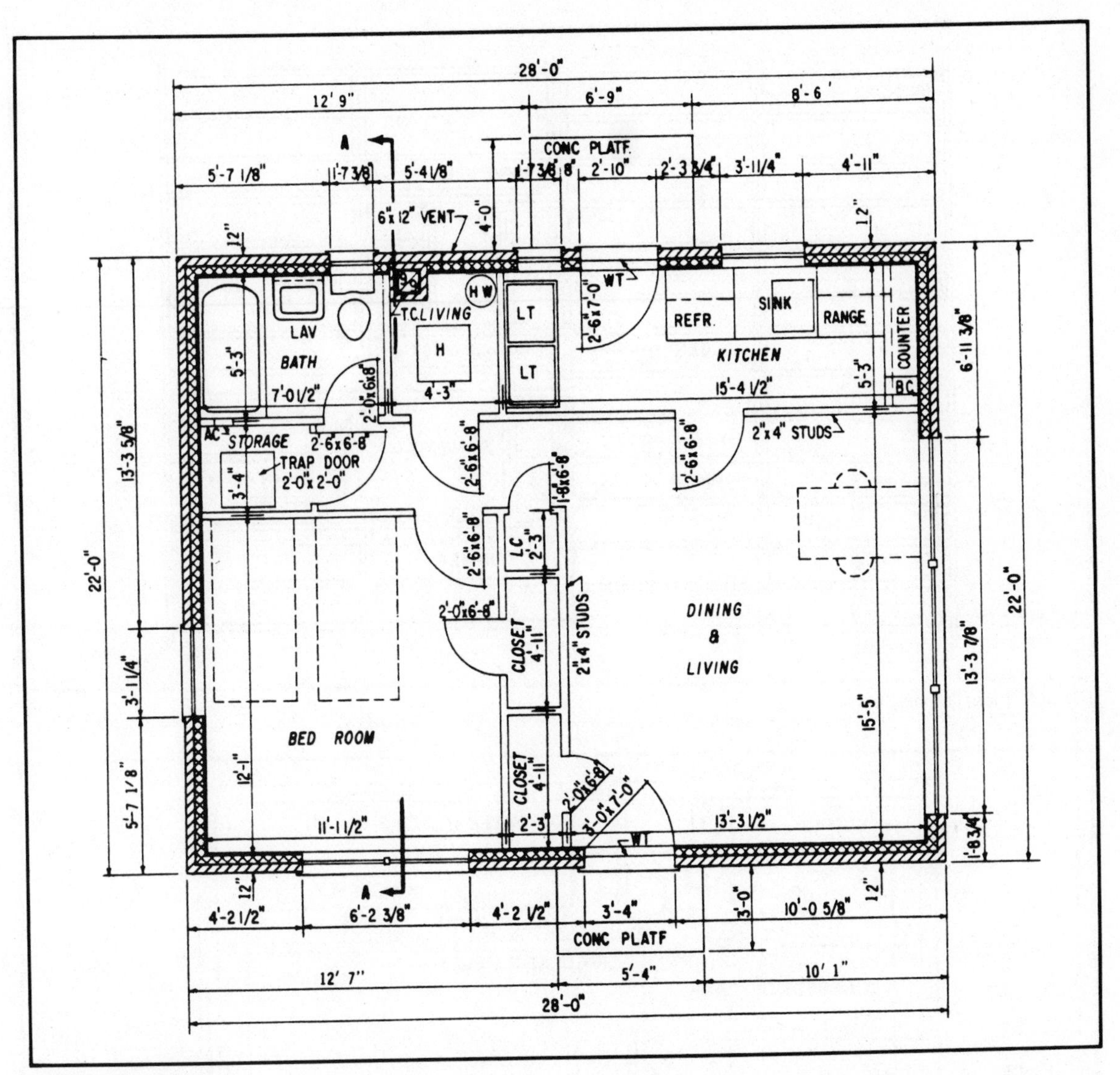

Fig. 2-8. Floor plan.

gives similar information about the rafters, ridge, purlins, and other structural members in the roof.

A *utililty plan* is a floor plan that shows the layout of a heating, electrical, plumbing, or other utility system. Figure 2-10 shows a heating plan and Fig. 2-11 offers an electrical plan for the sample home. Figure 2-12 illustrates a drain plan for a typical home and lot.

Elevations and Details

Elevations show the front, rear, and sides of a structure projected on vertical planes parallel to the planes of the sides. As you can see in Fig. 2-13, the elevations give you a number of important vertical dimensions, such as the perpendicular distance from the finish floor to the top of the rafter plate and from the finish floor to the tops of door and window finished openings.

A *section view* (Fig. 2-14) is a cross-sectional view cut by vertical planes. The most important sections are the *wall sections* (Fig. 2-15). Figure 2-16 shows the material symbols commonly used for sec-

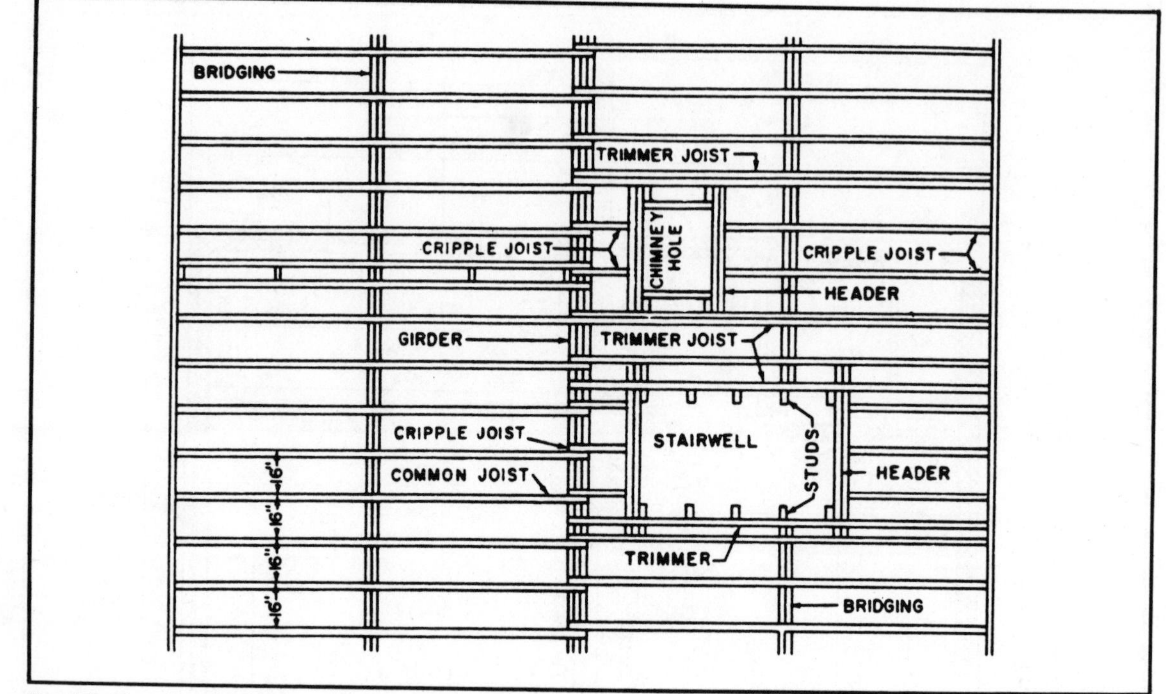

Fig. 2-9. Floor framing plan.

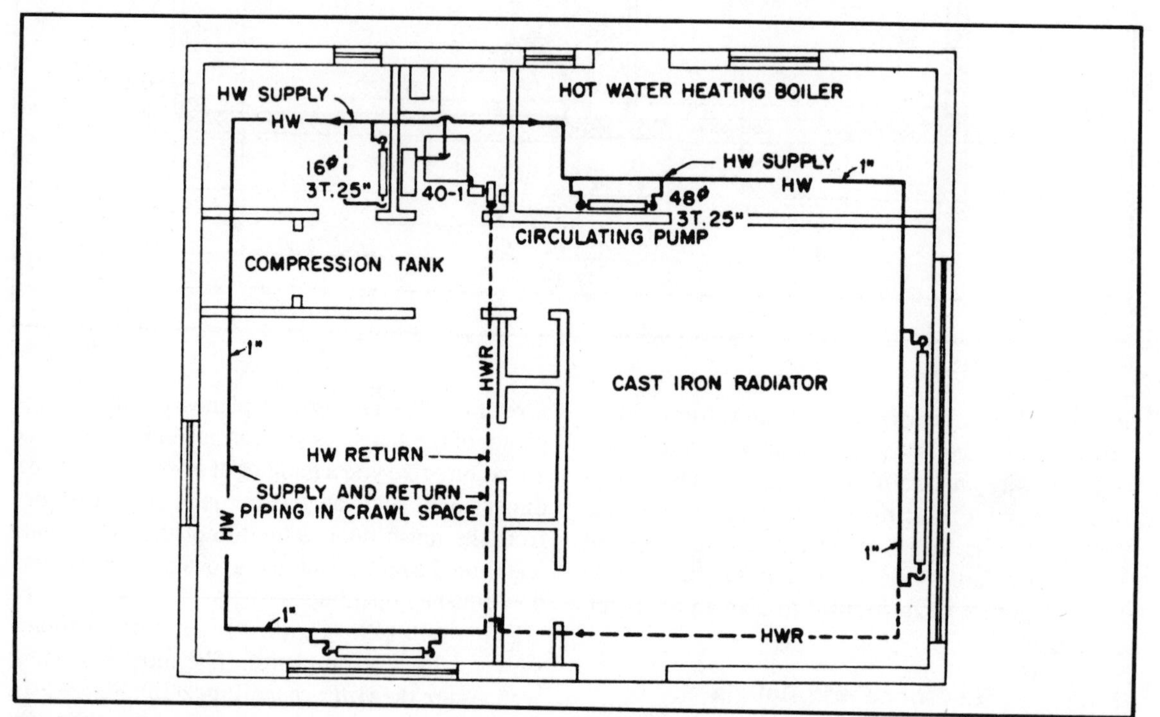

Fig. 2-10. Heating plan.

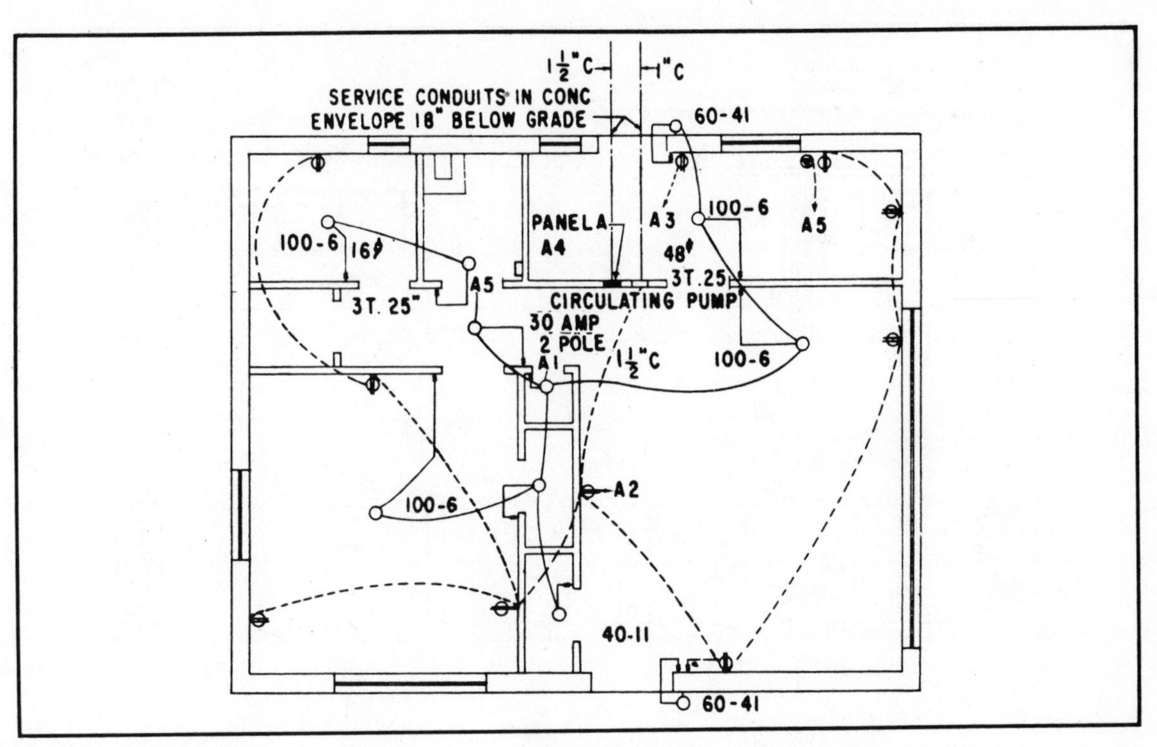

Fig. 2-11. Electrical plan.

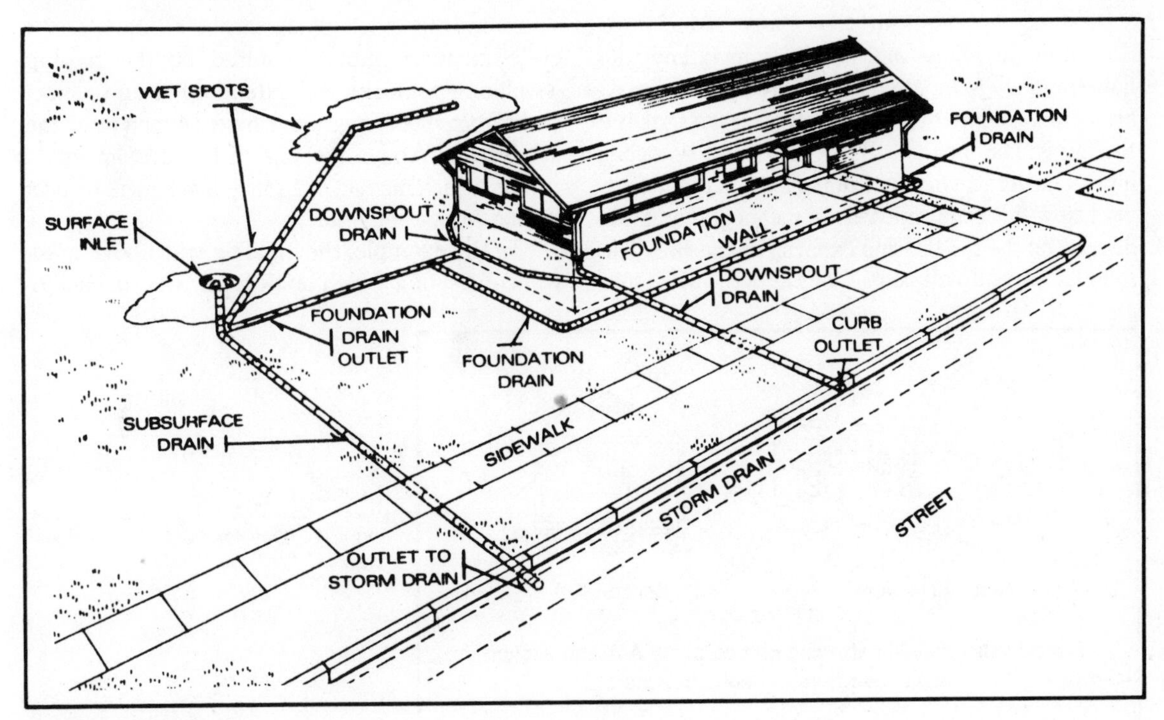

Fig. 2-12. Lot drain plan.

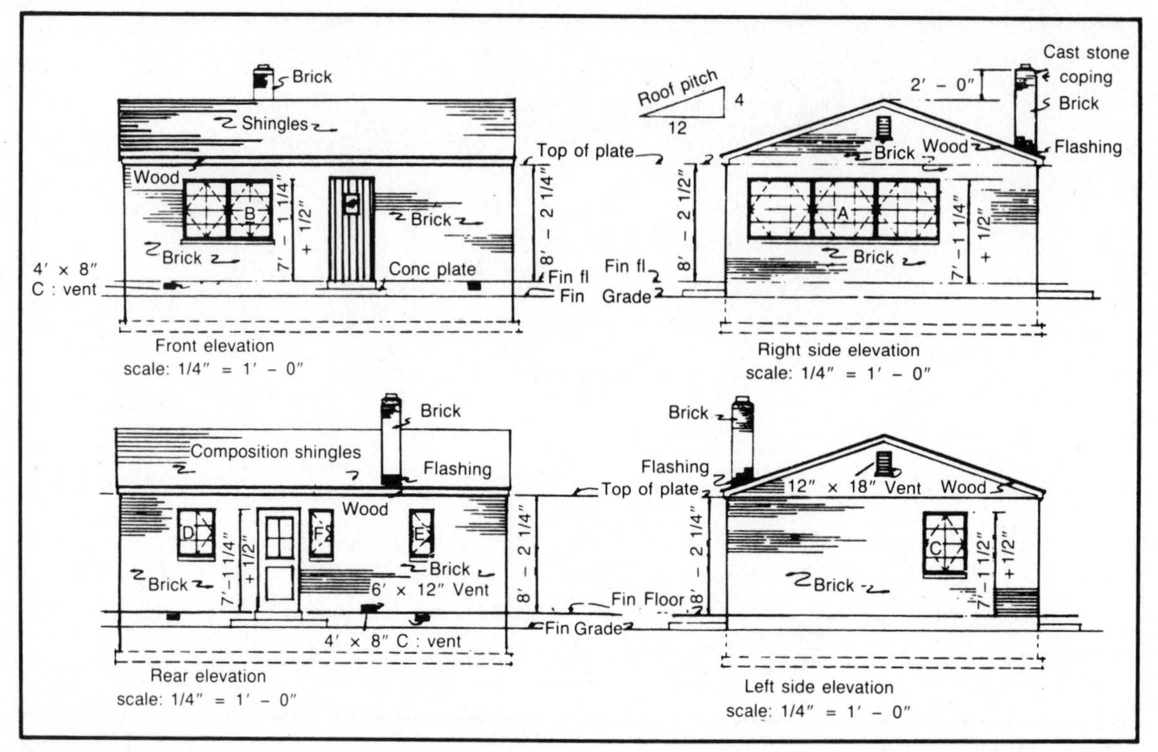

Fig. 2-13. Elevation plan.

tion drawings. The wall section shows how the house and the foundation work together. In the example on the left of Fig. 2-15, you learn that the footing will be concrete, 2 feet wide, and 10 inches high. The vertical distance of the bottom of the footing below finished grade varies, meaning that it will depend on the soil bearing capacity at the site. The foundation wall will consist of 12-inch CMU (concrete masonry units) centered on the footing. Twelve-inch blocks will extend up to an unspecified distance below grade, where a 4-inch brick facing begins. Above the line of the bottom of the facing, 8-inch instead of 12-inch blocks will be used in the foundation wall.

In the example, the building wall above grade will consist of a 4-inch brick facing tier, backed by

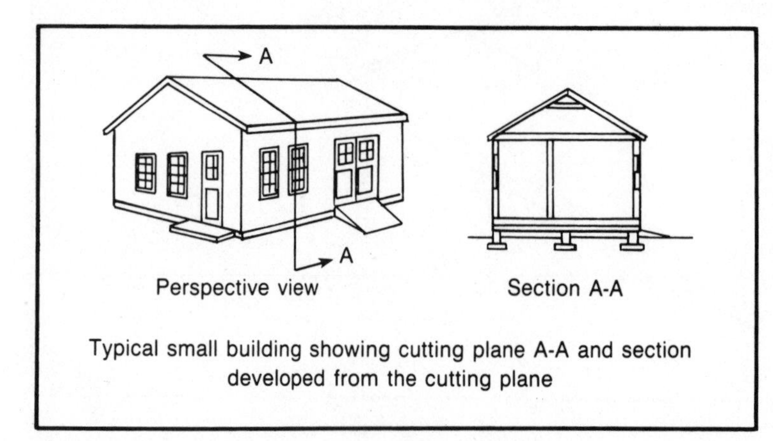

Fig. 2-14. Development of a section view.

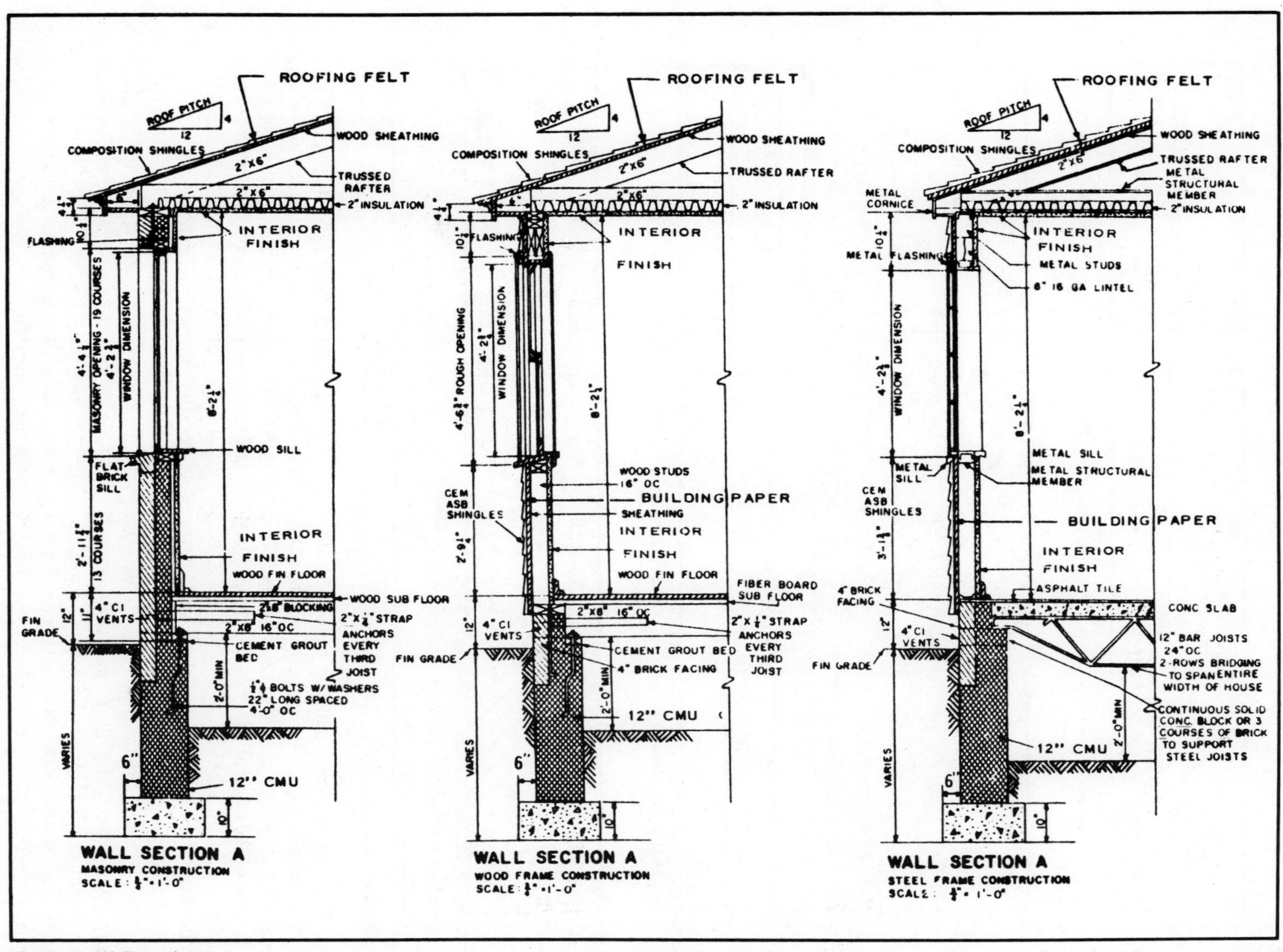

Fig. 2-15. Wall sections.

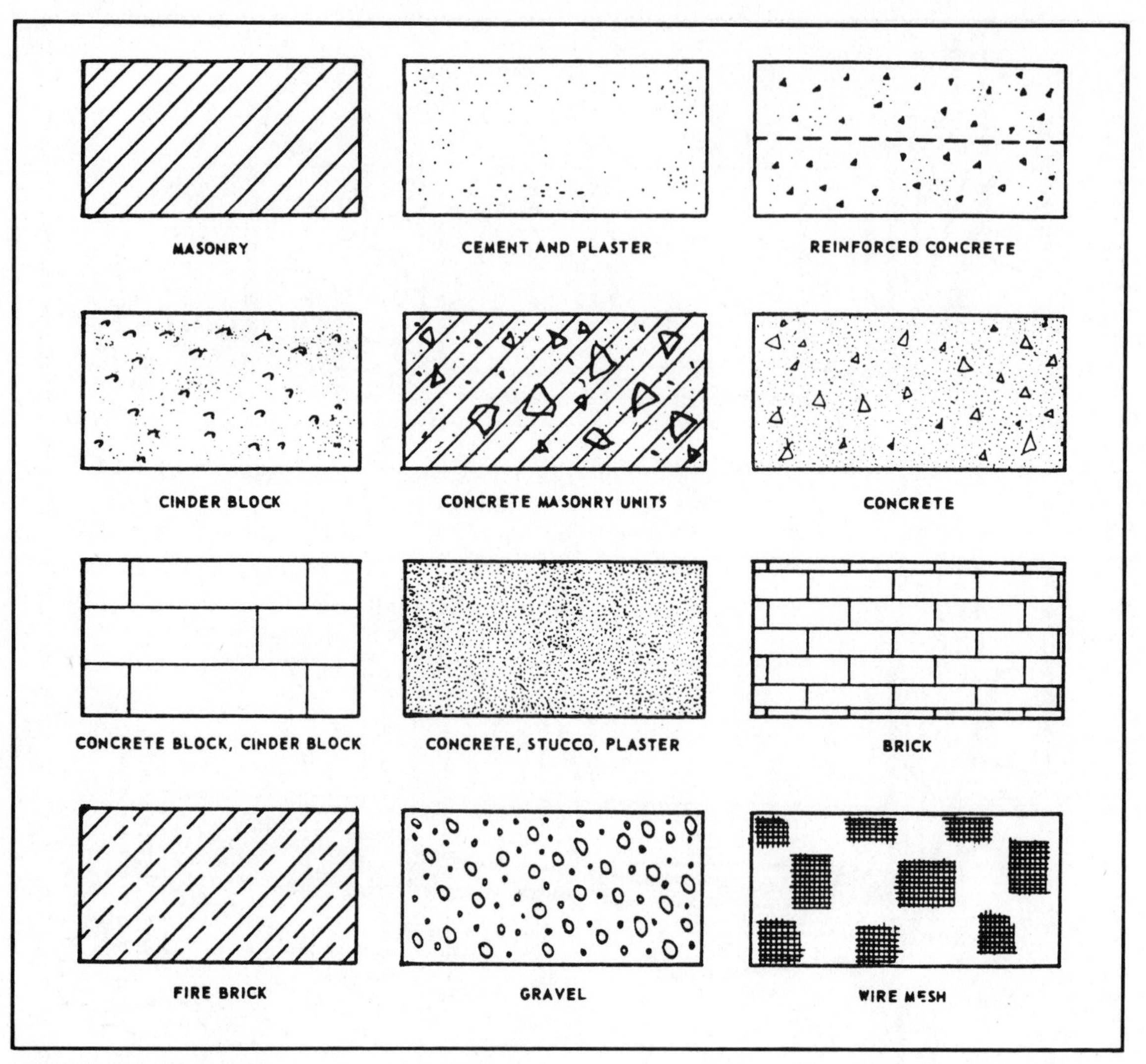

Fig. 2-16. Material symbols.

a backing tier of 4-inch cinder blocks. The floor joists, consisting of 2 × 8s placed 16 inches O.C. (on center), will be anchored on 2- × -4 sills bolted to the top of the foundation wall. Every third joist will be additionally secured by 2- × -1/4-inch strap anchor embedded in the cinder block backing tier of the building.

Detail drawings are drawings that are done on a larger scale than that of the general drawings and show features not clear on the general drawings. Figure 2-17 shows detail drawings for a typical home.

USING A SURVEYOR'S LEVEL

The surveyor's level is important in locating and laying out building lines and establishing elevations for foundations. Whether you do your own surveying or have it done, you should understand the principles and equipment involved.

The *elevation* of any object is its vertical distance above or below an established height on the earth's surface. This established height is referred to as a *reference plane,* or simply the *reference.* The most commonly used reference plane for elevations is *mean* (or average) *sea level,* which is assigned to

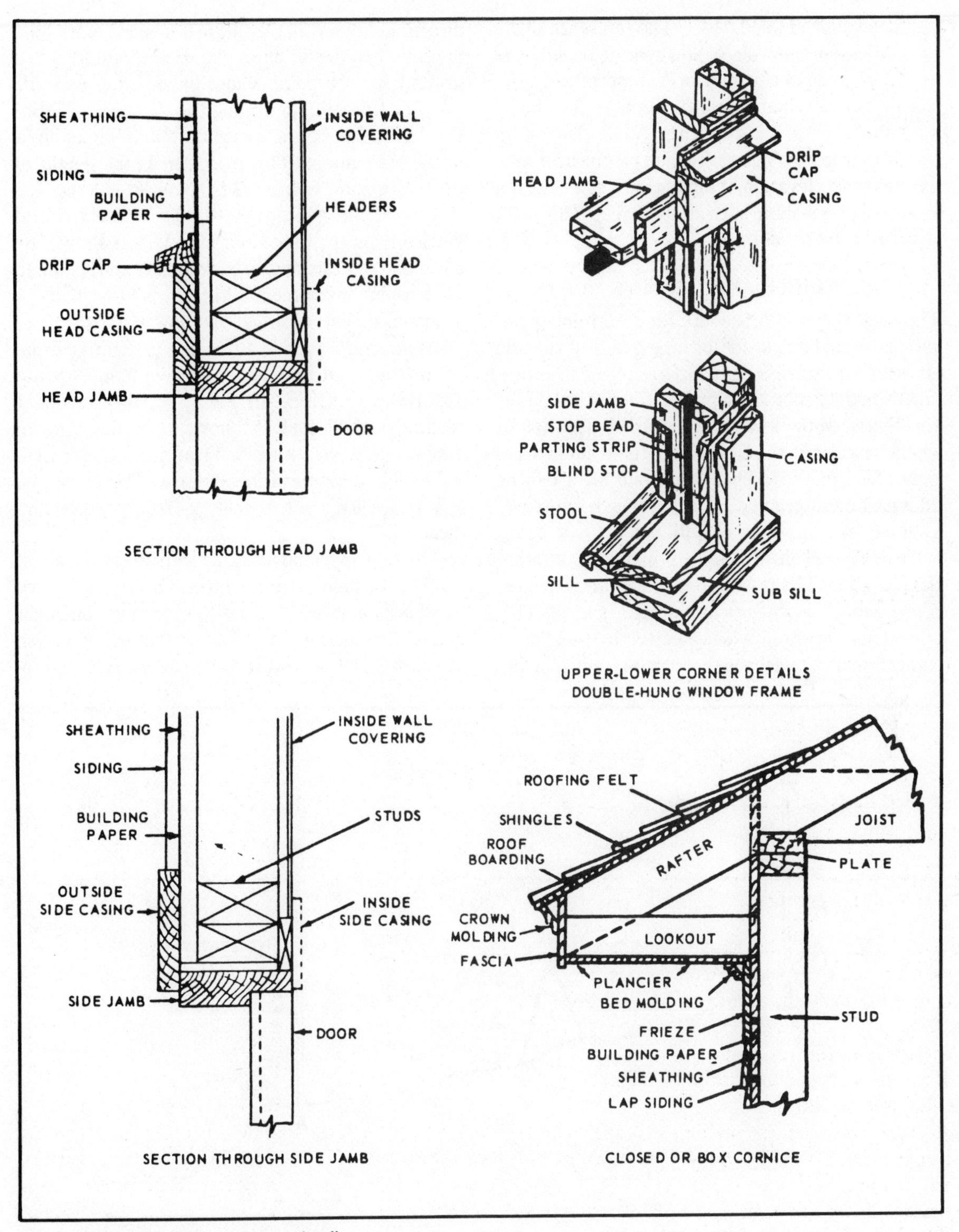

Fig. 2-17. Door, window, and eave detail.

be an elevation of 0000.0 feet. The reference plane for a construction project, however, is usually the height of some permanent or semipermanent object in the immediate vicinity, such as the rim of a manhole cover, a road, or the finish floor of an existing structure. This object can be given its relative sea-level elevation, if known, or it can be given an assumed elevation, such as 100.0 feet. This point is called a *bench mark.*

Differential Leveling

The most common procedure for determining elevations in the field, or for locating points at specific elevations, is called *differential leveling.* This procedure is nothing more than finding the vertical difference between the known or assumed elevation of a bench mark and the elevation of the point in question. Once the difference is measured, it can be added to or subtracted from the bench mark elevation to determine the elevation of the new point.

Figure 2-18 illustrates the principle of differential elevation. The instrument shown in the center represents an engineer's or surveyor's level. This optical instrument provides a perfectly level line of sight through a telescope that can be trained in any direction. Point **A** in the figure is a bench mark having a known elevation of 365.01 feet. Point **B** is a ground surface point whose elevation is desired.

The first step in finding the elevation of Point **B** is to determine the elevation of the line of sight of the instrument. This is known as the height of the instrument, written *H.I.* To determine the H.I. you would take a backsight (B.S.) on a level rod held vertically on the bench mark (B.M.), as shown, by a rodman or assistant. A backsight is always taken after a new instrument setup by sighting back to a known elevation to get the new H.I.

The *leveling rod* is a rod that is graduated upward from 0 (at its base) in feet with appropriate subdivisions of feet. In Fig. 2-18, the backsight reading is 11.56 feet. It follows, then, that the elevation of the H.I. must be 11.56 feet greater than the bench mark elevation, point **A**. Therefore, the H.I. is 365.01 + 11.56 feet, or 376.57 feet, as indicated.

Next, you would train the instrument ahead on another rod held vertically on **B**. This step is known as *taking a foresight* (F.S.). After reading a foresight of 1.42 feet on the rod, it follows that the elevation at point **B** is 1.42 feet lower than the H.I. There-

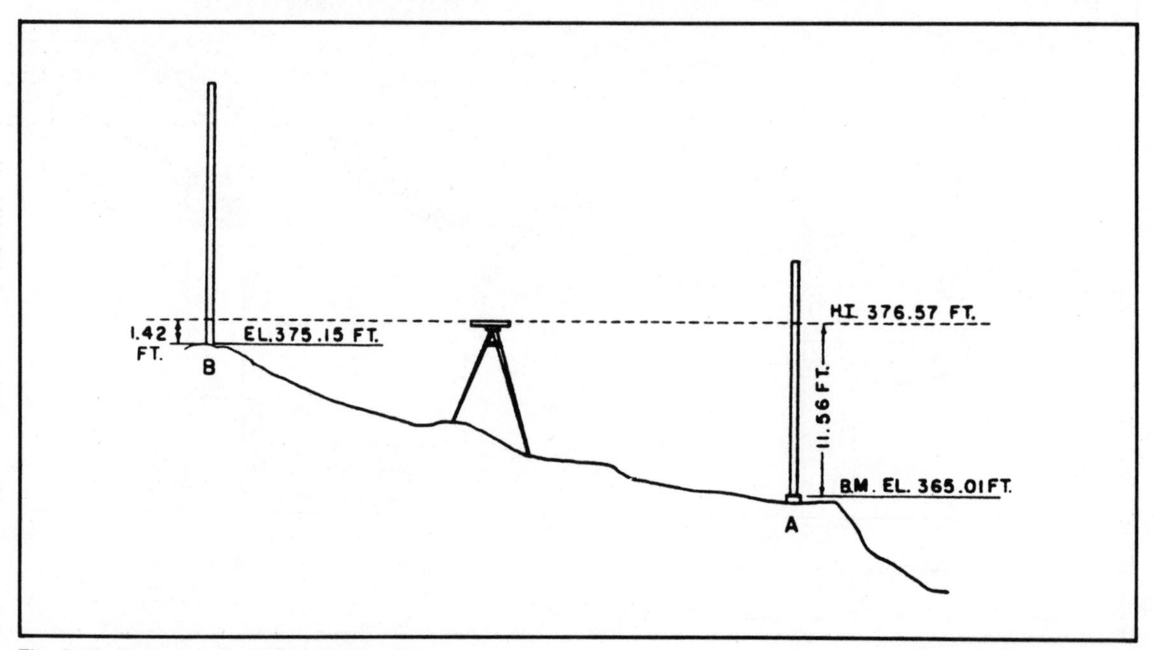

Fig. 2-18. Procedure for differential leveling.

fore, the elevation of point **B** is 376.57 minus 1.42, or 375.15 feet. (Refer to Fig. 2-19.)

Figures 2-20 and 2-21 show what the surveyor will see on a typical rod. Figure 2-22 illustrates a surveyor's notebook.

GRADING

Building a foundation typically requires that earth be moved. The grade, or contour of the earth at the site, is excavated. The term *grade* is used several different ways. In one sense it refers to the steepness of a slope. For example, a slope that rises 3 vertical feet for every 100 horizontal feet has a grade of 3 percent. A more accurate term for the steepness of a slope, however, is *gradient.*

In another sense, *grade* simply means surface. On a foundation wall section, for example, the line that indicates the ground surface level outside the building is marked grade or *grade line.*

Grade Elevations

The elevation of a surface at a particular point is a *grade elevation.* A grade elevation might refer to an existing, natural earth surface or a hub or stake used as a reference point, called *existing grade.* It also might refer to a proposed surface to be created artificially, called the *prescribed grade, plane grade,* or *finished grade.*

Grade elevations of the surface area around a structure are indicated on the plot plan. Because a natural earth surface is usually irregular in contour, existing grade elevations on such a surface are indicated by *contour lines* on the plot plan; that is, by lines that indicate points of equal elevation on the ground.

Contour lines that indicate existing grade are usually dotted lines. Existing contour lines on maps, however, are sometimes represented by solid lines. If the prescribed surface to be created artificially will be other than a horizontal plane surface, prescribed grade elevations will be indicated on the plot plan by solid contour lines.

On a level, horizontal plane surface, the elevation is the same at all points. Grade elevation of a surface of this kind cannot be indicated by contour lines because every contour line indicates an elevation different from that of every other contour line. Therefore, a prescribed level surface, to be artificially created, is indicated on the plot plan by outlining the area and inscribing inside the line the prescribed elevation, such as "First Floor Elevation 127.50."

Grade Stakes

The first earth-moving operations for a structure usually involve the artificial creation of a level area

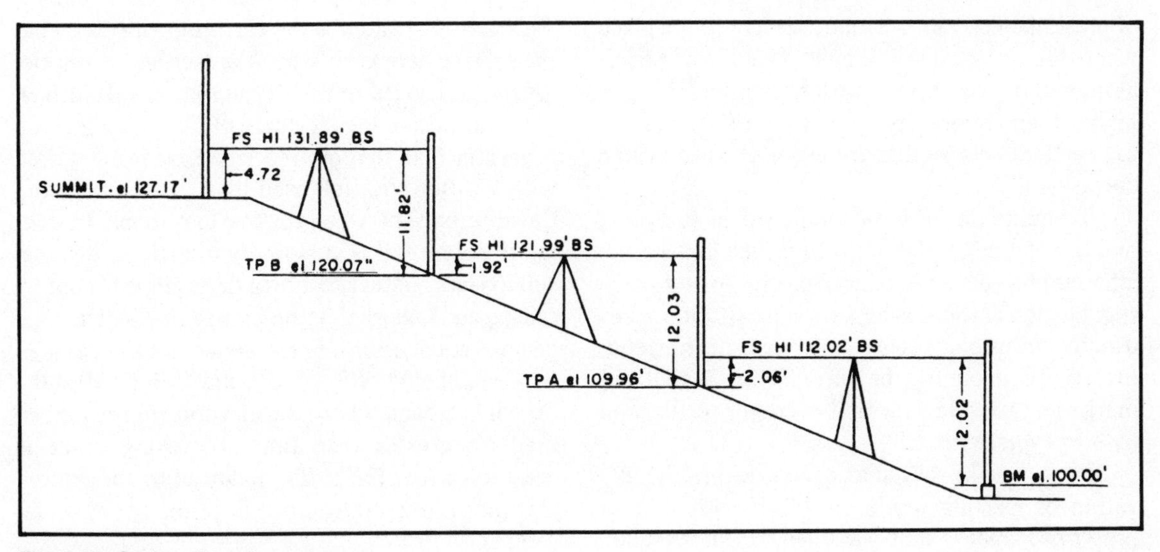

Fig. 2-19. Series of turning points.

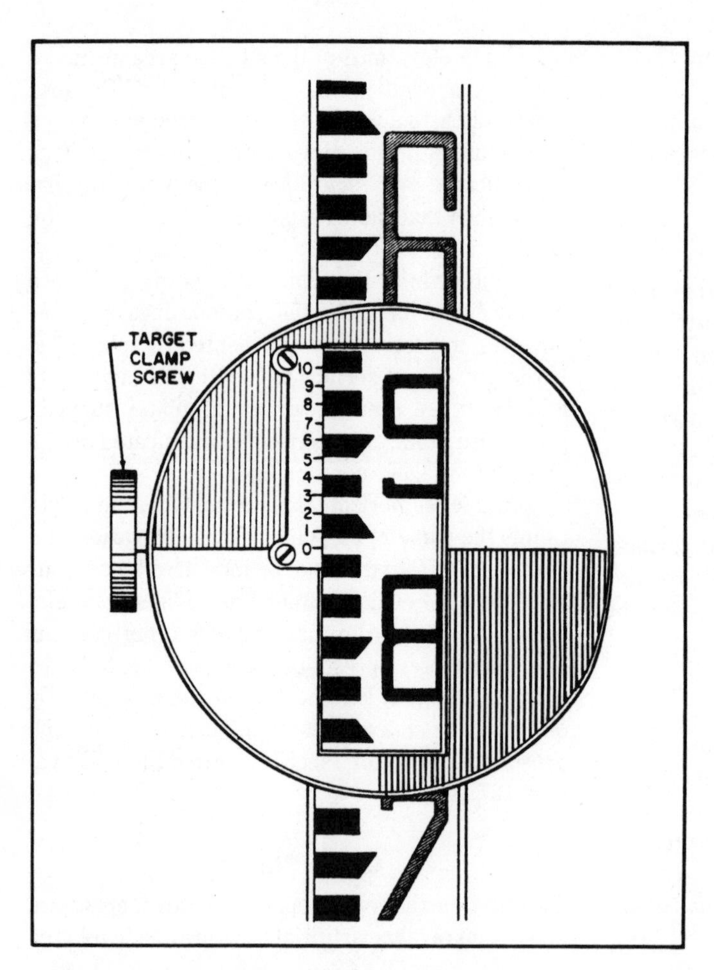

Fig. 2-20. Philadelphia rod set for target reading of less than 7000 feet.

of prescribed elevation at and adjacent to the place where the structure will be built. This grading operation involves removing earth from areas that are higher than the prescribed elevation (*cut*) and filling earth into areas that are below the prescribed elevation (*fill*).

To guide the earth-moving crew, a sufficient number of grade stakes must be driven in the area. The number depends mainly on the extent of irregularities in the existing surface. Grade stakes usually consists of about 18-inch lengths (depending on the amount to be cut) of 1 × 2s that are marked on the side with lumber crayon (called *keel* by surveyors).

First, a stake driven at a point where the elevation of existing grade coincides with that of prescribed grade is simply marked *GRD* (for grade). This point indicates to the earth-moving crew that the surface here is already at prescribed grade elevation, and no cut or fill is required. A stake driven at a point where the elevation of the existing grade is greater than that of prescribed grade is marked with a **C** (for cut), followed by a figure indicating the difference between the two elevations. In writing this figure, it is customary to indicate decimal subdivisions of feet, not by a decimal point, but by raising and underlining the figures that indicate the decimal subdivisions. For example, for a cut of 6.25 feet you should write $C\ 6^{\underline{25}}$, not C 6.25. A stake driven at a point where the elevation of prescribed grade is greater than that of existing grade is marked with an **F** (for fill), followed by the figures that indicate the difference between the two elevations.

Grade Rod and Ground Rod

Obtain the elevation of prescribed grade at each point where a grade stake will be driven from the plot plan. Once you know this elevation, all you need to know to mark the stake correctly is the elevation of existing ground at the point. You've already learned how to use the engineer's level. In setting grade stakes, however, you often will set a number of stakes from a single instrument setup. In such a case, you speed up the computational procedure by applying values called grade rod and ground rod.

Grade rod is simply the rod reading you get on a particular point. When you know the grade rod (difference between plan elevation and H.I.) and the *ground rod* (read on a rod set on the point) for a particular point, you can rapidly determine the mark for the stake by applying the following rules.

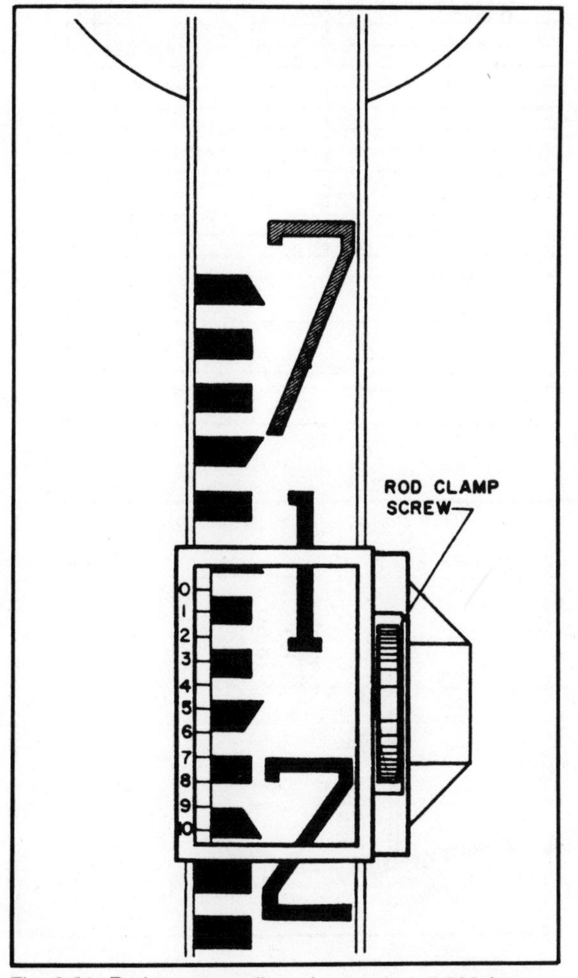

Fig. 2-21. Rod target reading of more than 7.000 feet.

The H.I. is always greater than the elevation of existing ground at the point; if it were less, you couldn't read a rod on the point from that particular setup. However, it might be greater or less than the elevation of prescribed grade. If the H.I. is less than the elevation of prescribed grade (Fig. 2-23), the difference in elevation between existing ground and prescribed grade (which is what you need to know to mark the stake) amounts to the sum of the grade rod and ground rod.

In the example in Fig. 2-23, the elevation of the prescribed grade is 131.12 feet; the H.I. is 127.62 feet. The grade rod is therefore 131.12 minus 127.62, or 3.50 feet, and the H.I. is less than prescribed grade elevation. You can see that fill must be added at this point and that the vertical depth of fill equals the sum of ground rod plus grade rod: 5.60 plus 3.50, or 9.10 feet. Therefore, you would mark the stake *G* 9^{10} feet.

If the H.I. is greater than the prescribed grade elevation, the difference in elevation between existing and prescribed grade (that is, the vertical depth of cut or fill) equals the difference between the ground rod and grade rod. Whether the stake should be marked with a **C** or an **F** depends upon which of the two is the larger.

If ground rod is larger than grade rod (Fig. 2-24), the stake takes an **F**. Here, the difference between existing ground level and prescribed grade elevation equals the difference between ground rod and grade rod: 5.60 minus 4.42, or 1.18 feet. Ground rod is larger than grade rod, and you can see that fill is required to bring the ground line up to grade. Therefore, you would mark this stake *F* 1^{18}.

If grade rod is larger than ground rod (Fig. 2-25), the stake takes a **C**. Here again, the difference between existing ground elevation and prescribed grade elevation equals the difference between grade rod and ground rod: 7.90 minus 5.60, or 2.30 feet. Grade rod is larger than ground rod, and you can see that cut is required to bring the

LEVELS FOR SUMMIT ELEVATION

Sta	(+) B.S.	H.I.	(−) F.S.	Elev.
B.M.1	12.02	112.02		100.00
T.P.1	12.03	121.99	2.06	109.96
T.P.2	11.82	131.89	1.92	120.07
Summit			4.72	127.17
	35.87		8.70	
	8.70			127.17
	27.17			100.00
				27.17
	Chk			

RETURN LEVEL RUN (FOR CHECK)

Sta	B.S.	H.I.	F.S.	Elev.
Summit	4.53	131.70		127.17
T.P.1			11.65	120.05
	1.88	121.93		
T.P.2			11.98	109.95
	2.10	112.05		
			12.03	100.02

10 JAN. 19—

Clear, cold	⊼ Johnson, BUI
Dumpy level #1	ϕ Jones, BUCN
Phila rod #2	
manhole rim	A & B streets

BM #1	100.02	CHECK ELEVATION
BM #1	100.00	TRUE "
	+0.02	DIFFERENCE

Fig. 2-22. Field notes for differential leveling.

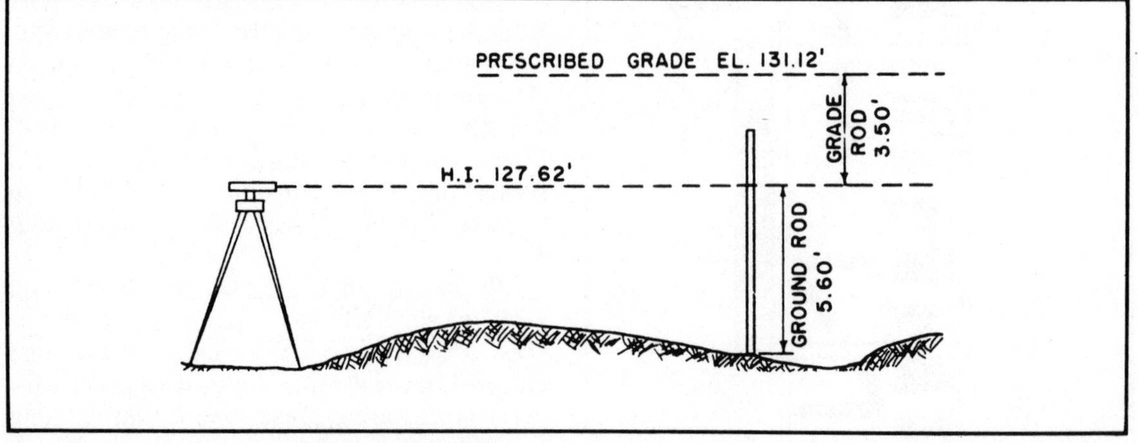

Fig. 2-23. H.I. less than prescribed grade elevation.

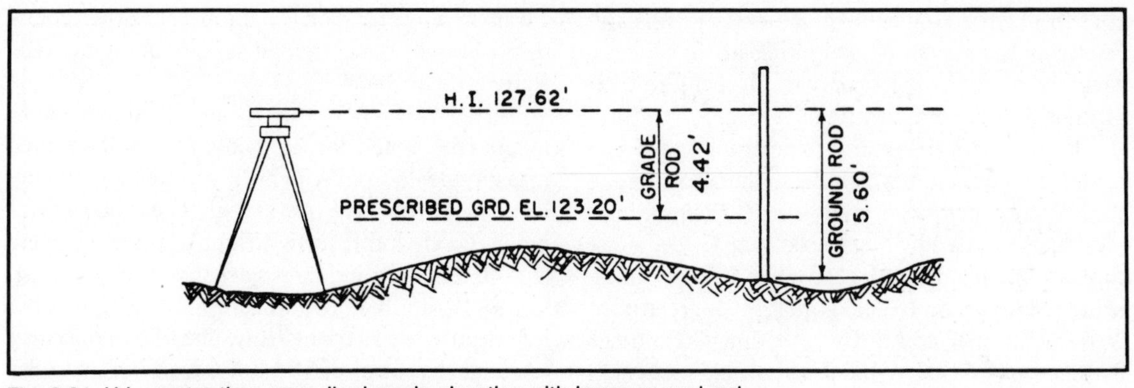

Fig. 2-24. H.I. greater than prescribed grade elevation with larger ground rod.

ground line down to grade. Therefore you would mark this *C* 2^{30}.

EXCAVATING

Grading means removing earth to create a surface of desired grade elevation at and adjacent to the place where a structure will be erected. After grading has been accomplished, further earth moving is usually required. If the structure is to have a below-grade basement, for example, earth lying within the building lines must be removed down to the prescribed finished basement floor elevation, less the thickness of the basement floor paving and subfill. After this earth is removed, further earth might need to be removed for footings under the foundation walls. This type of earth removal is generally known as *excavating*.

Laying Out Building Lines

Before foundation and footing excavation for a building can begin, the building lines must be laid out to determine the boundaries of the excavations. Points shown on the plot plan, such as building corners, are located at the site from a system of horizontal control points. This system consists of a framework of stakes, driven pipes, and other markers located at points of known horizontal location. A point in the structure, such as a building corner, is located on the ground by reference to one or more nearby horizontal control points.

An example of using horizontal control points is shown in Fig. 2-26. This figure shows two horizontal control points consisting of monuments A and B. The term *monument*, doesn't necessarily mean an elaborate stone or concrete structure. In

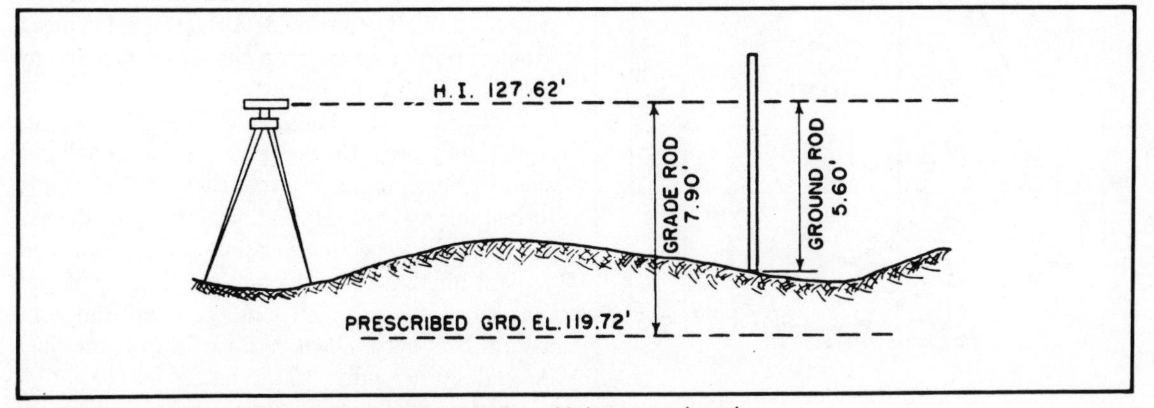

Fig. 2-25. H.I. greater than prescribed grade elevation with larger grade rod.

structural horizontal control, it simply means any relatively permanent object, either artificial (such as a driven length of pipe) or natural (such as a tree) of known horizontal location.

In Fig. 2-26, the straight line from **A** to **B** is a control baseline from which the building corners of the structure can be located. Corner **E**, for example, can be located by first measuring 15 feet along the baseline from **A** to locate point **C**; then measuring off 35 feet on **CE**, laid off at 90 degrees (that is, perpendicular) to **AB**, By extending **CE** another 20 feet, you can locate building corner **F**. Corners **G** and **H** can be similarly located along a perpendicular run from point **D**, which is itself located by measuring 55 feet along the baseline from **A**.

Checking Perpendicular

The easiest and most accurate way to locate points on a line or to turn a given angle, such as 90 degrees, from one line to another is by using a surveying instrument called a *transit.* If you don't have a transit, however, you can locate the corner points using tape measurements by applying the Pythagorean theorem.

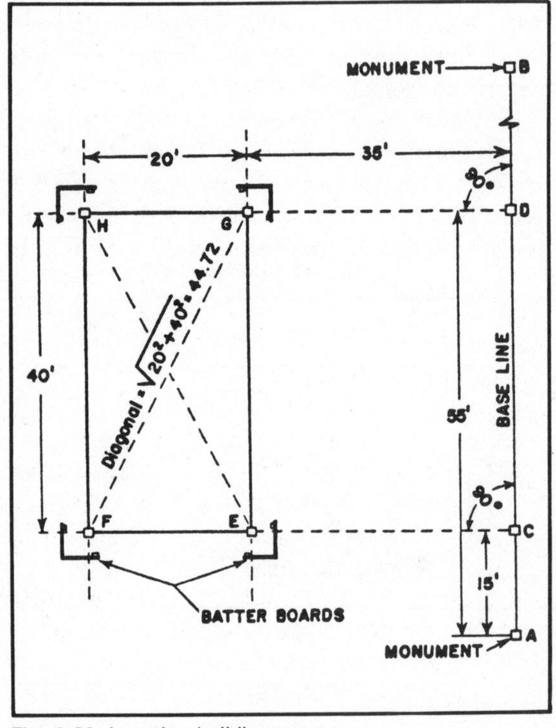

Fig. 2-26. Locating building corners.

First, stretch a cord from monument **A** to monument **B** and locate points **C** and **D** by tape measurements from **A**. Now, if you examine the figure, you will observe that straight lines connectings points **C**, **D**, and **E** form a right triangle with one side 40 feet long and the adjacent side 35 feet long.

By the Pythagorean theorem, the length of the hypotenuse of this triangle (the line **ED**) would equal the square root of $35^2 + 40^2$, which is about 53.1 feet. Because the figure **EGCD** is a rectangle, the diagonals both ways (**ED** and **CG**) are equal; therefore, the line from **C** to **G** also should measure 53.1 feet.

Checking Rectangular Layout

You always check a rectangular layout for its accuracy by checking the diagonals. The diagonals of any rectangle are equal. You check the layout by measuring the diagonals. If the layout is correct, the two diagonals will measure the same (or very nearly the same) distance. If you wish to know the value of the correct diagonal length, you can compute it by using the Pythagorean theorem.

Batter Boards

Hubs driven at the exact locations of building corners will be disturbed as soon as excavation for the foundation begins. To preserve the corner locations, and also to provide a reference for measurement down to prescribed elevations, erect batter boards (Fig. 2-27).

Nail each pair of boards to three 2- × -4 corner stakes, as shown. Drive the stakes far enough outside the building lines so that they will not be disturbed during excavation. Locate the top edges of the boards at a specific elevation, usually some convenient number of whole feet above a significant prescribed elevation, such as the top of the foundation. Nail cords, located directly over the lines through corner hubs, to the batter boards.

Figure 2-27 shows how a corner point can be located in the excavation by dropping a plumb bob from the point of intersection between two cords. You should always make sure that you have complete information as to exactly what lines and elevations are indicated by the batter boards.

EXCAVATION DIMENSIONS

With regard to the dimensions of cellar or basement excavations, the specifications usually say something like the following: "Excavations shall extend 2′0″ outside of all basement wall planes and to 9″ below finished planes of basement floor levels."

The 2-foot space is the customary allowance made for working space outside the foundation walls. It is a space that must be backfilled after the foundations have set. The 9 inches below finished planes of basement floor levels is the allowance for basement floor thickness (usually about 3 inches) plus the thickness of cinder or other fill placed under the basement floor (usually about 6 inches).

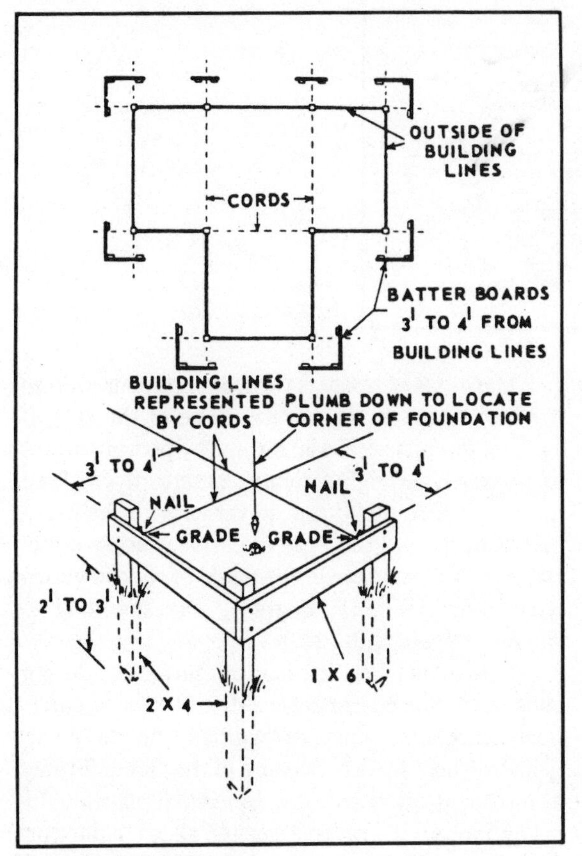

Fig. 2-27. Batter boards.

The actual depth below grade to which a basement excavation must be carried is determined by studying a wall section like the one shown in Fig. 2-28. This section shows that the depth of the basement excavation below grade would, in this case, equal 8 feet 0 inches (vertical distance between basement and first floor finished planes) minus 1 foot 6 inches (vertical distance between surface grade and first floor finished plane) plus 9 inches (3 inches of pavement floor plus 6 inches of cinder fill) for a total of 7 feet 3 inches.

The top of the footing comes level with the top of the 6-inch cinder fill. The footing is 2 inches deeper than the fill, however. Therefore, the footing excavation would be carried 2 inches lower than the basement floor elevation, or to 7 feet 5 inches below grade.

If a specified elevation were prescribed for the finished first-floor-line, then the basement floor and footing excavation must be carried down to the corresponding elevation, without reference to surface grade. Suppose, for example, that the specified elevation for the finished first-floor line is 163.50 feet. The elevation to which the basement floor elevation must be carried is then 163.50 − 8 feet + 3 inches + 6 inches, or 154.75 feet. The elevation to which the footing excavation must be carried is 154.58 feet (163.50 − 8 + 0.25 + 0.67 feet). Suppose the batter board cords were at elevation 165.00 feet. Then the vertical distance from the cords to the bottom of the basement floor excavation must be 165.00 − 154.75, or 10.25 feet. The vertical distance from the cords to the bottom of the footing excavation is 165.00 − 154.58, or 10.42 feet (10 feet 5 inches).

Excavations should never be carried below the proper depths. If a basement floor or footing excavation is carried too far, the error should not ordinarily be corrected by refilling. It is almost impossible to attain the necessary load-bearing density by compacting the refill unless special, carefully controlled procedures are used. For a basement floor excavation, you can correct a relatively small error by increasing the vertical dimen-

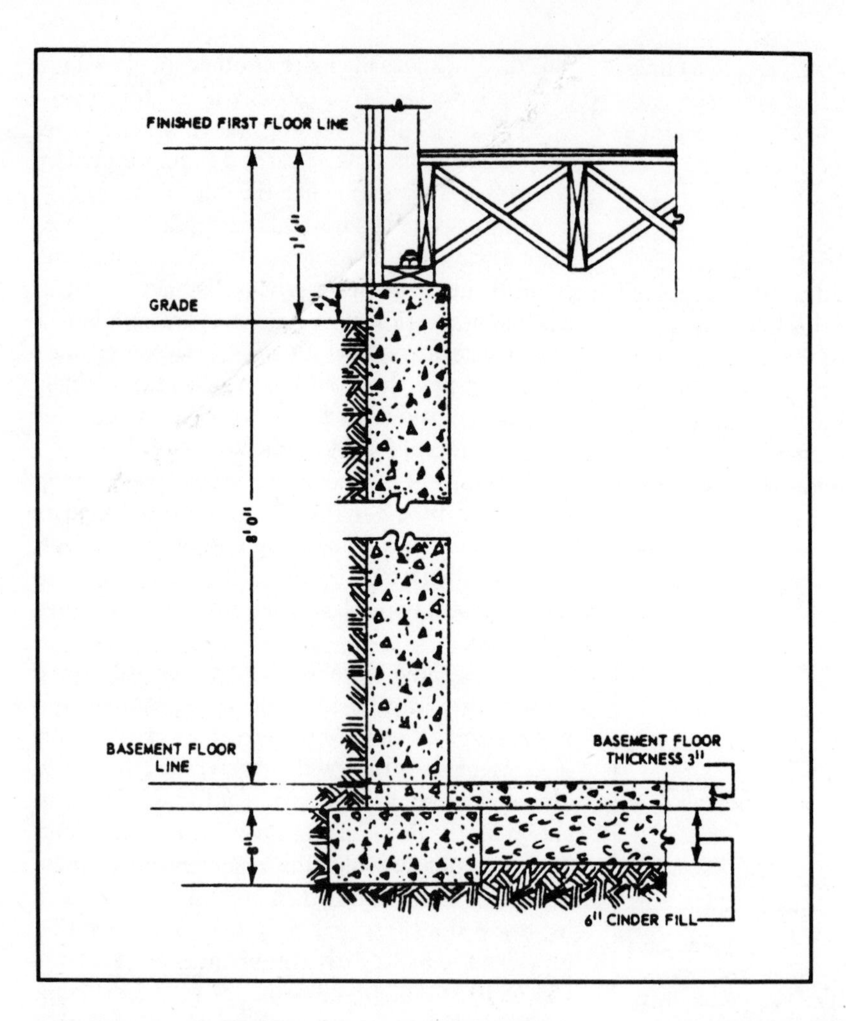

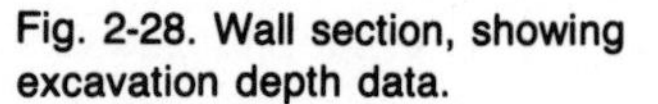
Fig. 2-28. Wall section, showing excavation depth data.

sion of the subfloor fill by the amount of the error. For a footing excavation, you can correct the error by increasing the vertical dimension of the footing by the amount of the error. Both of these methods mean additional and unnecessary expense for the extra material.

EXCAVATION SAFETY

When your work involves excavation, there are definite precautions you should observe to prevent accidents. To avoid slides or cave-ins, support the sides of excavations 4 feet or more in depth by substantial and adequate sheathing, sheet piling, bracing, or shoring, or slope the sides to the angle of repose.

The *angle of repose* is the angle, measured from the horizontal, of the natural slope of the side of a pile of granular material formed by pouring grains or particles through a funnel, practically without impact. The angle of repose varies with the moisture content and the type of earth or other material. For ordinary earth, the angle of repose varies from about 20 to 45 degrees, corresponding to slopes of about 2.8:1 to 1:1.

The sides of an excavation, however, do not consist of poured particles. Many types of earth, because of their cohesive qualities, will stand vertically without failure. Because of the nonuniformity of most soils, however, the times and places of local and intermittent cave-ins and slides cannot ordinarily be predicted. Therefore, it is conservative and safe to require the bank to be laid back to the angle of repose, or natural slope, of the material being excavated.

It is seldom practical to slope the sides of foundation and footing excavations to the angle of repose. Therefore, any such excavation to a depth of 4 feet or more must be supported as specified. The reason for the 4-foot figure is that a person needs to be buried only to chest level to suffocate in a cave-in. The pressure against the chest makes breathing impossible, and if a person's chest isn't freed within a minute or two, he will suffocate, even though his head and shoulders are out in the air. Don't take chances; use correct sheathing, piling, and shoring methods.

Sheathing

Sheathing consists of wooden planks placed edge to edge either horizontally or vertically (Fig. 2-29). Horizontal planking is used for excavations with plane faces; vertical planking, when it is necessary to follow curved faces. Sheathing is supported by longitudinal wales or rangers that are nailed to the sheathing and bear against transverse shores or braces. Sheathing must be progressively installed and braced as every 4-foot stage of depth is reached.

Sheet Piling

Drive sheet piling before the excavation begins. Wooden sheet piling (Fig. 2-30) consists of 2-, 3-, or 4-inch planks that are beveled at the lower end to facilitate penetration of the soil. Also, the lower ends are cut at an angle so the edge of a pile being driven bears against the edge of one previously driven. Care must be taken to place a pile with this angle inclined in the proper direction. The bevel on the lower end edge, too, must face toward the excavation. As a pile is driven, it tends to slant off in the direction away from the bevel. If the bevel is incorrectly turned away from the excavation, the excavation will progressively narrow, as shown in (5) of Fig. 2-30.

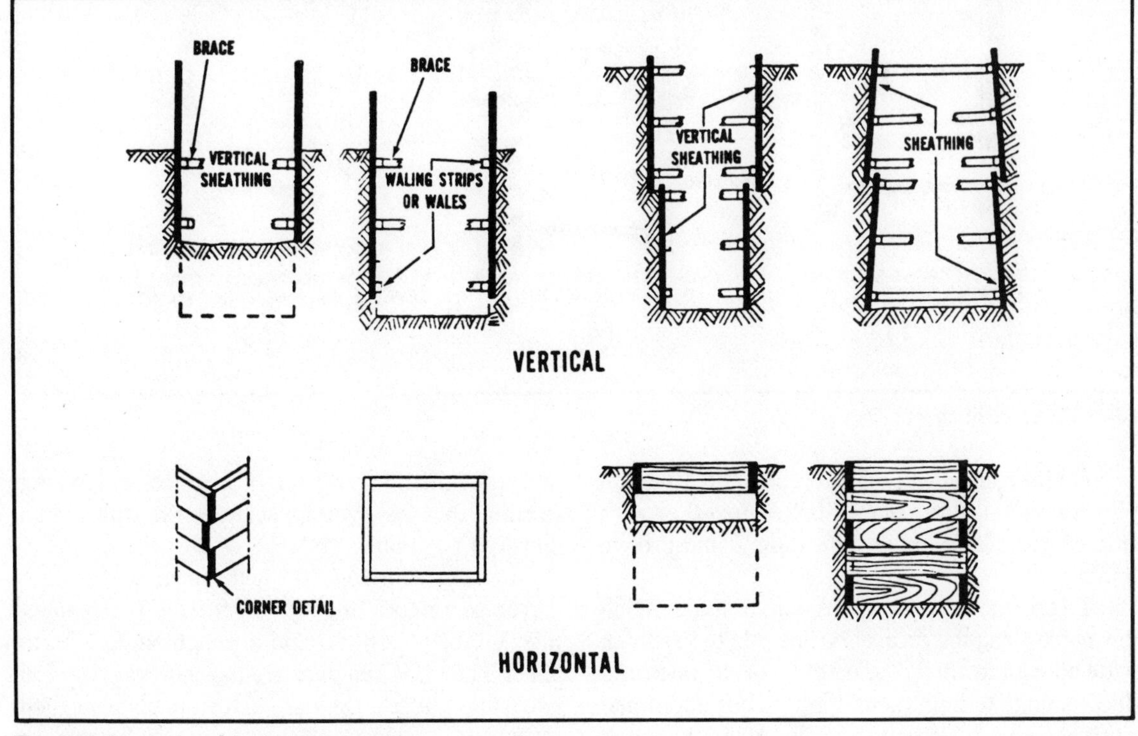

Fig. 2-29. Vertical and horizontal sheathing.

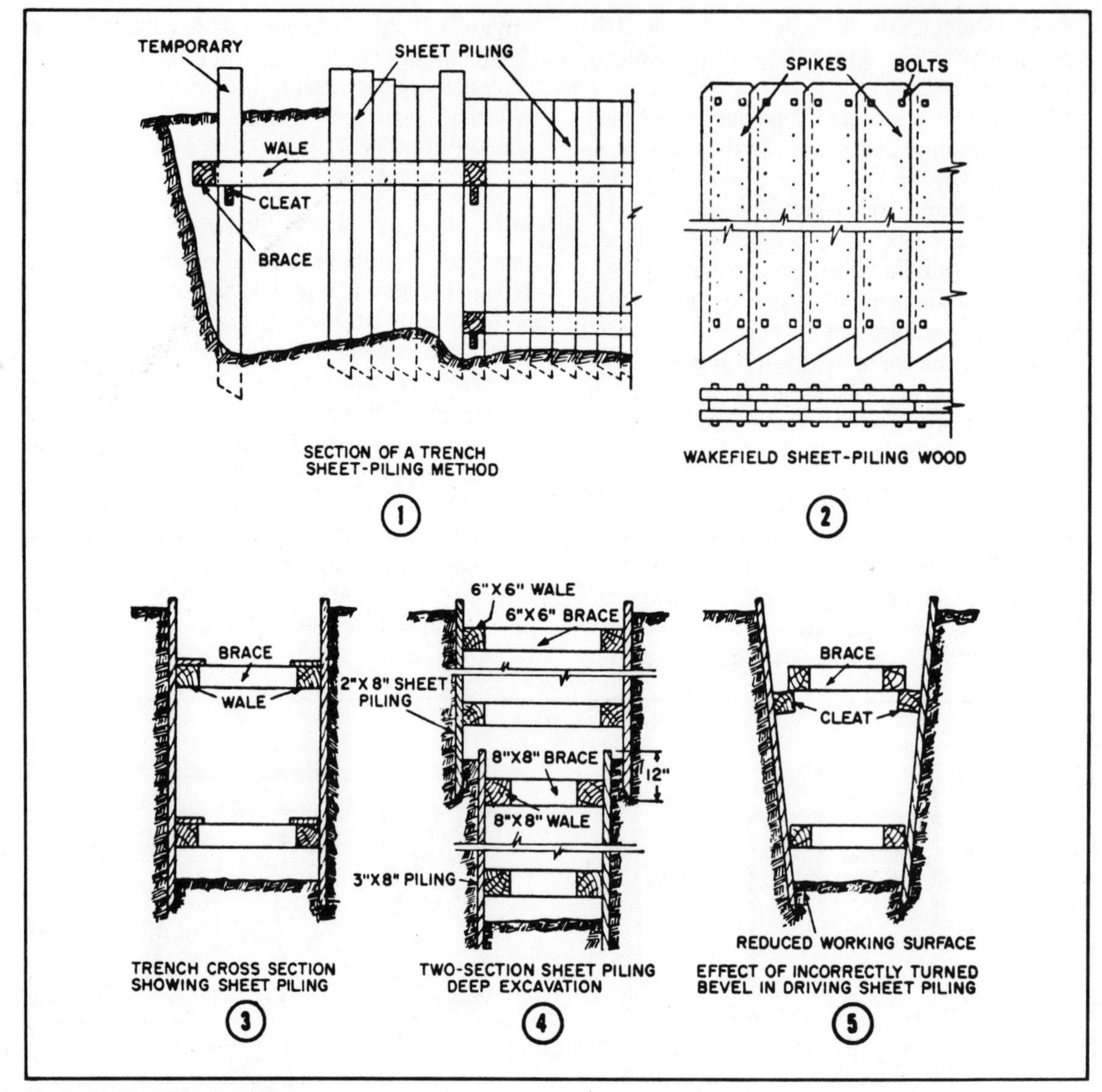

Fig. 2-30. Wooden sheet piling.

A Wakefield sheet piling is shown in (2). Each pile consists of three planks bolted together with the center plank offset for tongue-and-groove joining.

To set wooden piles for driving, first dig a shallow notch along the excavation line. Make the notch wide enough to admit the bottoms of the piles and deep enough to hold them upright. For additional upright support, lay pairs of wales along the notch and stake them in position. If the piles are too long to stand this way, you must brace the wales in an elevated position.

In favorable soil, a 2-inch sheet pile can be driven to a depth of about 16 feet; a 3-inch sheet pile, to about 24 feet; and a 4-inch sheet pile, to about 32 feet. When piles are too short to cover full excavation depth, they are driven in stages called *sections* (4).

Don't forget to install the dry wall when appropriate (Fig. 2-31).

SUPPORTING ADJACENT STRUCTURES

The removal of material near the foundations of a structure might threaten the stability of the foundations. When this situation is a possibility, you must provide temporary supports before excavation reaches the dangerous stage.

Shoring

A common method of providing support is by using up to 12-×-12 timbers, called *shores.* They should incline against the wall to be supported and extend across the excavation to a temporary footing consisting of a framework or mat of timbers laid on the ground. You can fit the upper ends of the shores into openings cut in the wall, or you can abut them to a timber bolted to the wall. You can place steel saddles in openings cut in concrete or masonry walls to support lifting or steadying shores.

It is good practice to set shores as nearly vertical as possible, in order to reduce lateral thrust against the wall. Whenever possible, locate heads of shores at floor levels to minimize the danger of pushing in the wall.

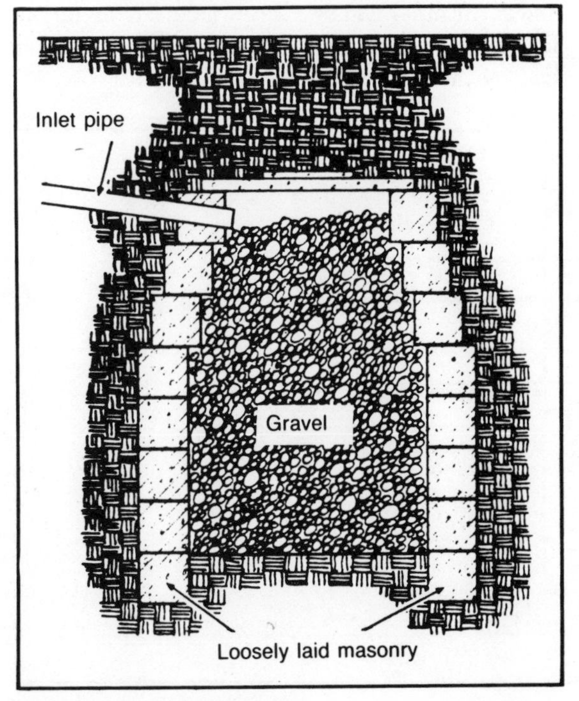

Fig. 2-31. Dry well installation.

Fig. 2-32. Standard steel screw jack.

Provision for inducing a lift or thrust in the shores is usually made by inserting jacks between the bases of the shores and the footing. Figure 2-32 shows a standard steel screw jack. One screw jack can apply a lift of as much as 100 tons. When you use a single screw jack with a shore, bore a hole in the base of the shore to admit the threaded portion of the jack. This arrangement is called a *pump.* For a larger lifting effect, attach a pair of jacks to a short timber called a *crosshead.* Pump and crosshead arrangements are illustrated in Fig. 2-33. An advantage of the crosshead arrangement is that after you have applied a lift, you can block the crosshead and remove the jacks for use elsewhere.

Hydraulic jacks provide a much stronger lift than screw jacks, but they cannot be used to support a load over a length of time. With a pair of hydraulic jacks in crosshead arrangement, however, you can set up and block as many shores as desired in a short time and with a minimum of labor.

Needling

Figure 2-34 shows a project that involves a construction procedure known as underpinning. The

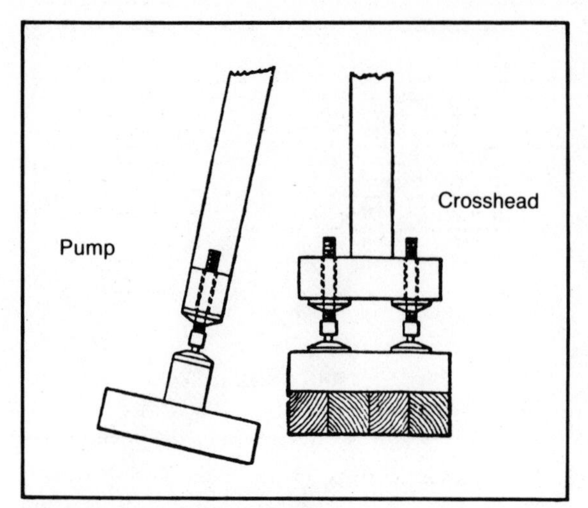

Fig. 2-33. Pump and crosshead.

feature marked **W** is a wall resting on a footing **BB**. Excavation is to be carried down to the level indicated by the horizontal dotted line, which means that the earth supporting the footing **BB** must be completely removed. The existing wall must be supported during this procedure, and subsequently, a new foundation wall extending upward from the footing as well. This process is called *underpinning.*

Support for the wall will be provided by a series of *needles,* or heavy timbers or steel beams, inserted horizontally through holes cut in the wall and wedged or jacked upward to assume the weight of the wall. The member marked **GG** in Fig. 2-34 is a needle. Carry out the needle and underpinning procedure in the following manner.

Before the actual underpinning operations are begun, complete preliminary investigations are made, especially regarding soil bearing capacity. Too much emphasis cannot be placed upon this point. The neglect of adequate study and proper interpretation can prove very costly. Soil samples must be properly taken and the results interpreted by trained engineers. The depth and character of the soil must be determined. It must be ascertained that a suitable stratum is not underlain by softer material.

After the preliminary investigations have been made, a pit **(DDDD)** is excavated down to the level of the top of the proposed new footing. At the bottom of this pit, place a layer of heavy timbers on a layer of thick planks. At grade level on the other side of the wall, lay a similar platform. Cut holes in the walls and insert the needles, each supported at the pit end by vertical timber and blocks **MN**, and at the other end by a screw jack. To obtain the lift at the pit end, drive wedges at **K**, at the other end by the jack.

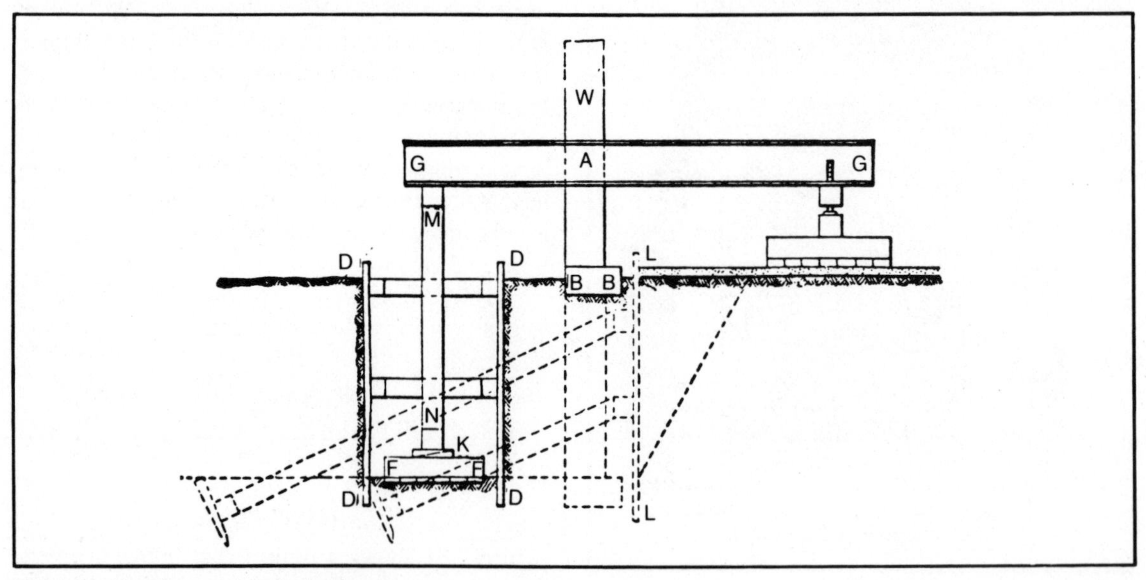

Fig. 2-34. Needling and underpinning.

Before you excavate the material under the wall, drive sheet piles at LL. As excavation proceeds downward, shore up these piles as shown to prevent a slide caused by the weight of the nearby grade-level platform.

Only that portion of the wall that is directly above a needle will receive direct support from the needle. Other portions of the wall above the needles will receive indirect support by oblique transfer of the needles' upward thrust through the material. Some parts of the wall above the needles will receive no support at all, however.

Consider, for example, the section of needled brick wall shown in Fig. 2-35. The oblique corbel outward of the upward thrust of the needles through the material is indicated by the line AAAAA. Only the portion of the wall above this line receives support from the needles. All of the wall below this line will hang when the support under the footing is removed. To some extent, cohesion will hold the hanging part to the supported part, but this is all the support it will have.

In a brick wall cohesion of this type will be slight; in a reinforced-concrete wall it will be much greater. Sometimes the hanging part of a needled wall is preserved by chaining or wiring it firmly up to the supported part.

When you have constructed the new footing and you have carried the new wall up to the old wall, you will need to join the wall sections. Joining the wall sections in a manner that will allow a transfer of the weight of the supported wall from the needles to the new wall without settlement presents a problem. In the case of a brick wall, wedging stones can be inserted in the new wall at needle level, as shown in Fig. 2-36. As the last course of brick between the old wall and the new is laid, compact the joint as tightly as possible by driving pieces of slate into the mortar, and by driving steel wedges between the wedging stones, to transfer most of the weight of the wall to the new underpinning. Then back off the jack to release the needle, and remove the needle.

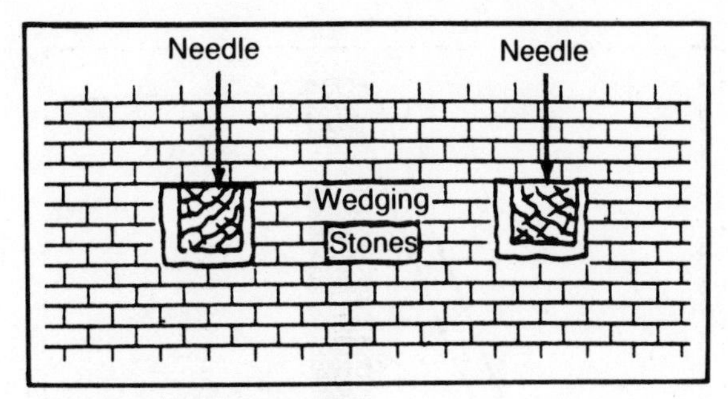

Fig. 2-36. Wedging stones.

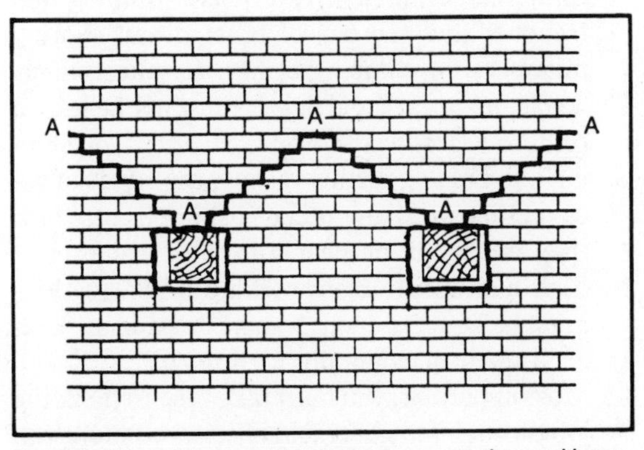

Fig. 2-35. Line of demarcation between supporting and hanging brick all in underpinning.

In the method of needling just described, the needle is used as a beam. It is supported at both ends and sustains a downward load in the middle. Figure 2-37 shows a situation in which this method of needling is not feasible, probably because erecting a supporting platform for the inner end of the needle inside the building is not feasible. In this case, the *figure-4 method* of needling is used. In this method, the needle serves as a cantilever rather than a beam.

Safe Excavation Practices

There are many safety precautions to follow as you work in and around foundation excavations.

- ☐ Burial alive in an excavation slide or cave-in is a terrible accident that occurs much too frequently. Application of the basic excavation safety rule—that any excavation 4 feet or more in depth must be protected by one of the prescribed methods—will dramatically reduce the chances of such an accident.
- ☐ Any trees, boulders, or other surface encumbrances located close enough to create a hazard must be removed before excavation begins.

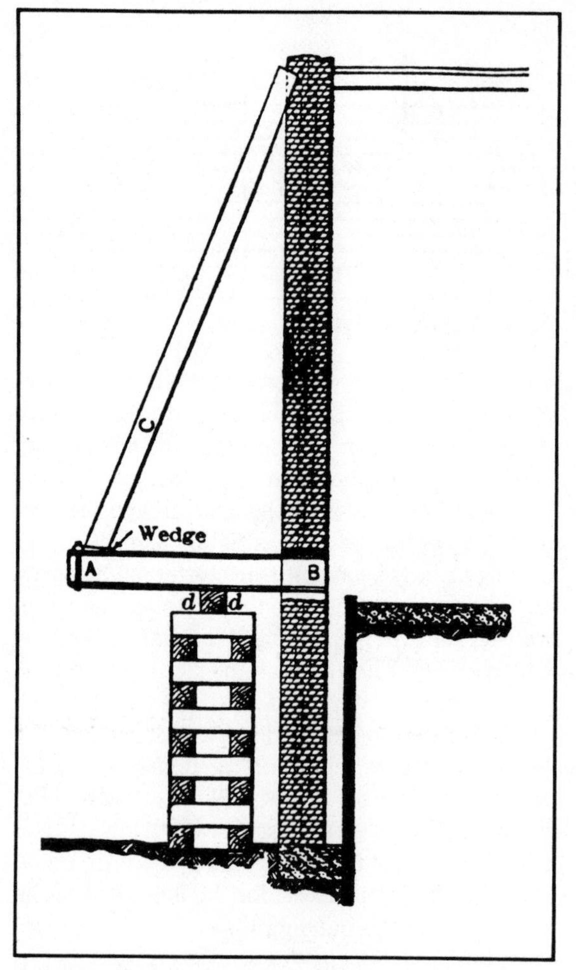

Fig. 2-37. Figure-4 method of needling.

☐ If it is necessary to bring power shovels, derricks, trucks, large quantities of supplies, or other heavy objects or materials near an excavation, the face of the excavation toward the object must be additionally shored and braced to resist the added pressure. No object or material should be placed within 2 feet of the edge of an excavation.

☐ When soil against the face of a masonry wall is excavated, do not presume that the wall will, of itself, provide sufficient support against lateral pressure from unexcavated material on the other face. The wall must be adequately shored and braced.

☐ Temporary sheet piling installed to permit construction of a retaining wall must not be removed until the wall has developed full strength. For a concrete wall, this is usually 28 days.

☐ Undercutting earth banks should be done only when it is unavoidable. It should then be done only if the overhang is kept progressively and adequately shored and braced.

☐ Excavations must be inspected after every rainstorm or similar hazard-increasing occurrence. The protection against slides or cave-ins must be increased if necessary.

☐ No sidewalk can be undermined without being progressively shored to carry a live load of 125 pounds per square foot. Excavated material must not be piled on sidewalks or walkways.

☐ All timber used for shoring, bracing, sheathing, and sheet piling must be sound, straight-grained timber of adequate strength and free from splits, shakes, large or loose knots, or other strength-impairing defects.

☐ Wooden sheet piling must not be less than 2 inches thick for a depth of up to 16 feet, not less than 3 inches for a depth of up to 24 feet, and not less than 4 inches for a depth of up to 40 feet.

☐ If pedestrains or vehicular traffic must be maintained over or near excavations, all proper safeguards, such as bridges, walkways, guardrails, barricades, warning flags, and lights, must be installed.

☐ Entrance to and exit from any excavation over 5 feet deep must be a properly constructed ramp, ladder, stairway, or hoist. Jumping into trenches and the use of bracing and shoring members for climbing can be very hazardous.

☐ Tools, materials, and debris must not be left on bridges or walkways over excavations, on shoring and bracing members, near the edges of excavations, or in any other position from which they might fall on people in the excavation.

☐ Pick and shovel men working in excavations must be kept far enough apart to avoid striking each other accidentally with tools.

☐ Most important, you must take time to think the job through for safety and efficiency before you start your foundation excavation.

Chapter 3

Concrete and Masonry

Concrete is a strong, economical building material that can be cast into practically any shape. Well-built concrete structures last indefinitely and require a minimum of maintenance.

WORKING WITH CONCRETE

Concrete (Fig. 3-1) has high compressive strength, weather resistance, and fire resistance. Its chief disadvantages are low tensile strength, high heat transmission, and water vapor permeability. These disadvantages can be offset, respectively, by proper use of steel reinforcement, insulation, and vapor barrier material.

Steel Reinforcement

Steel reinforcement increases the *tensile strength* of concrete; that is, its strength against pulling and bending forces and the effects of temperature and moisture changes. The steel is placed in the forms and the concrete is cast around it.

Reinforcing steel consists of round bars and wire mesh. Table 3-1 gives data on standard reinforcing bars, which are designated by numbers. Steel mesh, which is used to reinforce concrete slabs, is sold in rolls and mats.

Insulation

If you are using concrete where *heat transmission* (the loss or gain of heat) is an important factor, you should insulate walls, floors, or ceilings. Insulation can be provided in different ways with concrete.

You can install an insulating core between concrete faces to form a *sandwich panel.* (If the two concrete faces are reinforced with steel and connected with steel or reinforced-concrete shear pins, the panel will have built-in strength comparable to that of an I beam.) Building sidewalls can be sandwich panels made of two concrete faces with 1 1/2 to 2 inches of a semirigid insulating material between them. Such construction provides a relatively inexpensive wall that is about as resistant to heat flow as an insulating frame stud wall.

When concrete does not need to be dense to resist the flow of water, or hard to resist wear and

Table 3-1. Sizes, Area, and Weight of Standard Steel Reinforcing Bars.

Bar number	Diameter	Area	Approximate weight of 100 feet
	Inches	*Square inches*	*Pounds*
2	1/4	0.05	17
3	3/8	0.11	38
4	1/2	0.20	67
5	5/8	0.31	104
6	3/4	0.44	150
7	7/8	0.60	204
8	1	0.79	267

weathering, you can use a lightweight type of insulating material as aggregate. Such material provides insulation and lightens the weight of the concrete. It also greatly reduces the compressive strength and weather resistance of the concrete, however.

If you are building a house upon a concrete slab, you should provide insulation between the slab and the foundation wall. The insulation prevents the flow of heat from the perimeter of the slab.

When the concrete slab in a house does not have heat in the floor (hot air ducts around the perimeter, or radiant heating water or electrical heat in the floor), the subgrade beneath the slab should be insulated. Insulation will keep the inner surface temperature of the slab more nearly equal to the room temperature, making the floor more comfortable and reducing the likelihood of condensation. Insulation of the subgrade might even reduce heat loss enough to justify insulating the subgrade of heated floors. Subgrades of gravel or structural tile provide some insulation, but not nearly as much as even a 1-inch layer of some semirigid insulating material. Semirigid insulating materials include expanding polystyrene, foamed glass, and impregnated insulation board.

You also can use insulation to prevent condensation. Condensation on the north walls of concrete masonry houses, beneath rugs on floors of slab-on-grade construction, and on basement walls can cause problems with mildew, foul odors, and mustiness.

Note: condensation should not be confused with the passage of moisture through concrete. Concrete can be made watertight, but condensation can still be a problem.

Vapor Barrier

You can prevent the passage of water vapor through concrete floors by placing a layer of polyethylene, 4 to 6 mils thick, under the slab. Protect the polyethylene when you are installing it and when you are placing the concrete because a puncture will ruin the vapor barrier at that point.

You can prevent the passage of water vapor through walls consisting of sandwich-type tile-up concrete panels by using expanded polystyrene as the insulating core. It is both an insulating and a vapor-barrier material.

READY-MIXED CONCRETE

Ready-mixed concrete, or concrete ordered from and delivered by a concrete plant, is usually better than job-mixed concrete. In ordering ready-mixed concrete, you need to specify two things that determine the quality of the concrete:

- ☐ The number of bags of cement to be used per cubic yard of concrete
- ☐ The number of gallons of water to be used per bag of cement

The quality of concrete needed depends on the specific job. Use Table 3-2 as a guide in ordering the concrete.

Ready-mixed concrete is sold by the cubic yard. Your dealer can help you estimate the amount needed. Order 5 to 10 percent extra to allow for waste or slight miscalculation in the amount needed. For work that will be exposed to freezing and thawing, order air-entrained concrete.

Be ready to place the concrete when it is delivered. For best results, place the concrete in one continuous operation. Delays can reduce the quality of the job.

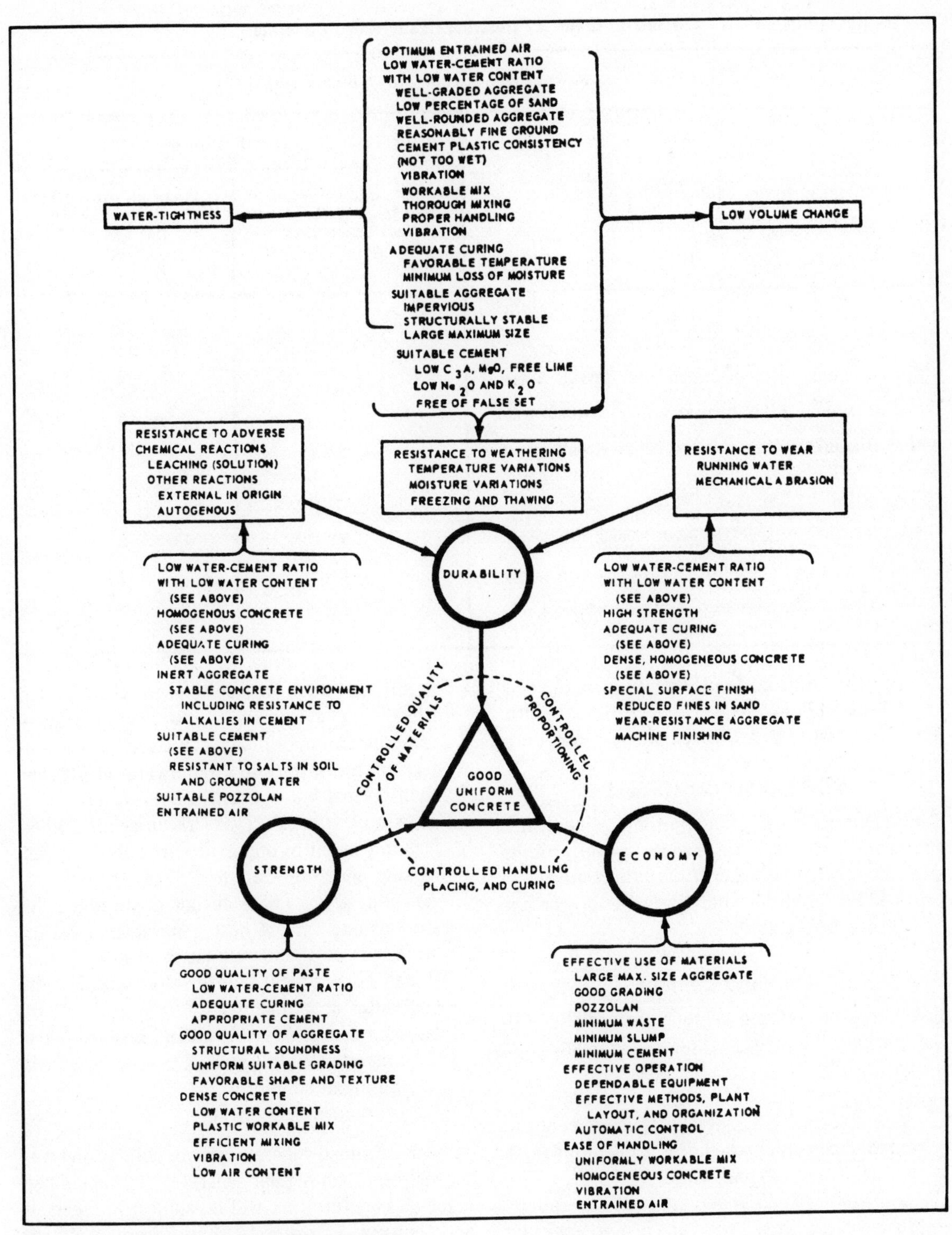

Fig. 3-1. Components of good uniform concrete.

Table 3-2. Guide for Ordering Ready-Mixed Concrete.

[Order medium-consistency concrete (3-inch slump)]		
Job	Specifications	
	Minimum number of bags of cement per cubic yard concrete	Maximum number of gallons of water per bag of cement
Flat work		
Severe exposure (garbage-feeding floors, floors in dairy plants)	7	5
Normal exposure (paved barnyards, floors for farm buildings, sidewalks)	6	6
Mild exposure (building footings, improvements in mild climates)	5	7
Formed work		
Severe exposure (managers for silage feeding, manure pits)	7¾	5
Normal exposure (reinforced walls, beams, tanks, foundations)	6½	6
Mild exposure (improvements in mild climates)	5½	7

Caution: never add water to ready-mixed concrete because doing so will reduce the strength, durability, and watertightness of the concrete.

JOB-MIXED CONCRETE

Concrete is made by mixing portland cement, fine aggregate (sand), coarse aggregate (gravel, crushed stone, or other), and water. Admixtures are sometimes added for various purposes, one of which is to improve workability.

Portland Cement

Five types of portland cement are available, although you might need to special order some from a local distributor:

- ☐ *Normal (Type I)* is a general-purpose cement and the most commonly used. It is available as regular, air-entrained, and white.
- ☐ *Modified (Type II)* generates heat less rapidly during curing than Type I and is more resistant to sulfate attack. It can be used where added precaution against sulfate attack is important—for example, in drainage structures where sulfate concentrations in groundwater are higher than normal but not unusually severe.
- ☐ *High-early strength (Type III)* can be used when it is important to obtain strength shortly after casting concrete—in 1 to 3 days.
- ☐ *Low-heat (Type IV)* is designed to reduce the amount and rate of heat generated in curing large masses of concrete (such as large dams). It has no practical use for the typical do-it-yourselfer or homebuilder.
- ☐ *Sulfate-resistant (Type V)* should be used when the concrete will be exposed to soils or waters of high alkali content.

Air-entrained concrete is more durable and water resistant than regular concrete. You should use it for all concrete that will be subject to freezing and thawing. In concrete, 5 to 7 percent of entrained air by volume is desirable. You can order air-entrained, ready-mixed concrete, or if you do your own mixing, you can use air-entrained portland cement or add an air-entrained admixture to the concrete.

Portland cement is sold in paper bags containing 94 pounds of cement, or 1 cubic foot. You must store the cement in a dry place until you use it. If it becomes damp, it can set in the bag. Never use cement that has set in the bag; discard it.

Stored cement can become lumpy. If the lumps can be readily pulverized between the thumb and forefinger, you can use the cement; if not, you should discard it.

Fine Aggregate

Fine aggregate, or sand, consists of all grains, small pebbles, or particles of crushed stone that will pass through a 1/4-inch mesh wire screen. The sand should be clean, hard, and well graded. *Well graded* means ranging in size from fine (excluding dust) to coarse. Well-graded aggregate makes stronger concrete than aggregate that is nearly uniform in size. More cement is required when the sand is fine.

Sand from saltwater beaches requires thorough washing to remove salt and other impurities. It is usually more economical to use sand from other sources.

The sand should be free from harmful amounts of vegetable matter, loam, or clay. You can test sand for excessive silt or clay content as follows:

- ☐ Put 2 inches of the sand in a quart fruit jar. Add water until the jar is three-quarters full. Screw on the cover and shake the jar vigorously until the sand is thoroughly washed. Let the contents settle. The silt will be deposited in a layer above the sand.
- ☐ Measure the layer of silt after the contents have settled for 24 hours. If it is more than 1/8 inch thick, the sand is not clean enough for concrete unless the silt is removed by washing.

You can test sand for harmful amounts of organic matter as follows:

- ☐ Dissolve a heaping teaspoonful of lye in 1/2 pint of clear water. Any household lye that consists of at least 94 percent sodium hydroxide is suitable.
- ☐ Pour the solution into a glass jar containing 1/2 pint of the sand. Cover the jar and shake it vigorously for 1 or 2 minutes.
- ☐ Let the contents settle for several hours. The color of the liquid then will indicate whether the sand contains harmful amounts of organic matter. A clear color indicates clean sand. A straw color indicates some organic matter, but not an objectionable amount. Darker colors indicate an excessive amount of organic matter, in which case the sand is unsuitable and must be washed and retested before use.

Caution: measure the materials carefully when you are making this test because variations in the concentration of the solution can alter the color. Avoid spilling the solution because lye can cause severe personal injury and can damage clothing.

Lignite or coal deposits in the soil can give the liquid a very dark color. Such impurities might not be present in sufficient quantity to reduce the strength of the concrete appreciably, however, and the sand might otherwise be acceptable. Laboratory tests should be made to determine their exact effect.

Coarse Aggregate

Gravel, crushed stone, and crushed slag are commonly used as coarse aggregate in concrete work. Coarse aggregate should be sound, hard, and free of the same impurities that are objectionable in sand. Use only a minimum of soft, flat, or elongated particles.

The particles should range in size from 1/4 inch up to 1 1/2 or 2 inches. The nature of the work determines the maximum size to use. The largest particles should not be more than one-fourth the thickness of the wall or slab.

You can greatly reduce the amount of concrete for foundations and walls having a large cross section by using large stones (about the size of a man's head) as aggregate. The stones should be sound and clean and at least 1 inch in the concrete so they do not show in the finished surface.

Bank-Run Gravel

Sometimes you can obtain bank or creek gravel, which can fill the requirement for both sand and gravel. It is frequently used in small concrete jobs just as it comes from the pit or creek.

Such material occasionally consists of nearly the right proportion of sand and gravel required for good concrete, but usually there is an excess of either fine or coarse material. Most gravel banks contain an excess of sand in proportion to coarse material. More cement paste is required to produce concrete of a given quality when there is a high proportion of fine aggregate.

To determine whether bank-run gravel contains the right proportion of sand and gravel, screen a representative sample of at least 2 cubic feet over a 1/4-inch mesh screen. Material that passes through is sand; that retained is gravel. The proportion should be comparable to that given in Table 3-3 for the specific job.

If the proportion is not approximately correct, separate the sand and gravel by screening and remixing them in the right proportion. The concrete will be stronger, and you might save enough cement to pay for the cost of screening.

Some commercial firms sell a mixed aggregate. The sand and gravel are separated and then recombined in the correct proportion for concrete.

Lightweight Aggregates

Concrete ordinarily weighs about 150 pounds per cubic foot. Lighter weight concrete can be made by using lightweight aggregates that are available in many parts of the country. Lightweight aggregates also increase the fire-resistant and insulating qualities of the concrete.

Lightweight aggregates consist of cinders or expanded materials, such as clay, shale, or slag. They produce concrete weighing from 100 to 130 pounds per cubic foot. Concrete weighing as little as 50 pounds per cubic foot can be made with very lightweight aggregates, such as pumice or expanded mica.

Concrete made with lightweight aggregates is especially useful for filling between floor sleepers, for making precast blocks and roof slabs, and for fireproofing. It must not be used for watertight structures or where it will be subject to abrasion or heavy loads. When structural strength is required, it should be used only by an experienced builder.

If you are using cinders in making concrete, they should be composed of hard, clean, vitreous clinkers free from sulfides, soot, and unburned coal or ashes. Soak them in water for 24 hours before use to remove any detrimental substances. When clean, they will not discolor the hands when rubbed between them. Ashes from cookstoves and domestic heaters are not suitable aggregate. Lignite ashes contain alkali, which disintegrates concrete.

Lava rock varies widely in chemical composition and physical qualities. Some lavas are so light and frothy or contain so much easily oxidizable material that they are unsuitable for concrete work. Lava rock found in Washington and Oregon is usually satisfactory. Rhyolite, which is a light-colored volcanic rock, and many of the darker basaltic lavas are suitable for concrete for buildings.

Water

Mixing water for concrete should be clean and free of strong acid, oil, alkali, and organic matter. Sea or brackish water should not be used because it might reduce the strength of the concrete. Alkali salts are destructive if they are present in excess of 0.5 percent.

Concrete Additives

Admixtures are sometimes put into concrete to improve workability, reduce segregation, entrain air, or accelerate setting and hardening. If improperly used, they can impair the quality of the concrete.

Fine materials, such as powdered pumice, fly ash, and hydrated lime, are sometimes added to improve workability. They usually reduce the strength of concrete. Better ways to improve workability include adding more cement to the mix, varying the mix proportions or the aggregate graduation, and placing the cement at a slower rate.

You must add air-entraining agents carefully. A small error in the amount added can make considerable difference in the amount of air in the con-

Table 3-3. Trial Concrete Mixtures for Various Kinds of Work.

Kind of work	Proportions			Water required per sack of cement when sand is—		
	Cement *Sacks*	Sand *Cubic feet*	Gravel *Cubic feet*	Wet *Gallons*	Moist *Gallons*	Dry *Gallons*
Very thin work—2 to 4 inches thick (fence posts, milk cooling tanks)	1	2	2	3½	3¾	4½
Exceptionally watertight and abrasion-resistant work—4 to 8 inches thick (tanks, corner posts, silos)	1	2	3	3¾	4½	5½
General reinforced and watertight work—8 to 12 inches thick (basement walls, pavements, steps)	1	2½	3½	4½	5	6½
Mass concrete work of moderate strength and not watertight (footings, foundation walls)	1	3	5	5	6	7

crete. Too much air is detrimental to the concrete. Follow the manufacturer's recommendation to obtain the desired air content.

To accelerate setting and hardening, you can add between 1 and 2 pounds of calcium chloride per sack of cement. Never add more than 2 pounds per sack of cement. You can add the chemical in solution in the mixing water or as crystals in the aggregate. Do not mix it with the cement until the materials are put in the mixer.

You can add finely ground mineral oxides to give concrete color; 3 to 6 pounds per sack of cement should be enough to develop the desired tone. Longer mixing time is necessary to obtain uniformity of color throughout the concrete.

FORMS

Fresh concrete is heavy and plastic. Forms for holding it in place until it hardens must be well braced and should have a smooth inside surface. Cracks, knots, or other imperfections in the forms can be permanently reproduced in the concrete surface.

Wood is commonly used for forms because of its light weight and strength. Because the cracks between boards can mar the concrete surface, plywood is often used. The special high-density overlay surface on plywood provides a smooth casting surface and facilitates removal of the forms for reuse.

If you are using unsurfaced wood for forms, oil or grease the inside surface to facilitate removal of the forms and to prevent the wood from drawing too much water from the concrete. Do not oil or grease the wood if the concrete surface will be painted or stuccoed.

Forms for flat work, such as pavements, can be 2-by-4-inch or 2-by-6-inch lumber; the size depends on the thickness of the slab. Stakes spaced 4 feet on center hold the forms in place.

Figures 3-2 and 3-3 show forms for straight-wall construction. Figure 3-4 illustrates a straight wall after the forms are removed. To prevent the forms from bulging, tie opposite studs together with 10- to 12-gauge wire, which should be twisted to draw the form walls tight against the wooden spacer blocks. Remove the blocks as you place the concrete.

Space the ties about 2 1/2 feet vertically on the studs. When you remove the forms, clip the wires close to the concrete and punch them back. Use mortar to point up any pit holes caused by punching back the wires.

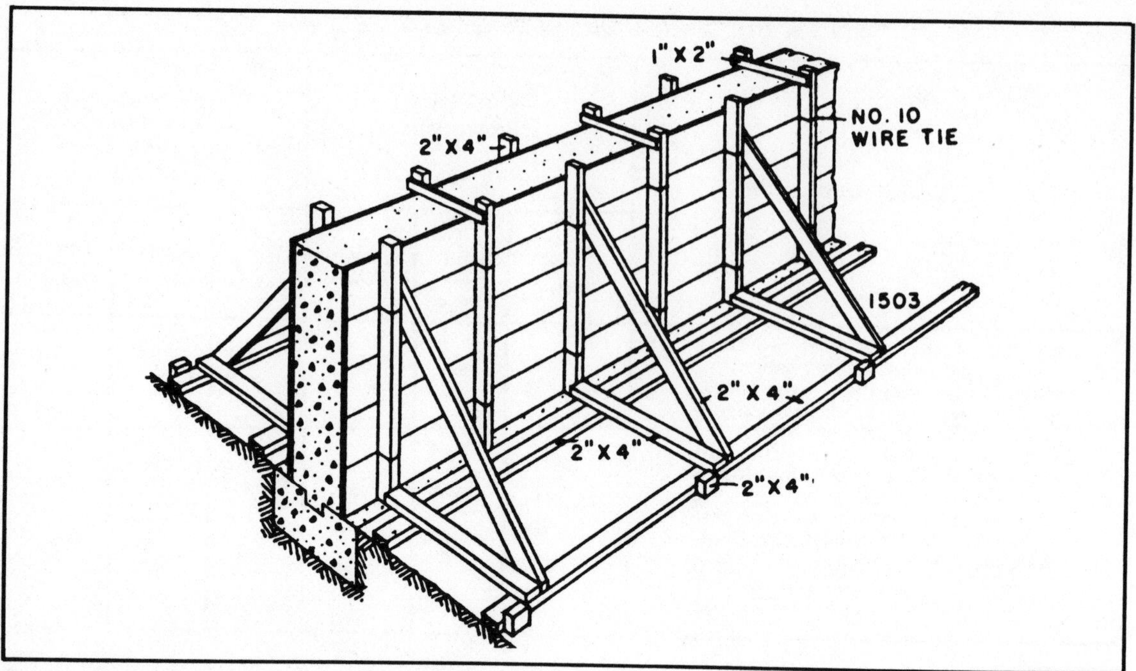

Fig. 3-2. Forms for a straight wall on level ground.

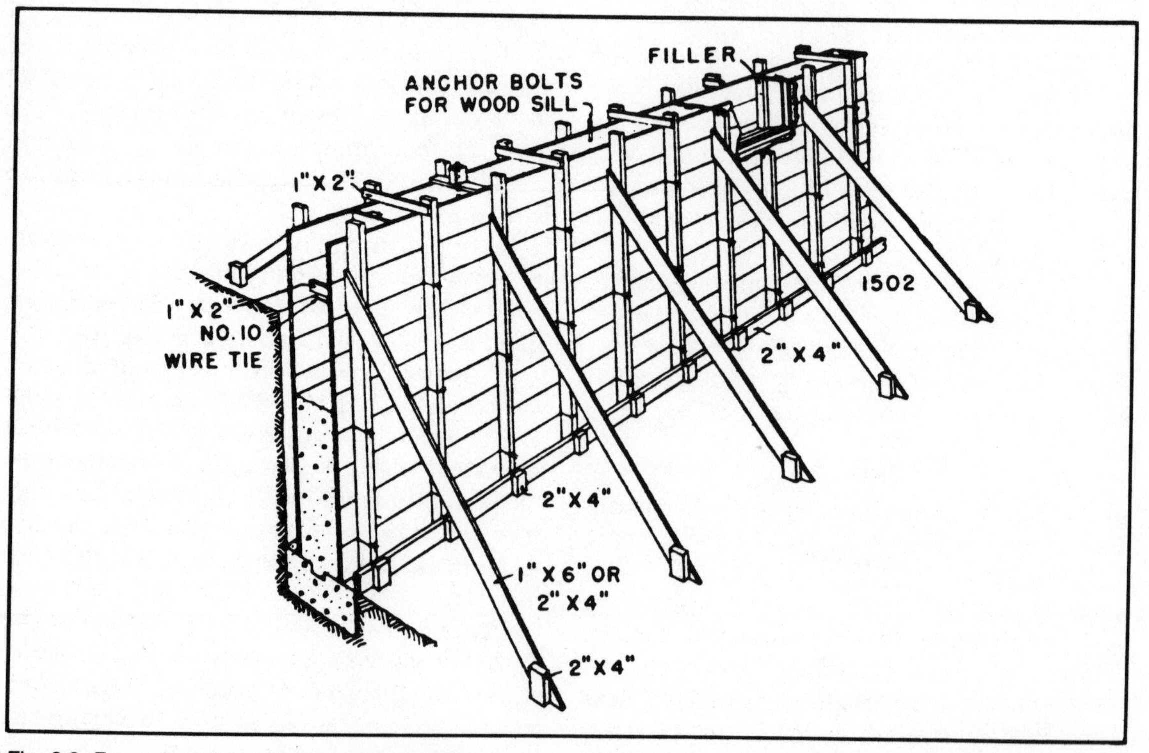

Fig. 3-3. Forms for a basement or cellar wall (C).

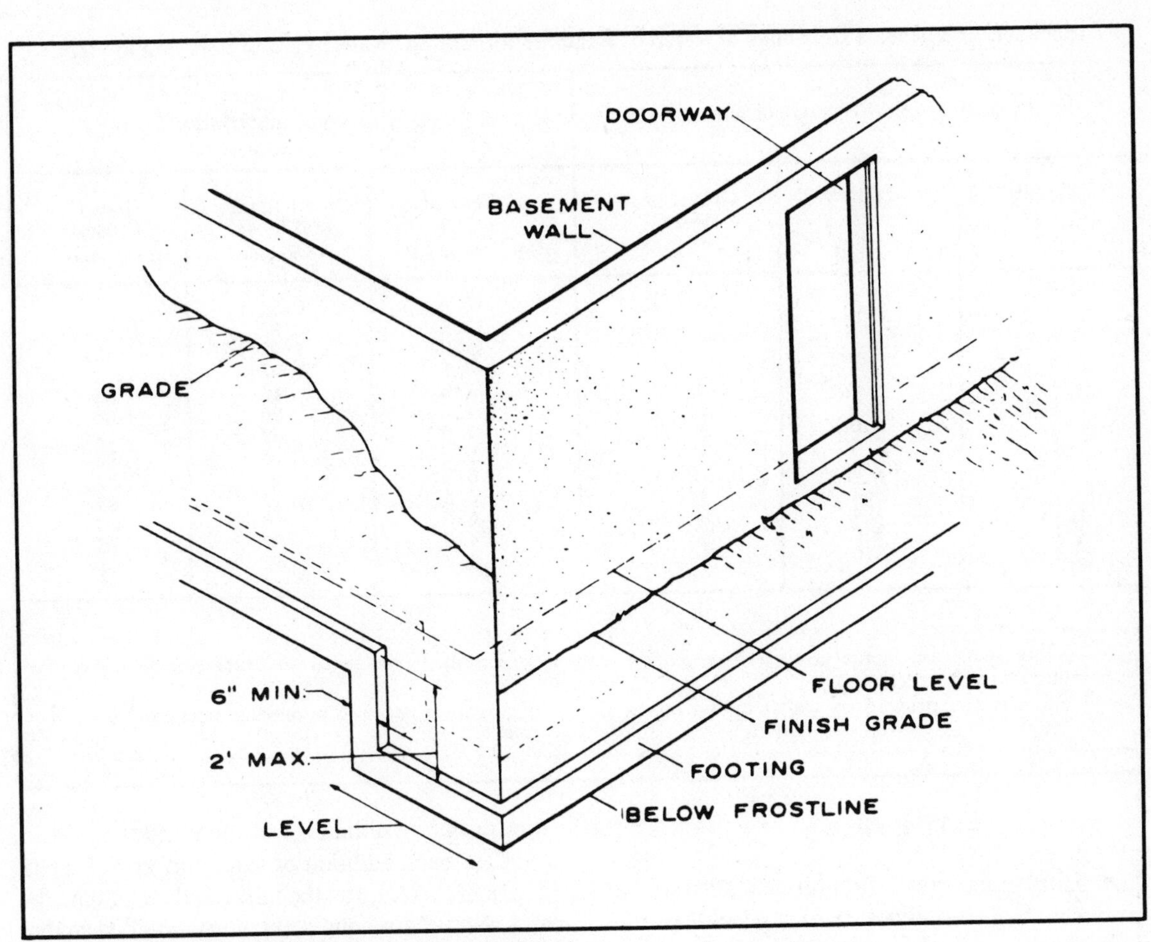

Fig. 3-4. Straight wall foundation after forms are removed.

You can obtain the finest natural finish on a concrete surface by casting on polyethylene. Sometimes polyethylene forms are used for decorative work, or a kraft paper with a polyethylene film surface is used as a form liner.

Use of forming materials that are left in place can be economical for reinforced floors, roof decks, and similar work. Such materials include steel, tile, and precast concrete unit masonry.

MIXING

Concrete mixes are designated by three numbers—for example: 1:2:3—indicating the proportion of cement, sand, and gravel (or other coarse aggregate) used. A 1:2:3 mix indicates 1 part cement, 2 parts sand, and 3 parts gravel. Different mixes are used for different kinds of concrete work.

Table 3-4 lists concrete mixes for various kinds of work. It also indicates the quantity of water required per bag of cement. The cement-sand-gravel proportion might need to be varied slightly to obtain a workable concrete mix, but the water-cement ratio should never be changed. The water-cement ratio determines the quality of the cement paste, which in turn determines the strength, durability, and watertightness of the concrete.

Note that the water-cement ratio varies according to the moisture content of the sand. You can determine the moisture content of sand by squeezing some in your hand. If the sand forms a firm ball, it is wet. If it forms a ball that tends to crumble, it is moist. If it falls free, it is dry.

Table 3-4. Approximate Quantities of Materials Required for Making 1 Cubic Yard of Concrete in Place.[1]

Proportions of the concrete or mortar [2]			Quantities of materials		
Cement	Sand	Gravel or stone	Cement *Sacks*	Sand (damp and loose) *Cubic yards*	Gravel (loose) *Cubic yards*
1	1. 5	----	15. 5	0. 86	----
1	2. 0	----	12. 8	. 95	----
1	2. 5	----	11. 0	1. 02	----
1	3. 0	----	9. 6	1. 07	----
1	1. 5	3	7. 6	. 42	0. 85
1	2. 0	2	8. 2	. 60	. 60
1	2. 0	3	7. 0	. 52	. 78
1	2. 0	4	6. 0	. 44	. 89
1	2. 5	3. 5	5. 9	. 55	. 77
1	2. 5	4	5. 6	. 52	. 83
1	2. 5	5	5. 0	. 46	. 92
1	3. 0	5	4. 6	. 51	. 85
1	3. 0	6	4. 2	. 47	. 94

[1] The quantities of materials required may vary as much as 10 percent, the variation depending on the aggregate used.
[2] The first four proportions are for mortar mixes. Coarse aggregate is not used in making mortar.

Trial Mixing

The cement-sand-gravel proportions given in Table 3-4 are trial proportions. It is not possible to give definite proportions because of the variation in aggregates.

Mix a batch of concrete using the appropriate trial proportion from the table and the correct water-cement ratio. Note that the amounts of material and the water-cement ratio are based on one-bag (of cement) batches. If you are using a smaller batch, reduce the amount of materials, maintaining the water-cement ratio.

Mix the concrete thoroughly. Then stop the mixer and examine the batch by reaching in and working the surface with a float. A workable mix should be smooth and plastic; not so wet that it will run or so stiff that it will crumble. If the mix is too wet, add some amounts of sand and gravel in the proper proportion until you obtain a workable mix. If the mix is too stiff, add some amounts of water and cement, maintaining the proper water-cement ratio, until you obtain a workable mix.

After each addition of sand and gravel, or of water and cement, run the mixer and reexamine the batch. Note the amounts of materials added so that you can initially proportion subsequent batches properly. Never add water to a mix without adding cement. Water dilutes and weakens the cement paste and thus reduces the strength of the concrete.

In some cases, the mix might be too sandy or too stony. In this situation, it is advisable to make a second trial batch and to vary the proportions of sand and gravel until you obtain the desired workability.

Be sure to measure concrete materials as accurately as possible. You can measure water in a pail marked off in gallons, half gallons, and lesser quantities. One bag of portland cement contains 1 cubic foot of cement. You can measure cement in quantities of less than one bag in a calibrated 1-cubic-foot box or pail. To measure sand and gravel, count the shovelsful required to fill a cubic foot box or pail. Measuring can then be done by shovelsful into the mixer. This method is sufficiently accurate for most concrete work.

Estimating Material Requirements

Table 3-4 shows the approximate amount of materials required to make 1 cubic yard of concrete of different mixes. To find the amount of materials required for a specific job:

- ☐ Determine the cubic yardage of the space inside the forms. This gives the number of cubic yards of concrete needed.
- ☐ Multiply the appropriate figures in the table by the number of cubic yards of concrete needed.

Example: You need 12 cubic yards of concrete of a 1:2:3 mix. The table shows that 7 bags of cement, 0.52 cubic yard of sand, and 0.78 cubic yard of gravel are required to make 1 cubic yard. Therefore, you need:

$7 \times 12 =$ 84 bags of cement
$0.52 \times 12 =$ 6 1/4 cubic yards of sand
$0.78 \times 12 =$ 9 1/2 cubic yards of gravel (approximate)

Order about 10 percent extra of each material to allow for waste or slight miscalculation in the amounts needed.

Table 3-5 shows normal concrete components.

Machine Mixing

Concrete should be thoroughly mixed. Thorough mixing increases the strength of the water-cement paste and improves the workability of the concrete. For machine mixing, allow 5 or 6 minutes mixing time after all the materials are in the drum.

PLACING CONCRETE

Place the concrete within 20 minutes after you have completed the mixing. In warm weather, initial set will occur in about 20 minutes, and if concrete is disturbed after the initial set, it loses strength. Never rewet and remix concrete that has set before it can be placed in the forms. Discard it.

Place concrete as near as practical to its final position in the forms. Honeycombing and segregation can occur if it is pushed or flowed for some distance into position. Spade the concrete as it goes into the forms.

If concrete is overworked in the form, the finer materials, including the cement paste, tend to work to the top, resulting in a nonhomogenous mixture of unequal density. Push a flat, spadelike tool down the wall of the form through the concrete to release air pockets against the wall (Fig. 3-5). Lightly rod the concrete throughout to relieve entrapped air and ensure intimate contact with the form.

Fresh concrete will not bond readily to hardened concrete, and the resultant seam can permit water to seep through. To bond new concrete to concrete that has been in place a short time, roughen the surface of the hardened concrete with a pick to expose the gravel or stone. Clean off loose particles.

To provide a bond for the next day's work, roughen the surface of the concrete just before it hardens. Before placing the new concrete, soak the hardened concrete with water, remove the excess water, and apply a coat of grout. If you will discontinue the pouring of a wall for some time, provide for bonding of future work by embedding short steel dowels in the concrete when it is poured or by cutting a rebate groove in the concrete. (Fig. 3-6, C). To bond a new wall to an old one, drill holes of dowels in the old wall and grout in the dowels. Then roughen, clean, and wet the old surface.

Placing concrete in freezing or very cold weather is not recommended. The expense and trouble of heating the mixing water and aggregates before use and of protecting the finished work (which might include supplying heat through some type of heating equipment) to prevent freezing is not warranted for most concrete jobs.

Hot, dry weather also presents special problems. You must start curing promptly to prevent too rapid drying, which can reduce the strength of the concrete and cause cracking. In hot weather, it is advisable to place concrete late in the afternoon when the temperature has dropped. Make sure, however, that you have enough time to complete the job before dark.

Table 3-5. Normal Concrete.

Class concrete (figures denote size of coarse aggregate in inches)	Estimated 28-day compressive strength, (pounds per square inch)	Cement factor, bags (94 pounds) of cement per cubic yard of concrete, freshly mixed	Maximum water per bag (94 pounds) of cement (gallons)	Fine aggregate range in percent of total aggregate by weight	Approximate weights of saturated surface-dry aggregates per bag (94 pounds) of cement	
					Fine aggregate (pounds)	Coarse aggregate (pounds)
(1)	(2)	(3)	(4)	(5)	(6)	(7)
B-1	1500	4.10	9.50	42-52	368	415
B-1.5	1500	3.80	9.50	38-48	376	498
B-2	1500	3.60	9.50	35-45	378	567
B-2.5	1500	3.50	9.50	33-43	373	609
B-3.5	1500	3.25	9.50	30-40	378	702
C-1	2000	4.45	8.75	41-51	329	387
C-1.5	2000	4.10	8.75	37-47	338	467
C-2	2000	3.90	8.75	34-44	338	529
C-2.5	2000	3.80	8.75	32-42	332	565
C-3.5	2000	3.55	8.75	29-39	334	648
D-0.5	2500	5.70	7.75	50-60	282	231
D-0.75	2500	5.30	7.75	45-55	288	288
D-1	2500	5.05	7.75	40-50	279	341
D-1.5	2500	4.65	7.75	36-46	287	413
D-2	2500	4.40	7.75	34-42	288	471
D-2.5	2500	4.25	7.75	32-40	287	509
D-3.5	2500	4.00	7.75	29-37	285	578
E-0.5	3000	6.50	6.75	50-58	238	203
E-0.75	3000	6.10	6.75	45-53	240	249
E-1	3000	5.80	6.75	40-48	233	297
E-1.5	3000	5.35	6.75	36-44	239	359
E-2	3000	5.05	6.75	33-41	241	410
E-2.5	3000	4.90	6.75	31-39	238	441
E-3.5	3000	4.60	6.75	28-36	237	503

Joints

Concrete expands and contracts with changes in temperature. You must provide joints to allow for the movement and thus prevent, reduce, or control cracking. Concrete slabs can be cast in 10- to 15-foot square or rectangular blocks. Place two layers of 15-pound builders' felt between the blocks (Fig. 3-6, A), or leave a 1/2-inch space between blocks. Later fill the spaces with asphalt.

Slabs also can be cast in 10- to 12-foot-wide strips. Provide joints between strips, as just indicated. Cut control joints, called *dummy* joints, at 10- to 15-foot intervals across each strip (Fig. 3-6, B). Cut the joints to one-fourth the thickness of the

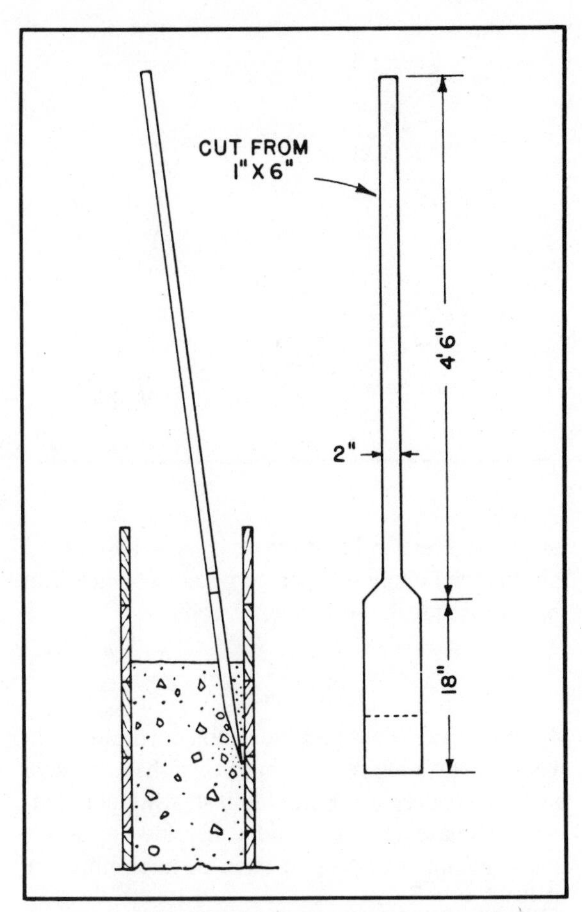

Fig. 3-5. Wooden spade for working concrete along the form.

slab with an ax or hoe and a straightedge.

Contractors often use concrete saws to cut control joints in concrete after it hardens. This method is satisfactory, but requires the proper tools and some experience.

Where a slab abuts a wall, provide a 3/4-inch joint between the slab and wall. Place two wedge-shaped boards along the wall (Fig. 3-6, C). After you have cast the slab, remove the boards and fill the joint with hot bitumen (asphalt).

Concrete floor slabs for houses are usually cast as one continuous unit. To control cracking, reinforce them with steel consisting of 10-gauge welding wire with a mesh size of 6 × 6 inches. The reinforcement will not prevent cracks, but will prevent them from enlarging. A general rule is to provide a steel area of 0.3 percent of the cross-sectional area of the concrete. This rule applies when you are casting ordinary slabs not more than 45 to 50 feet in length.

FINISHING

After the concrete is placed, *screed,* or level off the concrete in the forms with a straightedge (Fig. 3-7). Then float the concrete with a long-handled float, called a *bull float,* to obtain a more even finish. Delay final surface finishing until the concrete has become stiff and the water sheen has disappeared.

The three general textures of final surface finish are the rough broom, the wood float, and the steel trowel.

- ☐ The *rough-broom finish* provides a slipproof surface and should be used on yards, floors, and other walking surfaces.

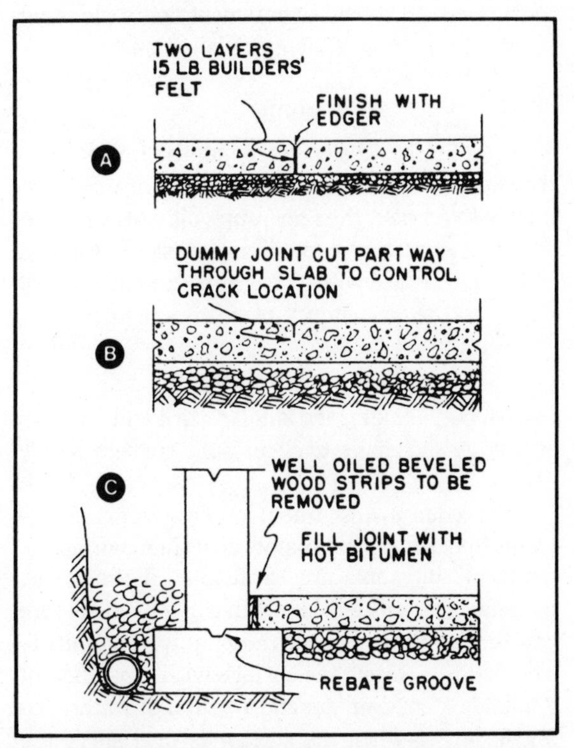

Fig. 3-6. Joints for concrete slabs: joint between concrete blocks or strips in slab (A); *dummy* joint to control cracking (B); joint between slab and wall.

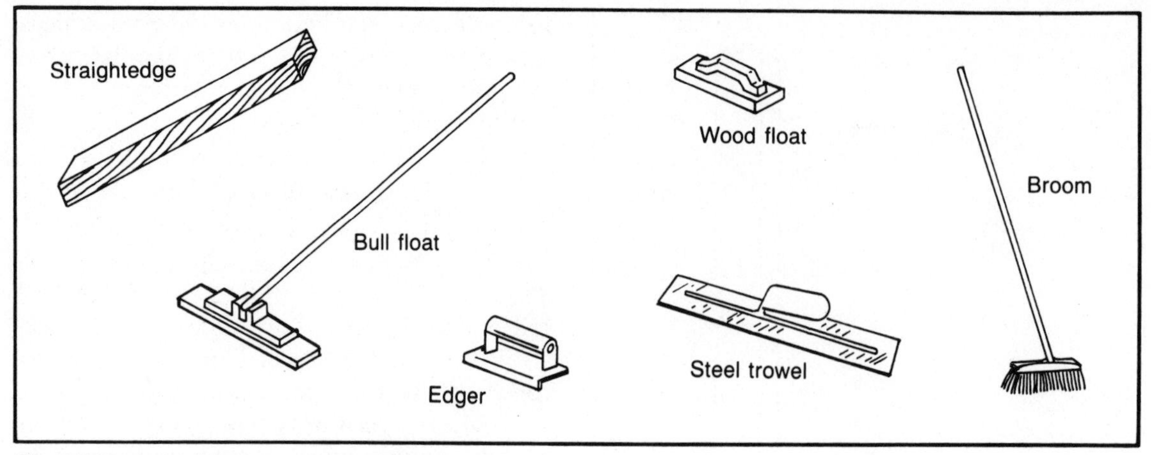

Fig. 3-7. Tools for finishing concrete surfaces.

- ☐ The *wood-flat finish* provides an even, gritty surface and is used on general work, such as building floors and sidewalks.
- ☐ The *steel-trowel finish* provides a smooth, dense surface and is used on basement floors and other work requiring a highly finished surface.

Walls

You can finish walls by removing the forms while the concrete is still partially set. Rub off irregularities with a brick, then point up voids with mortar or *neat cement* (a mix containing no sand). Finally, brush the surface with a coat of water and cement mixed to the consistency of heavy cream.

Special Treatments

Patios, porch floors, and similar work will be more attractive and easier to clean if the surface is clay tiled.

For a dense, dust-free floor, consider concrete paint. Special concrete paints containing abrasion-resistant pigment are available. Follow the manufacturer's directions when using them. Before you paint a floor, clean it thoroughly. To neutralize alkalinity, brush the surface with a solution of 4 pounds of zinc sulfate and 1 gallon of water. After 48 hours, when the surface is dry, apply the paint.

White portland cement mixed with water and applied with a stiff-bristled brush makes a good finish for concrete walls. If the walls need a gloss finish, you can apply a concrete paint. You also can install glazed tile on concrete walls.

CURING

Proper curing is essential for good concrete. Thin work requires more care in curing than massive work. If concrete dries out too fast, it might not attain adequate strength, and excessive shrinkage and cracking can occur. It must be kept moist for at least 4 or 5 days.

You can cure slabs and other flat work by several methods:

- ☐ Cover the concrete with burlap, straw, or similar material and keep the material wet.
- ☐ Cover the concrete with a watertight cover to seal in the water and prevent evaporation.
- ☐ Flood or pond the surface, retaining the water by means of an earth dike around the slab.
- ☐ Spray a commercial curing compound on the surface to seal it.

Polyethylene is one of the most effective watertight covers. As soon as possible after final finishing of the surface, wet it and cover it with a polyethylene sheet. If the sun beats down on the job, throw some straw on top of the polyethylene to reflect some of the heat.

You can cure walls by leaving the forms in place and keeping them wet. If it is necessary to remove the forms, cover the concrete surface with canvas or burlap, and keep that material wet.

You can prevent cracking in small precast units, such as fence posts and foundation piers, by *prestressing* the reinforcing steel. Release the steel after the concrete has cured from 30 to 60 hours, but while it is still "green." This method requires careful judgment as to the elasticity of the steel and the time to release the steel. Several trials might be necessary before it can be done successfully. Prestressing the steel reinforcement helps to achieve a good cure without cracks. It should not be used in place of curing by moisture retention, but only in conjunction with that method.

WATERTIGHT CONCRETE

First-class workmanship is essential for watertight concrete. Other important requirements are:

- ☐ Use a fairly rich mixture with a low water-cement ratio. Mix the concrete to a sluggishly flowing consistency. Maintain the same proportion of materials and the same mix consistency for each batch of the concrete.
- ☐ Reinforce the concrete.
- ☐ Place the concrete in one continuous operation. If that is not possible, provide watertight joints between hardened concrete and new concrete.
- ☐ Provide contraction and expansion joints.
- ☐ Cure the concrete properly.

You can take additional measures to ensure dry basement or cellar walls. These are:

- ☐ Grade the ground around the building to provide surface drainage away from the walls.
- ☐ Install drain tile around the outside of the footings.
- ☐ Apply a 1/2-inch-thick coat of portland cement and two coats of hot coal-tar pitch to the outside of the walls.

CONCRETE WALKS

Concrete walks may be 1 1/2 feet wide or more. Figure 3-8 shows a method of setting up the forms. Expansion joints are usually provided at 4- or 5-foot intervals.

To prevent water from standing on the surface, the walk can be built a little higher than the surrounding ground and crowned, or it can be built with one side slightly higher than the other. Water from downspouts should be diverted so that it will not flow across the walk and become an ice hazard in winter.

Ordinarily, a 4-inch slab of one-course construction is sufficient. If heavy vehicles will be driven over the walk, however, it should be 6 inches thick.

A broomed or float finish is desirable for level or slightly inclined walks. For steeper grades, a coarse, scored surface is advisable. This type of surface is made by running a stiff broom crosswise to the direction of travel.

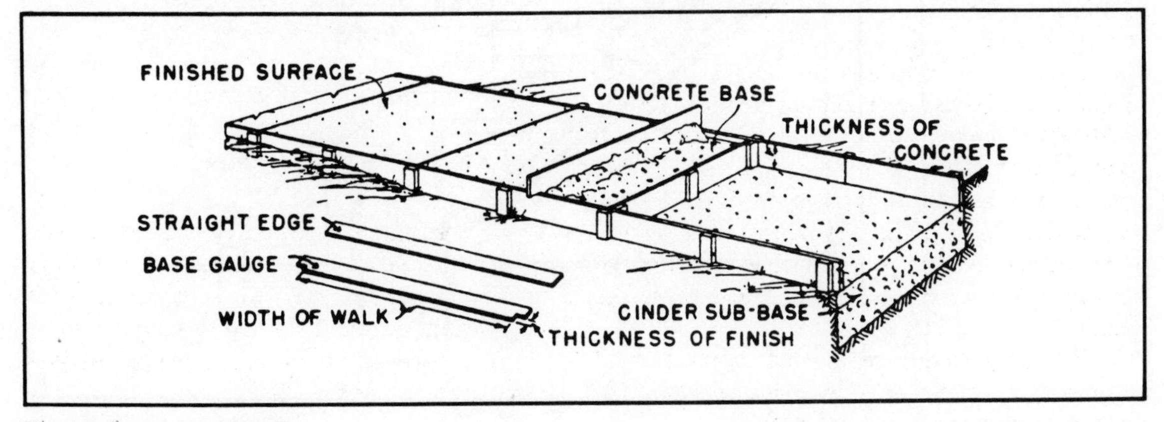

Fig. 3-8. Forms for sidewalks.

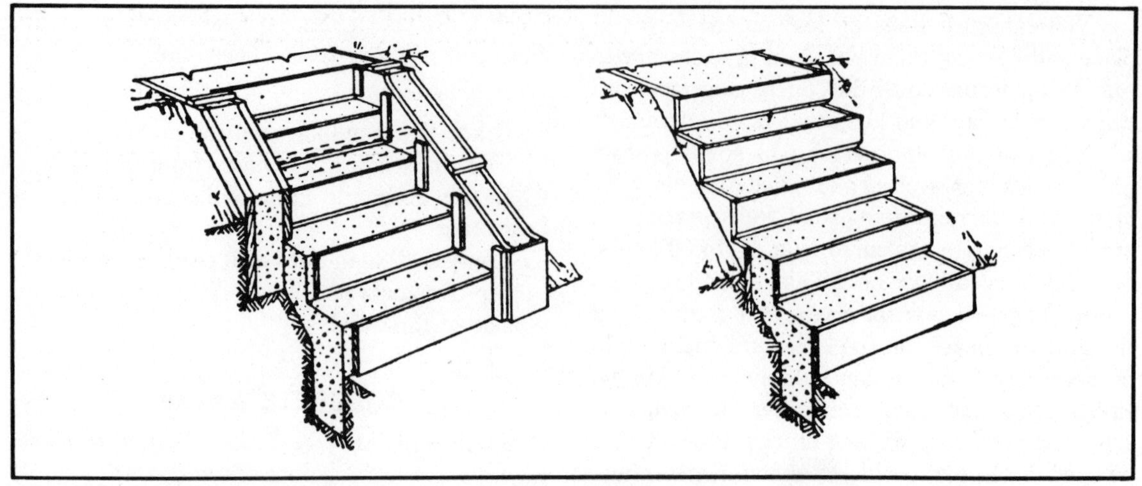

Fig. 3-9. Forms for earth-supported steps with cutaway to show construction.

CONCRETE STEPS

Concrete steps are built by casting the *risers* (vertical part) and *treads* (horizontal part) on an inclined slab. The thickness depends on the span or method of support.

When the slab rests on solid earth or on earth fill and there are only three or four steps from 3 to 4 feet wide, a 4-inch slab is sufficient. For wider and longer flights, however, the slab should be 6 or more inches thick. Figure 3-9 shows two arrangements of forms for earth-supported steps.

Steps that do not rest on solid earth or on earth fill must be self-supporting and, therefore, reinforced (Fig. 3-10). Table 3-6 gives the reinforce-

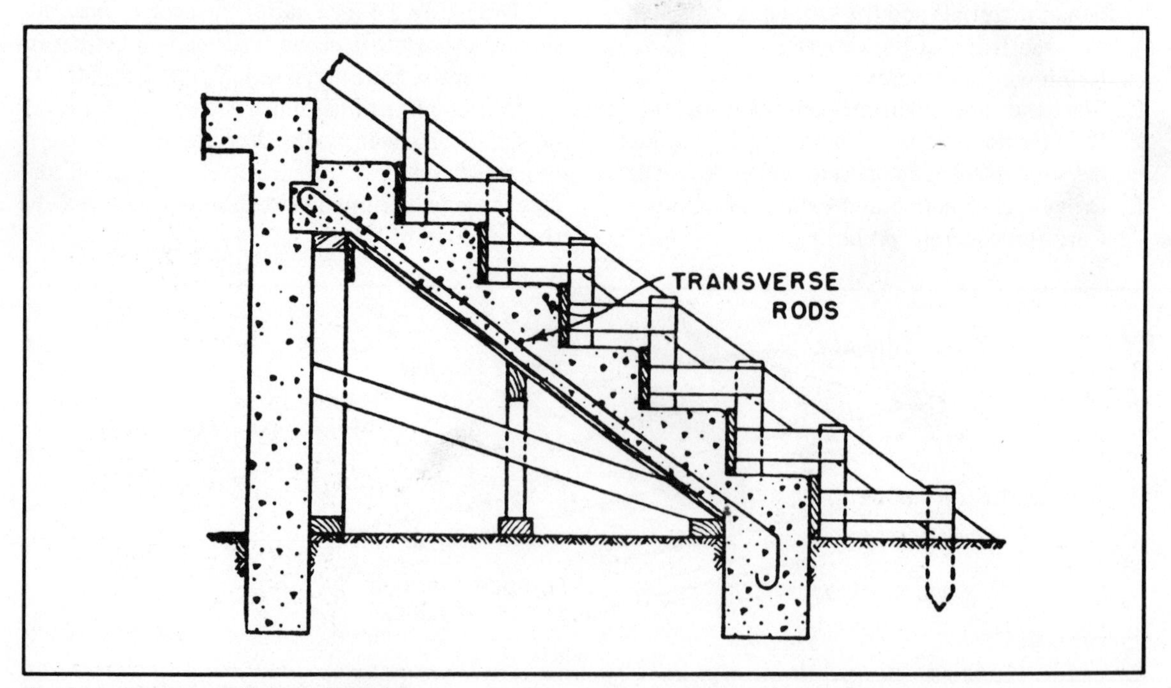

Fig. 3-10. Forms for self-supporting steps.

Table 3-6. Diameter and Spacing of Round Rods Required for Reinforcing Concrete Step Slabs.

Slab dimensions		Longitudinal rods		Transverse rods	
Length (feet)	Thickness inches	Diameter inches	Spacing inches	Diameter inches	Spacing inches
2 to 3	4	1/4	10	1/4	12 to 18
3 to 4	4	1/4	5 1/2	1/4	12 to 18
4 to 5	5	1/4	4 1/2	1/4	18 to 24
5 to 6	5	3/8	7	1/4	18 to 24
6 to 7	6	3/8	6	1/4	18 to 24
7 to 8	6	3/8	4	1/4	18 to 24
8 to 9	7	1/2	7	1/4	18 to 24

ment required for slabs of different lengths and thicknesses. Place the longitudinal rods lengthwise, from top to bottom, 1 inch up from the underside of the slab. Place the transverse rods in the opposite direction. They should extend across the width of the slab. Wire the longitudinal and transverse rods together where they intersect. Self-supporting steps must be firmly supported at the head, such as by a concrete porch or a masonry wall. You can secure the foot as in Fig. 3-10.

Low risers and wide treads are preferable for outdoor steps. Probably the safest and easiest steps to climb have risers 6 to 8 inches high and treads 10 to 11 inches wide. A good formula is: "Twice the height of the riser plus the width of the tread equals 25."

To provide more toe space and, therefore, easier and safer climbing, project the treads 3/4 inch beyond the risers. Treads should slope about 1/16 inch toward the front in order to shed water.

PORCH FLOORS

Porch floors can be laid on an earth fill or supported above the ground. When a porch is laid on earth fill, the earth must be well settled. Provide a porous subbase (gravel or crushed rock), and build an apron (Fig. 3-11) under the edges of the slab. The slab should slope about 1/4 inch per foot to drain off water. If the slab rests on earth fill more than 12 inches thick, you should support it like a porch built above ground.

You can support porches built above the ground with two walls built of 8-inch concrete block. Figure 3-12 shows construction details, and Table 3-7

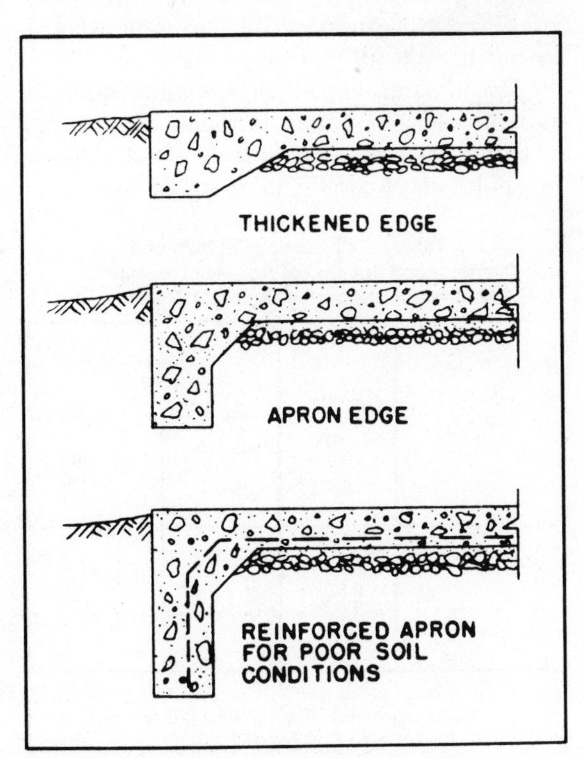

Fig. 3-11. Edge construction for concrete slabs.

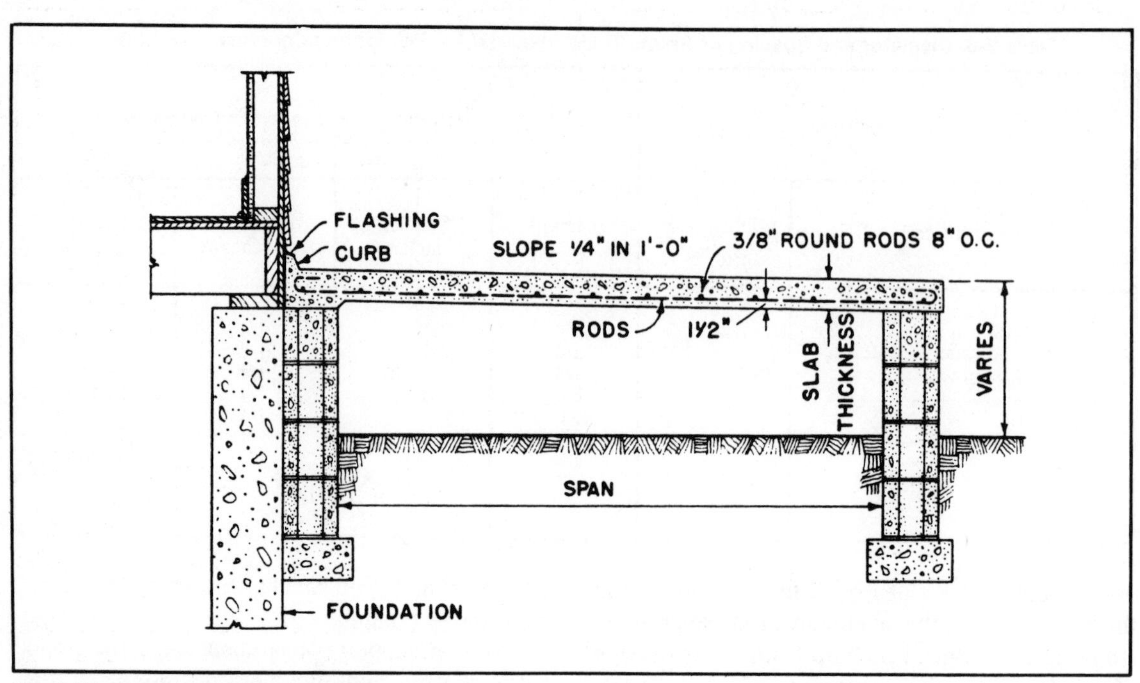

Fig. 3-12. Concrete porch construction.

gives the slab thickness and reinforcement required for porches 4 to 10 feet wide.

You also can support porches with reinforced girders resting on concrete piers. The "wall" method is less expensive, however, and is easier for adding small porches to existing buildings.

Table 3-7. Thickness of Slab and Diameter and Spacing of Rod Reinforcement Required for Supported Concrete Porches.

Width of porch (feet)	Slab thick-ness inches	Reinforcing rods[1]	
		Diam-eter inches	Spacing inches
4	5	3/8	7 1/2
6	5	3/8	6
8	5 1/2	1/2	9 1/2
10	6	1/2	8

[1] The transverse reinforcing rods are 3/8-inch round rods spaced 8 inches on center.

RETAINING WALLS

Two types of retaining walls are commonly built: gravity and reinforced cantilever. Figure 3-13 shows construction of a gravity-type wall. The footing should be below frost depth. Proper width for the footing depends on the height of the wall. A wall 3 feet high should have a footing 2 1/2 feet wide. A 4-foot wall needs a footing 3 feet wide. A 6-foot wall requires a 4-foot wide footing. An 8-foot high wall requires a 5 1/2-foot wide footing.

You can use clean, sound field stones no larger than one-half the thickness of the wall to reduce the amount of concrete needed. Do not place backfill until the concrete has thoroughly hardened, usually after 27 days.

The weep holes, which drain water from the back of the wall, consist of agricultural tile spaced 6 to 8 feet apart. They should drain 6 inches above the lower grade.

For retaining walls higher than 8 feet, the reinforced cantilever type might be less expensive (Fig. 3-14). It should be designed by an engineer. Reinforcement steel is attached as in Figs. 3-15 and 3-16.

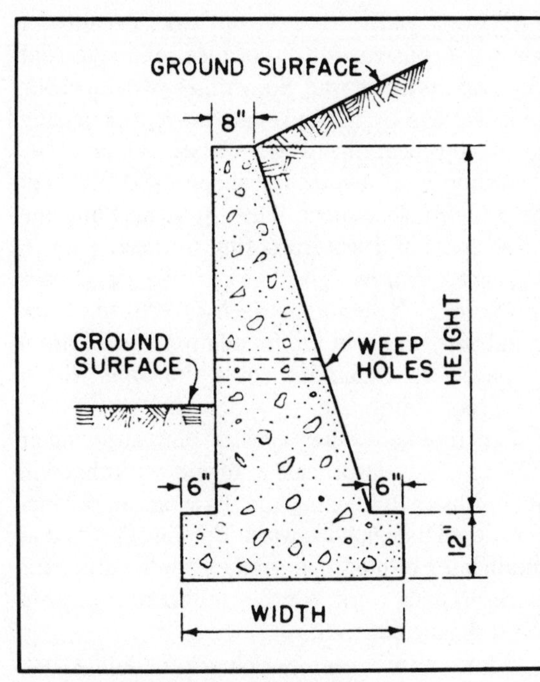

Fig. 3-13. Gravity-type retaining wall.

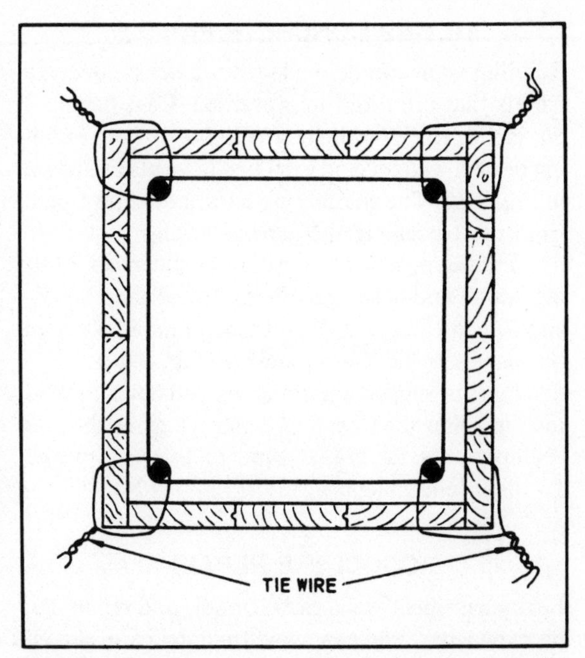

Fig. 3-15. Securing column reinforcing steel against displacement.

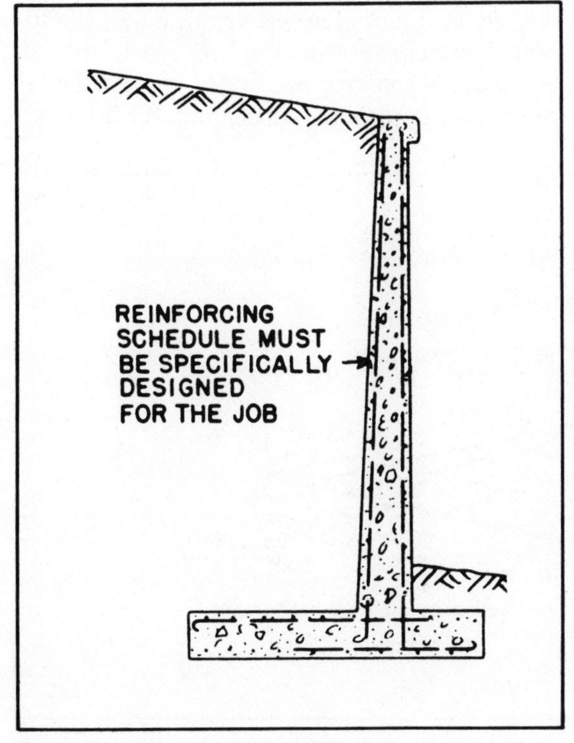

Fig. 3-14. Cantilever-type retaining wall.

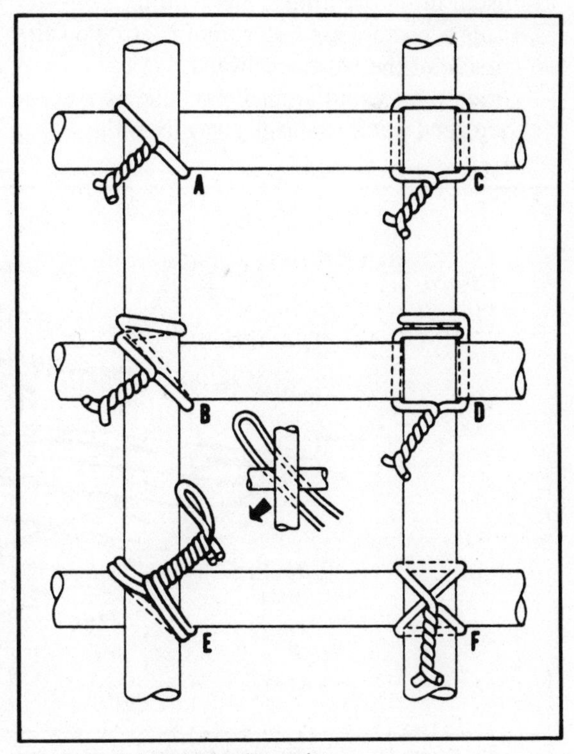

Fig. 3-16. Tying reinforcing rod.

TILT-UP CONCRETE PANELS

Building walls can be made from precast concrete panels that are tilted into position. Cast panels 3 5/8 inches thick on the concrete floor of the building or on a well-leveled sand bed in forms as shown in Fig. 3-17. The reinforcing bars should be 1 inch from the bottom of the panel.

Tilt the panels into position by means of a lifting frame and hoist (Fig. 3-18). Brace them in position, and then cast reinforced columns between the panels to tie them together (Fig. 3-19).

Piers to support the panels extend down to solid footing below the frost line. Table 3-8 gives the pier requirements for walls made of 8-×-8-foot and 10-×-10-foot panels.

CONCRETE REPAIR

Leaky basement walls cause trouble and repair can be expensive. You can avoid them by careful construction. Three basic preventive measures are:

- ☐ Install drain tile around the footing.
- ☐ Place a continuous waterproof coating on the outside of the basement walls.
- ☐ Grade the ground around the building to provide good water drainage away from the walls.

Existing walls can develop leaks because of poor initial construction or because soil movement causes cracks in the wall. Sometimes you can check minor leakage by inside repair work, but usually you must repeat the work. Chisel out cracks—preferably in an inverted V shape—and fill them with a hydraulic cement. Then give the walls one or two coats of a waterproofing masonry paint.

Inside repair work seldom controls major leakage for long. It might be necessary to excavate around the outside of the foundation, install drain tile, parge the walls, and cover the walls with a waterproof membrane.

Concrete floor repair consists of resurfacing or patching holes or worn spots. Inside work that will not be subject to freezing and thawing can be thinner in depth than outside work. Bonding agents that contain latex or modified epoxies might suffice for thin resurfacing work, but they are more expensive than ordinary concrete.

When you are patching holes or regrading shallow sinks, trim the old concrete away until you reach sound material, or at least trim it deep enough to allow for 1 inch of new material. Edges of adjacent good concrete should be kept nearly vertical. Soak the old concrete with water, remove the excess water, and apply a coat of grout. While the sur-

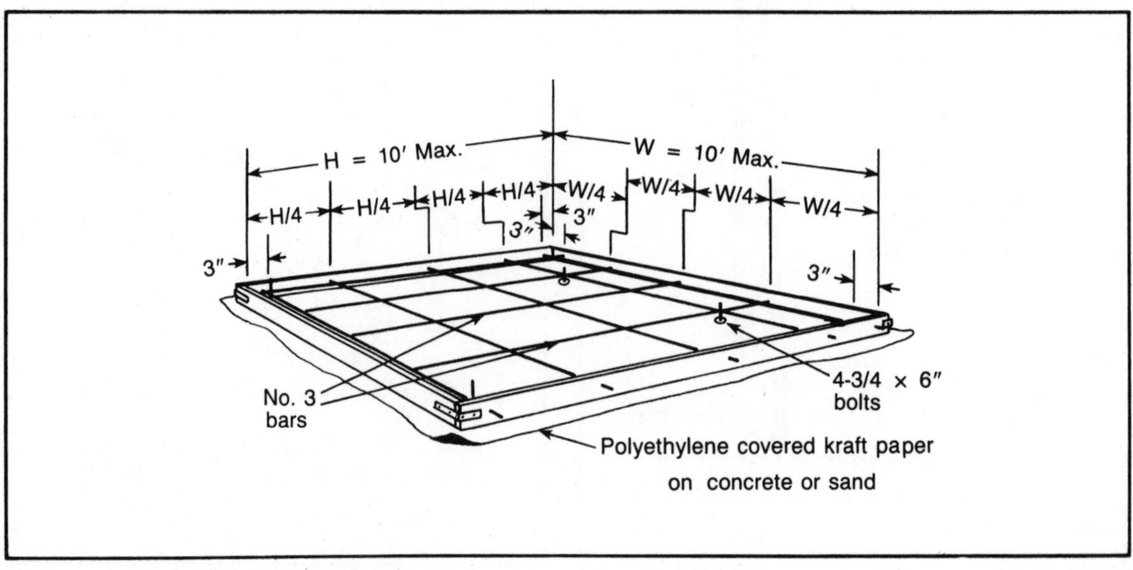

Fig. 3-17. Forms for tilt-up concrete panels.

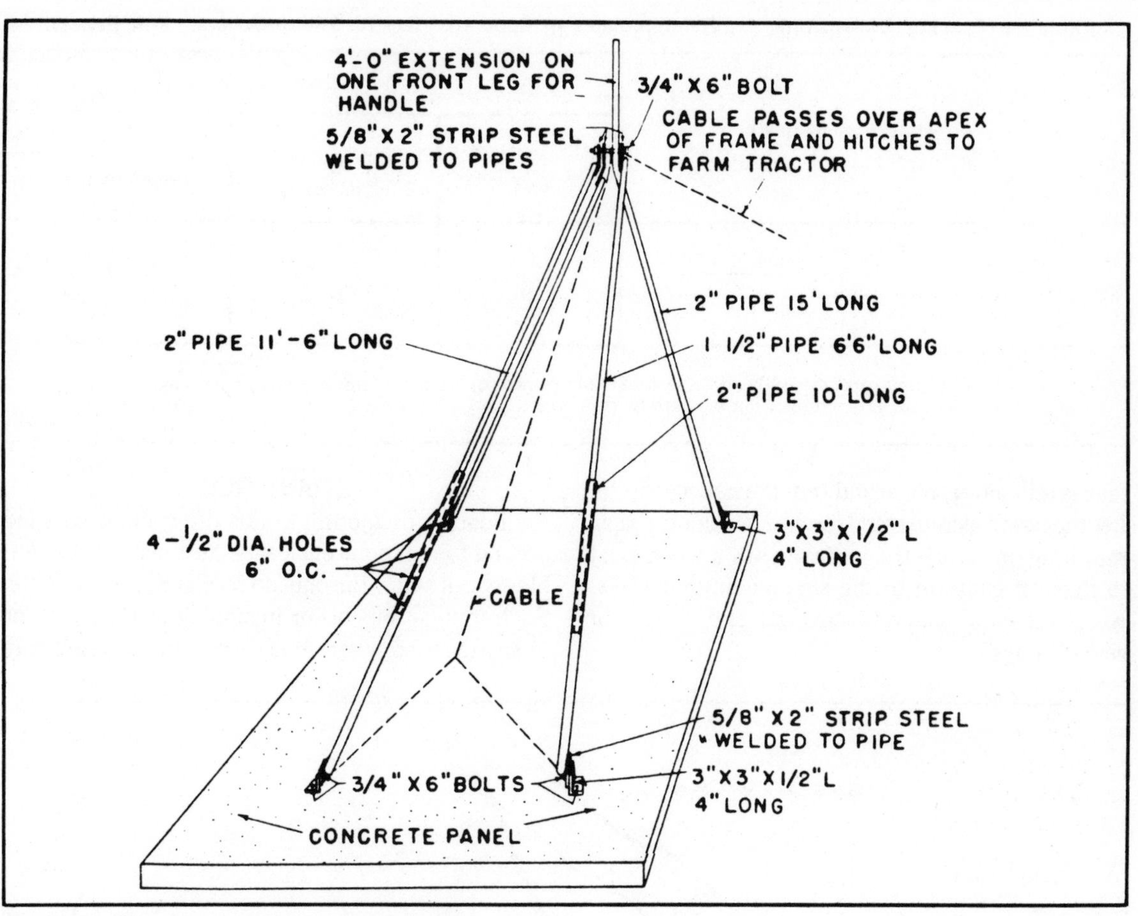

Fig. 3-18. Lifting frame for tilt-up concrete panels.

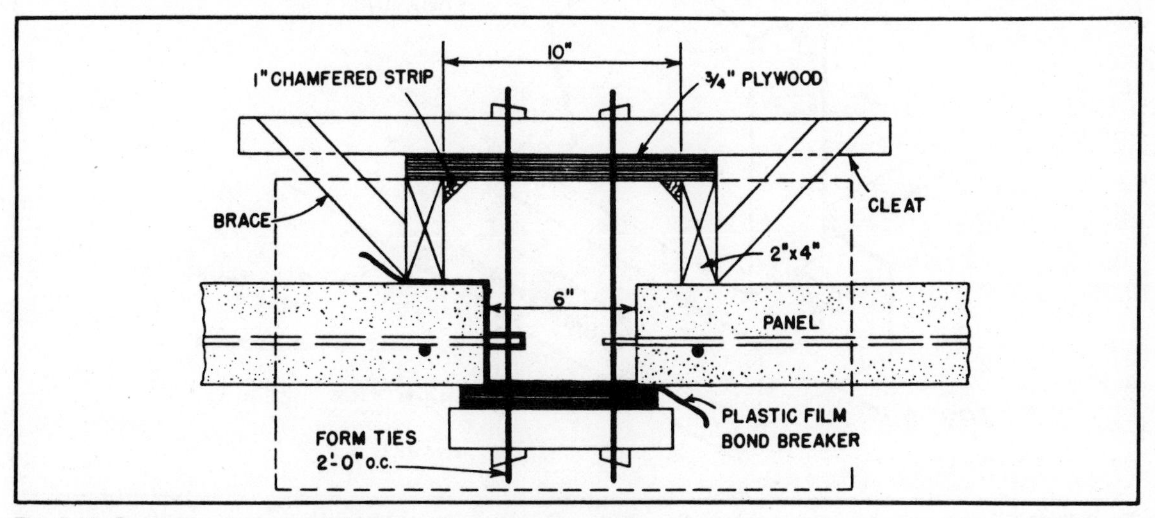

Fig. 3-19. Concrete column forms between tilt-up concrete panels.

Table 3-8. Spacing, Dimensions, and Reinforcement of Concrete Piers for Tilt-up Concrete Wall Panels.

Panel dimensions (feet)	Piers		
	Spacing feet	Dimensions [1] inches	Vertical reinforcement
8 by 8	8	12 by 24	3 No. 4 bars
10 by 10	10	12 by 32	3 No. 6 bars

[1]Soil-bearing capacity assumed to be 4,000 pounds per square foot. For clear-span roof construction, increase either dimension of piers by 50 percent.

face is still moist, place and ram the new concrete. Let the new concrete stand for 5 to 20 minutes, then ram it again. Work the surface with a wood float to make it conform to the surrounding concrete. Keep the new concrete covered and moist for several days.

DRAIN TILE

Foundation or footing drains often must be used around foundations enclosing basements or habitable spaces below the outside finish grade (Fig. 3-20). Such foundations occur in sloping or low areas or in any location where it is necessary to drain away

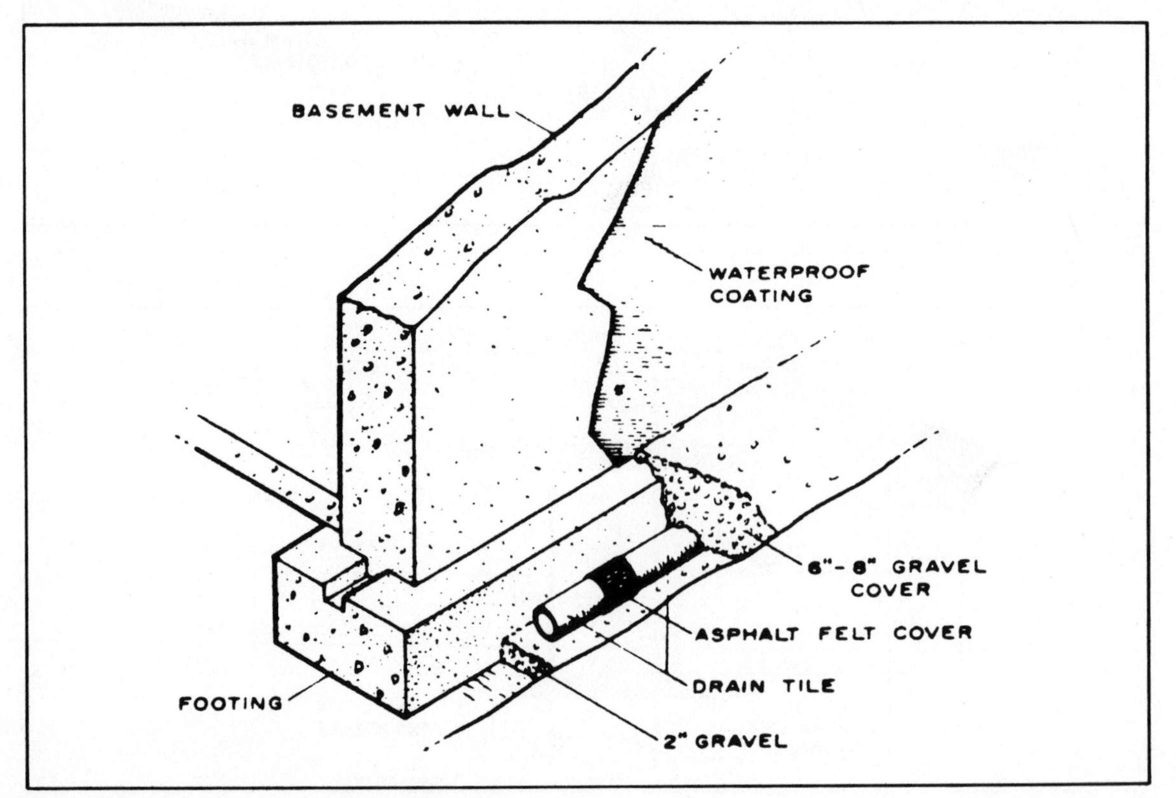

Fig. 3-20. Drain tile for soil drainage at outer wall.

subsurface water. This precaution prevents damp basements and wet floors. Drain tile is often necessary where habitable rooms are developed in the basement or where houses are located near the bottom of a long slope subjected to heavy runoff.

Install drains at or below the area to be protected. They should drain toward the outfall or ditch, or into a sump where the water can be pumped into a storm sewer. Place clay or concrete drain tile, 4 inches in diameter and 12 inches long, at the bottom of the footing level on top of a 2-inch gravel bed. Place tiles end to end and space them about 1/8 inch apart. Cover the top of the joint between the tiles with a strip of asphalt felt or similar paper; use 6 to 8 inches of gravel over the tile.

Dry wells for drainage water are used only when the soil conditions are favorable for this method of disposal. Local building regulations vary somewhat and should be consulted before you begin construction of a drainage system.

Chapter 4

Foundation Walls and Piers

Foundation walls form an enclosure for basements or crawl spaces and carry wall, floor, roof, and other building loads. The two types of walls most commonly used are poured concrete and concrete block. Treated wood foundations (Chapter 6) also might be used when accepted by local codes.

Wall thicknesses and types of construction are ordinarily controlled by local building regulations. Thicknesses of poured concrete basement walls can vary from 8 to 10 inches and concrete block walls from 8 to 12 inches, depending on story heights and length of unsupported walls.

Clear wall height should be no less than 7 feet from the top of the finish basement floor to the bottom of the joists; greater clearance is desirable to provide adequate headroom under girders, pipes, and ducts. Many contractors pour 8-foot-high walls above the footings, which provide a clearance of 7 2/3 feet from the top of the finish concrete floor to the bottom of the joists. Concrete block walls, 11 courses above the footings with a 4-inch solid cap block, produce about a 7 1/3-foot height to the joists from the basement floor.

POURED CONCRETE WALLS

Poured concrete walls (Fig. 4-1) require forming that must be tight and also braced and tied to withstand the forces of the pouring operation and the fluid concrete. Poured concrete walls should be *double-formed* (formwork constructed for each wall face). Reusable forms are used in the majority of poured walls. Panels consist of wood framing with plywood facings and are fastened together with clips or other ties. Wood sheathing boards and studs with horizontal members and braces are sometimes used in the construction of forms in small communities. As in reusable forms, formwork should be plumb, straight, and braced sufficiently to withstand the pouring operations.

Set in place frames for cellar windows, doors, and other openings as you erect the forming. Also set in place forms for the beam pockets, which are located to support the ends of the floor beam.

Reusable forms usually require little bracing other than horizontal members and sufficient blocking and bracing to keep them in place during pouring operations. Forms constructed with vertical

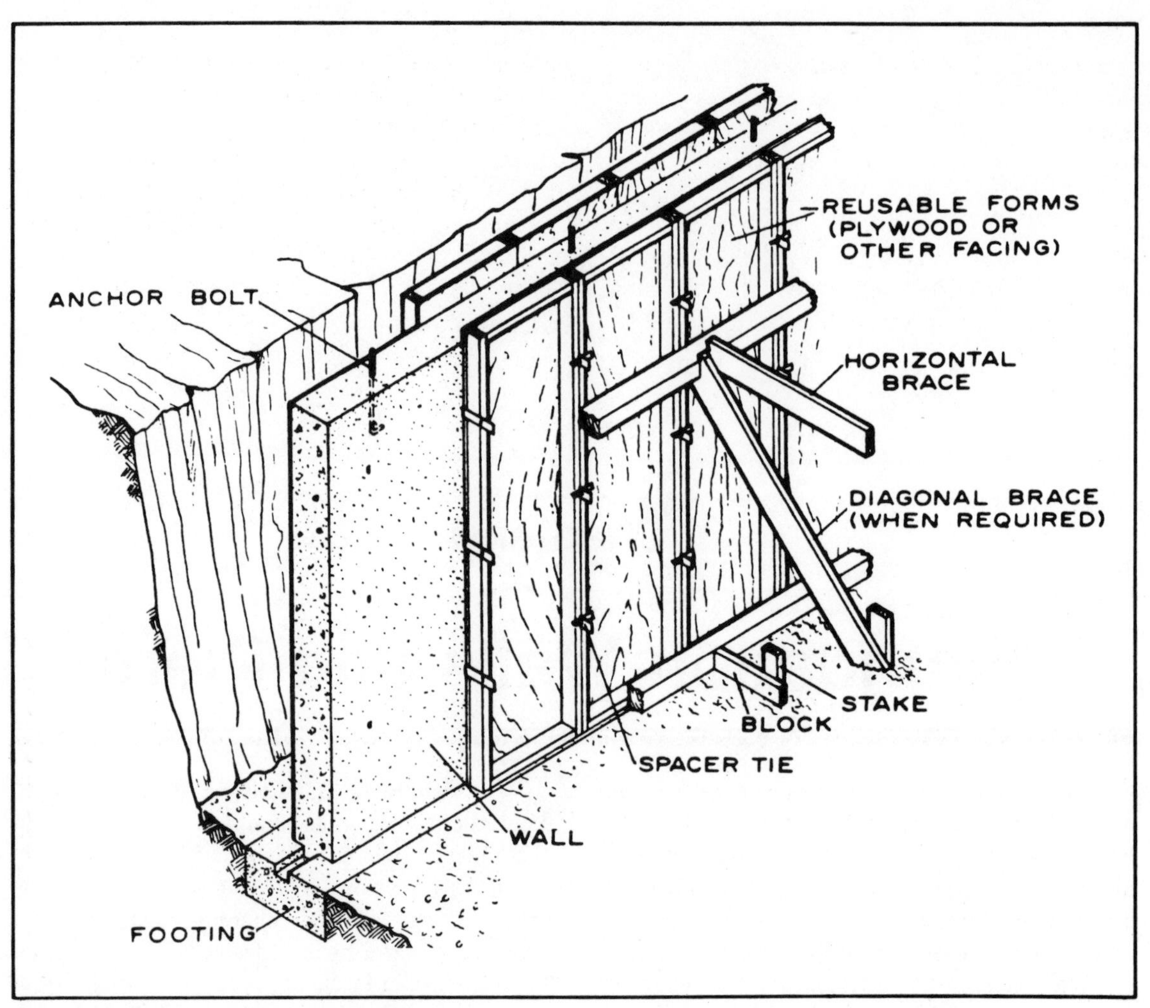

Fig. 4-1. Framing for poured concrete walls.

studs and waterproof plywood or lumber sheathing require horizontal whalers and bracing.

Use level markers of some type, such as nails along the forms, to ensure a level foundation top. This provides a good level sill plate and floor framing.

Pour concrete continuously, without interruption, and constantly puddle it to remove air pockets. Work the material under window frames and other blocking. If you have used wood spacer blocks, remove them; do not permit them to become buried in the concrete. Place anchor bolts for the sill plate while the concrete is still plastic. Always protect concrete when temperatures are below freezing.

Do not remove forms until the concrete has hardened and acquired sufficient strength to support loads imposed during early construction. At least two days, and preferably longer, are required when temperatures are well above freezing, and perhaps a week when outside temperatures are below freezing.

You can make poured concrete walls dampproof with one heavy cold or hot coat of tar or asphalt. Apply the tar or asphalt to the outside from the footings to the first grade line. Such coatings are usually sufficient to make a wall watertight against ordinary seepage (such as occurs after a rainstorm), but do not apply them until the surface of the concrete has dried enough to ensure good adhesion. In poorly drained soils, you can use a membrane.

CONCRETE BLOCK WALLS

Concrete blocks are available in various sizes and forms, but those generally used are 8, 10, and 12 inches wide. Modular blocks allow for the thickness and width of the mortar joint, so are usually about 7 5/8 inches high by 15 5/8 inches long. These dimensions result in blocks that measure 8 inches high and 16 inches long from centerline to centerline of the mortar joints.

Concrete block walls require no framework. Start block courses at the footing and lay them up with about 3/8-inch mortar joints, usually in a common bond (Fig. 4-2). Tool joints smooth to resist water seepage. Use a full bed of mortar in all contact surfaces of the block. When pilasters (column-like projections) are required by building codes or

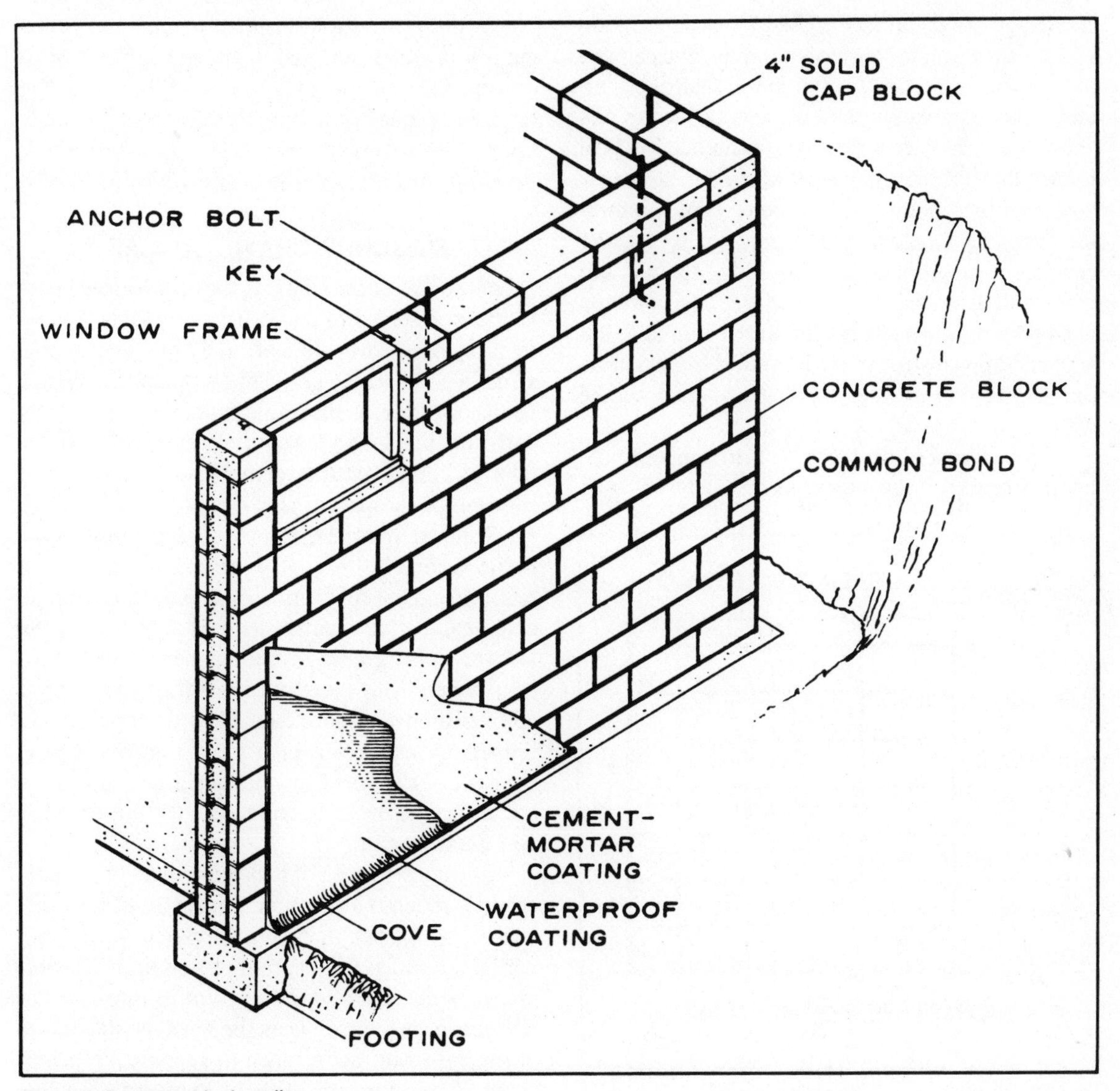

Fig. 4-2. Concrete block walls.

to strengthen a wall, place them on the interior side of the wall and terminate them at the bottom of the beam or girder supported. Set basement door and window frames with keys for rigidity and to prevent air leakage.

Cap block walls with 4 inches of solid masonry or concrete reinforced with wire mesh. Place anchor bolts for sills through the top two rows of blocks and the top cap. Anchor them with a large plate washer at the bottom and fill the block openings solidly with mortar or concrete.

If you are using an exposed block foundation as a finished wall for basement rooms, you can use the *stack bond pattern* (Fig. 4-3) for a pleasing effect. This pattern consists of placing blocks one above the other, resulting in continuous vertical mortar joints. When you use this system, however, you must incorporate some type of joint reinforcing every second course. This reinforcing usually consists of small-diameter, steel longitudinal and cross rods arranged in a grid pattern. The common bond does not normally require this reinforcing, but when additional strength is desired, it is good practice to incorporate this bonding system into the wall.

Protect freshly laid block walls in temperatures below freezing. If the mortar freezes before it sets, it can result in low adhesion, low strength, and joint failure.

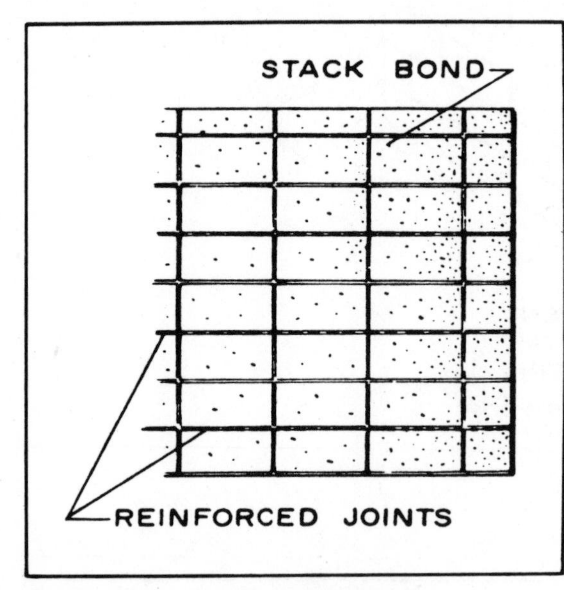

Fig. 4-3. Stack bond with reinforced joints.

To provide a tight, waterproof joint between the footing and wall, an elastic caulking compound is often used. Waterproof the wall by applying a coating of cement mortar over the block with a cove formed at the juncture with the footing (refer again to Fig. 4-2). When the mortar is dry, a coating of asphalt or other waterproofing normally will ensure a dry basement.

For added protection against wet soil conditions, you can mop a waterproof membrane of roofing felt or other material over the cement mortar coating. Use shingle-style laps of 4 to 6 inches. Hot tar or hot asphalt is commonly used over the membrane. This covering prevents leaks if minor cracks develop in the blocks or joints between the blocks.

MASONRY CRAWL SPACES

In some areas of the country, the crawl-space house is often preferred to one constructed over a basement or on a concrete slab. It is possible to construct a satisfactory crawl-space house by using a good soil cover, a small amount of ventilation, and sufficient insulation to reduce heat loss. These details will be covered later in this chapter.

One of the primary advantages of the crawl-space house over the full-basement house is the reduced cost. Little or no excavation or grading is required except for the footings and walls. In mild climates, the footings are located only slightly below the finish grade. In the northern states, however, where frost penetrates deeply, the footing is often located 4 or more feet below the finish grade, thus requiring more masonry work and increasing the cost. Always pour the footings over undisturbed soil and never over fill, unless you use special piers and grade beams.

The construction of a masonry wall for a crawl space is much the same as that required for a full basement, except that no excavation is required within the walls. Waterproofing and drain tile normally are not required for this type of construction. The masonry pier replaces the wood or steel posts of the basement house used to support the center beam. Footing size and wall thicknesses vary somewhat by location and soil conditions. A common minimum thickness for walls in single-story frame houses is 8 inches for hollow concrete block and 6 inches for poured concrete. The minimum footing thickness is 6 inches; the width is 12 inches for concrete block and 10 inches for the poured foundation wall for crawl-space houses. In well-constructed houses, however, it is common practice to use 8-inch walls and 16- × -8-inch footings.

Poured concrete or concrete block piers often are used to support floor beams in crawl-space houses. They should extend at least 12 inches above the ground line. The minimum size for a concrete block pier is 8 × 16 inches with a 16- × -24- × -8-inch footing. A solid cap block is used as a top course. Poured concrete piers should be at least 10 × 10 inches in size with a 20- × -20- × -8-inch footing.

Unreinforced concrete piers should be no greater in height than 10 times their least dimension. Concrete block piers should be no higher than four times the least dimension. The spacing of piers should not exceed 8 feet on center under exterior wall beams and interior girders set at right angles to the floor joists, and 12 feet on center under exterior wall beams set parallel to the floor joists. Exterior wall piers should not extend above grade more than four times their least dimension unless they are supported laterally by masonry or concrete walls. The size of the pier footings for walls should be based on the load and the capacity of the soil.

SILL PLATE ANCHORS

In wood-frame construction, the sill plate is anchored to the foundation wall with 1/2-inch bolts hooked and spaced about 8 feet apart (Fig. 4-4). In some areas, sill plates are fastened with masonry nails, but such nails do not have the uplift resistance of bolts. In high-wind and storm areas, well-anchored plates are very important. A sill sealer is often used under the sill plate on poured walls to

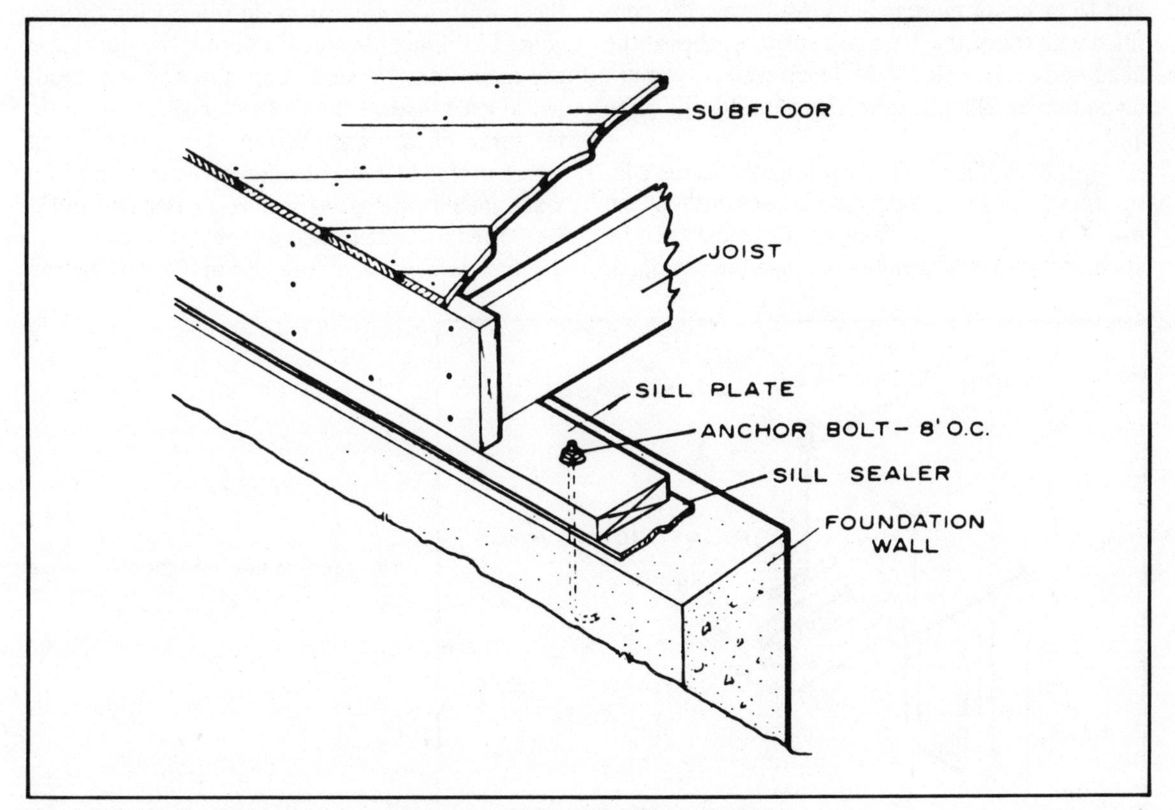

Fig. 4-4. Anchoring floor system to concrete or masonry walls with a sill plate.

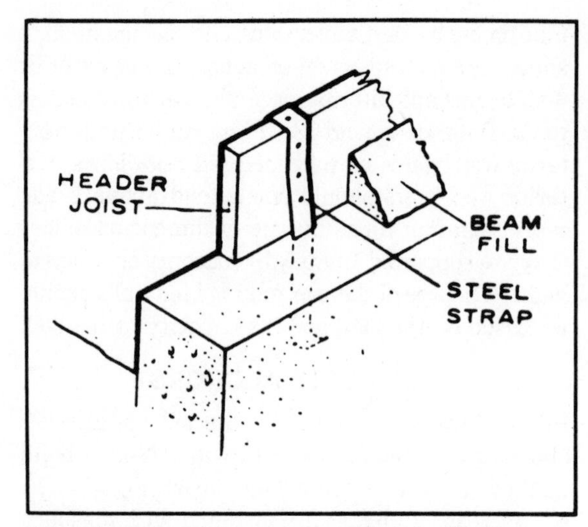

Fig. 4-5. Anchoring floor system without a sill plate.

take care of any irregularities that might have occurred during curing of the concrete. Embed anchor bolts 8 inches or more in poured concrete walls and 16 inches or more in block walls with the core filled with concrete. Use a large plate washer at the head end of the bolt for the block wall. If you are using termite shields, install them under the plate and sill sealer.

Although it is not the best practice, some contractors construct wood-frame houses without the use of a sill plate. The floor system must then be anchored with steel strapping. Place the strapping during the pouring or between the block joints. Bend strap over the joist or the header joist and fasten it by nailing (Fig. 4-5). The use of a concrete or mortar beam fill provides resistance to air and insect entry.

REINFORCING POURED WALLS

Poured concrete walls normally do not require steel reinforcing except over window or door openings located below the top of the wall. This type of construction requires that a properly designed steel or reinforced-concrete lintel be built over the frame (Fig. 4-6). For poured walls, lay the rods in place while you are pouring the concrete so that they are about 1 1/2 inches above the opening. Use paint primer on frames or treat them before installation. For concrete block walls, use a similar reinforced poured concrete or a precast lintel.

Where concrete work includes a connecting porch or garage wall not poured with the main basement wall, you must provide reinforcing-rod ties (Fig. 4-7). Place these rods during the pouring of the main wall. Depending on the size and depth, use at least three, 1/2-inch deformed rods at the intersection of each wall. You can use keyways in addition to resist lateral movement. Such connecting walls should extend below normal frost line and be supported by undisturbed ground. Wall extensions in concrete block walls are also of block. Construct

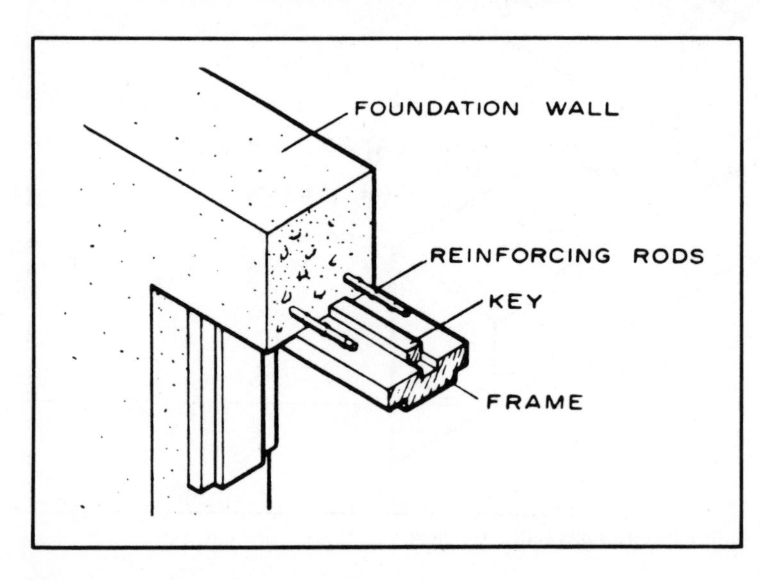

Fig. 4-6. Rod ties used for door frames.

them at the same time as the main walls over a footing placed below the frost line.

MASONRY VENEER

If you are using masonry veneer for the outside finish over wood-frame walls, the foundation must include a supporting ledge or offset about 5 inches wide (Fig. 4-8). This offset results in a space of about 1 inch between the masonry and the sheathing for ease of laying the brick. Use a base flashing at the brick course below the bottom of the sheathing and framing. You should lap the flashing with sheathing paper. Locate weep holes at this course to provide drainage. Form the weep holes by eliminating the mortar in a vertical joint. Use corrosion-resistant metal ties—spaced about 32 inches apart horizontally and 16 inches vertically—to bond the brick veneer to the framework. If you are using other than wood sheathing, secure the ties to the studs.

Lay brick and stone in a full bed of mortar; avoid dripping mortar into the space between the veneer and sheathing. Tool outside joints to a smooth finish to get the maximum resistance to water penetration. Protect masonry laid during the cold weather from freezing until after the mortar has set.

BEAM NOTCHES

When basement beams or girders are wood, the wall notch or pocket for such members must be large enough to allow at least 1/2 inch of clearance at the sides and ends of the beam for ventilation (Fig. 4-9). Unless the wood is treated, there is danger of decay where beams and girders are so tightly set in wall notches that moisture cannot readily es-

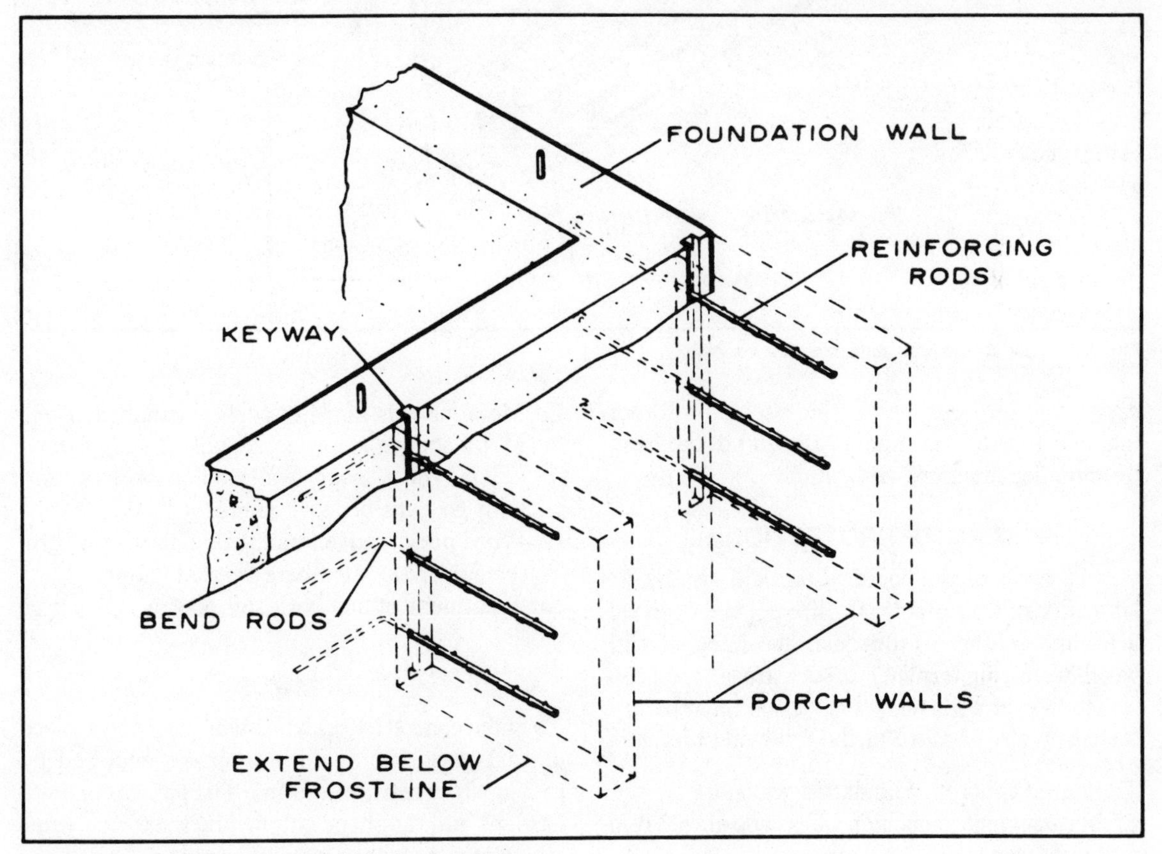

Fig. 4-7. Using reinforcing rods within the foundation.

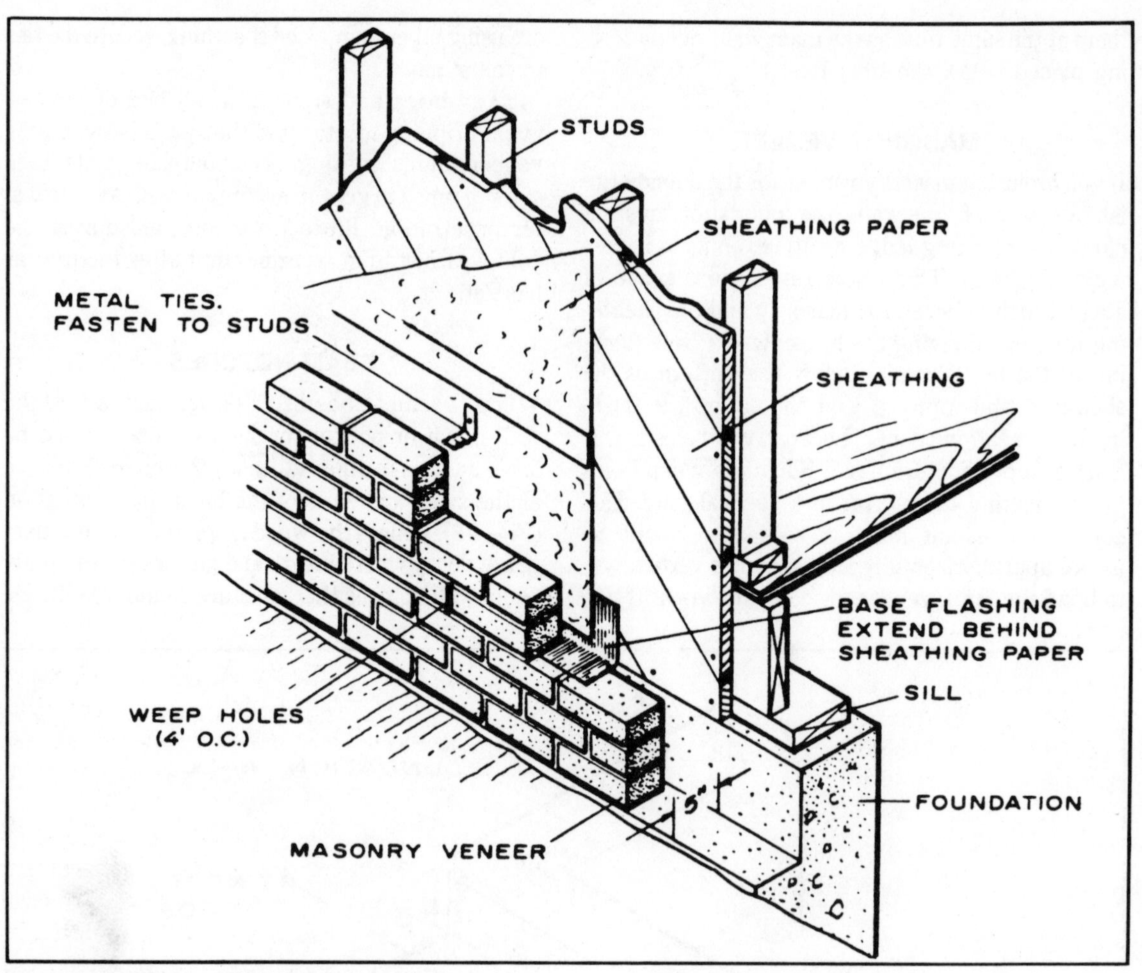

Fig. 4-8. Wood-frame wall with masonry veneer.

cape. A waterproof membrane, such as roll roofing, is commonly used under the end of the beam to minimize moisture absorption.

TERMITE PROTECTION

Certain areas of the country, particularly the Atlantic Coast, Gulf States, Mississippi and Ohio valleys, and southern California, are infested with wood-destroying termites. In such areas, wood construction over a masonry foundation must be protected by one or more of the following methods:

- ☐ Poured concrete foundation walls.
- ☐ Masonry unit foundation walls capped with reinforced concrete.
- ☐ Metal shields made of rust-resistant material. Metal shields are effective only if they extend beyond the masonry walls and are continuous, with no gaps or loose joints.
- ☐ Wood-preservative treatment. This method protects only the members treated.
- ☐ Treatment of soil with soil poison.

FORMWORK

To make concrete foundation walls and piers, place or cast plastic concrete into spaces enclosed by previously constructed forms. The plastic concrete hardens into the shape outlined by the forms, after which the forms are usually removed.

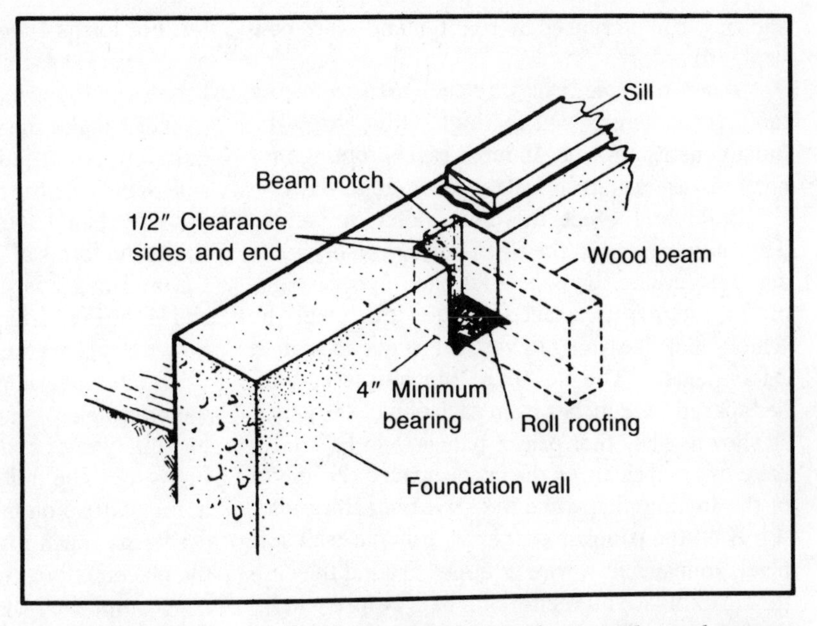

Fig. 4-9. Notch for wood beam.

Forms for concrete structures must be tight, rigid, and strong. If the forms are not tight, there will be a loss of paste. A loss of paste might cause weakness or a loss of water, which might cause sand streaking. The forms must be strong enough and braced well enough to resist the high pressure exerted by the concrete.

Forms, or parts of forms, are often omitted when a firm earth surface exists that is capable of supporting or molding the concrete. In most footings, for example, the bottom of the footing is cast directly against the earth, and only the sides are molded in forms. Many footings are cast with both bottom and sides against the natural earth. In this case, however, the specifications usually call for larger footings. A foundation wall is often cast between a form on the inner side and the natural earth surface on the outer side.

Form Materials

Form materials can be of wood, plywood, steel, or other approved material. Forms for concrete pavement, other than on curves, should be metal. On curves, flexible or curved forms of metal or wood can be used. Wood forms, for surfaces exposed to view in the finished structure and requiring a standard finish, should be tongue-and-groove boards or plywood. For exposed surfaces, undressed square-edge lumber can be used. Forms for surfaces requiring special finishes can be plywood or tongue-and-groove boards or lined with plywood, a nonabsorptive hard-pressed fiberboard, or other approved material.

Tongue-and-groove boards must be dressed to a uniform thickness, evenly matched, and free from loose knots, holes, and other defects that would affect the concrete finish. Plywood, other than for lining, should be concrete-form plywood not less than 5/8 inch thick. Surfaces of steel forms must be free from irregularities, dents, and sags.

Formwork Terms

Strictly speaking, it is only those parts of the formwork that directly mold the concrete that can be correctly referred to as *forms.* The rest of the formwork consists of various bracing and tying members used to strengthen the forms and to hold them rigidly in place.

Wall, column, and floor slab forms were formerly built by joining boards edge to edge, but built-up forms have been largely replaced by plywood forms. Plywood forms are tighter, more warp-resistant, and easier to construct than board forms, and they can be reused more often and more conveniently.

When possible, excavate the earth to form a mold for concrete wall footings. Otherwise, you must construct forms. In most cases, footings for columns are square or rectangular (Fig. 4-10).

Build and erect the four sides in panels. Thoroughly moisten the earth before you place the concrete. Make the panels for the opposite sides of the footings to exact footing width. Nail the 1-inch-thick sheathing to vertical cleats spaced on 2-foot centers. The cleats are 2-inch dressed lumber spaced 2 1/2 inches from each end of the panel, as shown. The other pair of panels (*b* in Fig. 4-10) have two end cleats on the inside spaced the length of the footing plus twice the sheathing thickness.

Hold the panels together with No. 8 or 9 soft black annealed iron wire wrapped around the center cleats. Place all reinforcing bars before you install the wire. The holes on each side of the cleat permitting the wire to be wrapped around the cleat should be less than 1/2 inch in diameter to prevent leakage of mortar through the hole.

Hold the panels in place with form nails until you install the tie wire. Drive all form (duplex) nails from the outside if possible to make stripping easier. For forms 4 feet square or larger, drive stakes as shown. These stakes and 1- × -6 boards nailed across the top are designed to prevent spreading. You can make the side panels higher than the required depth of footing and mark them on the inside to indicate the top of the footing. If the footings are less than 1 foot deep and 2 feet square, construct the forms of 1-inch sheathing without cleats. Cut and nail boards for the sides of the form as shown in Fig. 4-11. The forms can be braced and no wire ties are needed.

Sometimes it might be necessary to place a footing and a small pier at the same time. The form for this type of concrete construction is shown in Fig. 4-12. The units are similar to the one shown in Fig. 4-10. You must provide support for the upper form in such a way that it doesn't interfere with the placement of concrete in the lower form. This is accomplished by nailing a 2×2 or 4×4 to the lower form, as shown. Then nail the top form to these supports.

Formwork for a wall footing is shown in Fig. 4-13, and methods of bracing the form are given in Fig. 4-14. The sides of the forms are made of 2-inch lumber having a width equal to the depth of the footing. These pieces are held in place with

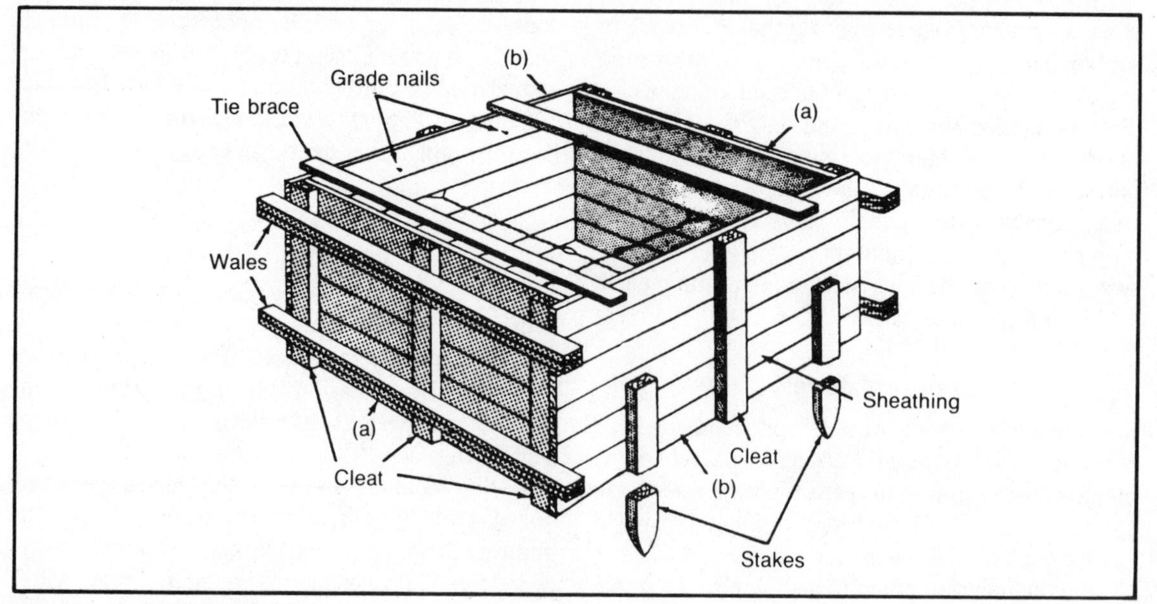

Fig. 4-10. Typical large footing form.

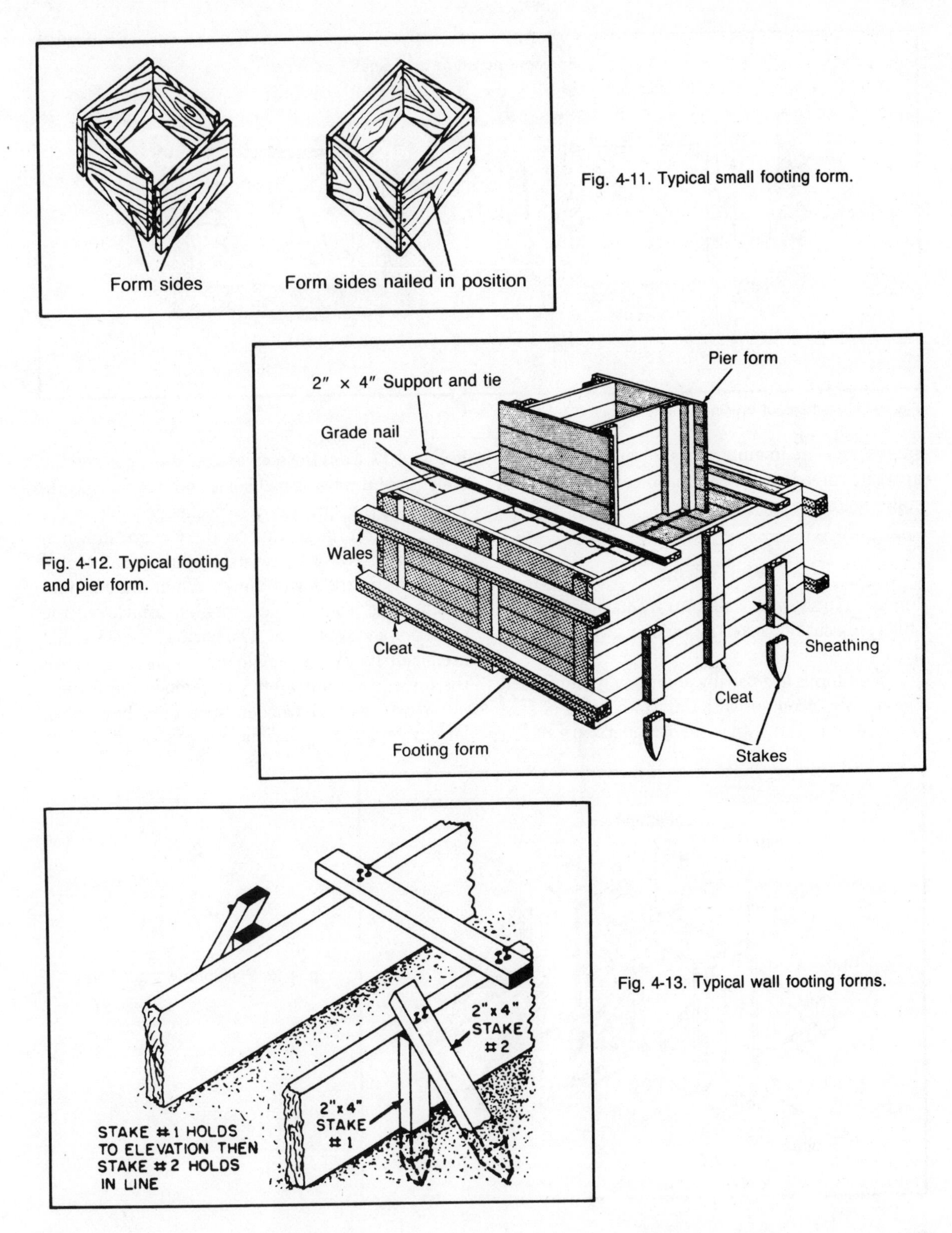

Fig. 4-11. Typical small footing form.

Fig. 4-12. Typical footing and pier form.

Fig. 4-13. Typical wall footing forms.

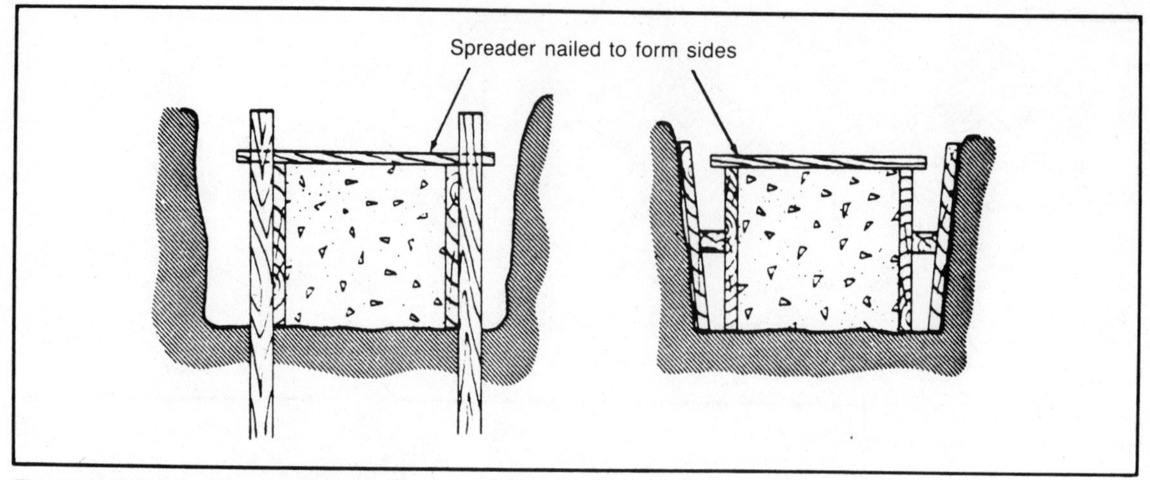

Fig. 4-14. Methods of bracing footing forms.

stakes and are maintained the correct distance apart by spreaders. The short brace shown at each stage holds the form in line.

Wall Forms

Figure 4-15 shows a wall form without wales. The studs are usually backed by wales, as shown in Fig. 4-16.

Wall forms are usually additionally reinforced against displacement with ties. Two types of simple wire ties, used with wood spreaders, are shown in Fig. 4-17. Pass the wire around the studs and the wales, and through small holes bored in the sheathing. Place the spreader as close as possible to the studs, and set the tie taut by the wedge (shown in the upper view) or by twisting with a small toggle (as shown in the lower view). When the concrete reaches the level of the spreader, knock out the spreader and remove it. The parts of the wire that are inside the forms remain in the concrete; cut off the outside surplus after you remove the forms.

Wire ties and wooden spreaders have been largely replaced by various manufactured devices

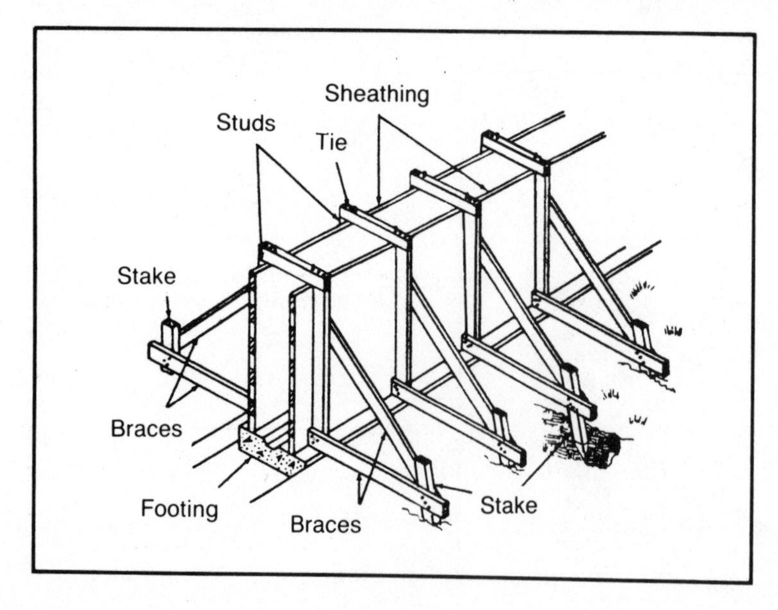

Fig. 4-15. Wall form without wales.

Fig. 4-16. Wall form with wales.

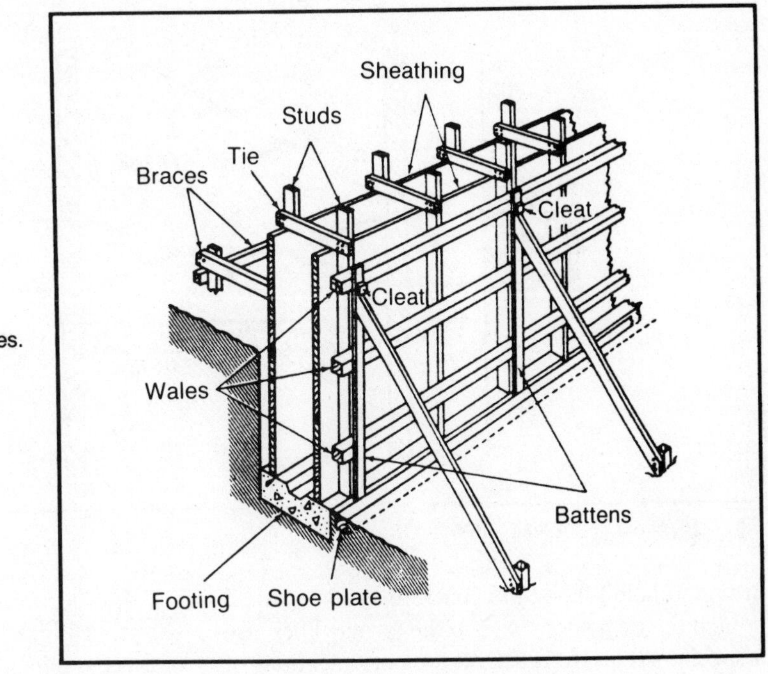

in which the functions of the tie and the spreader are combined. Figure 4-18 shows one of these devices, called a *snap tie.* These ties are made in various sizes to fit various wall thicknesses. The tie holders can be removed from the tie rod. The rod goes through small holes bored in the sheathing and also through the wales, which are usually doubled for that purpose.

Tapping the tie holders down on the ends of the rod brings the sheathing to bear solidly against the spreader washers. (To prevent the tie holder from coming loose, drive a duplex nail in the

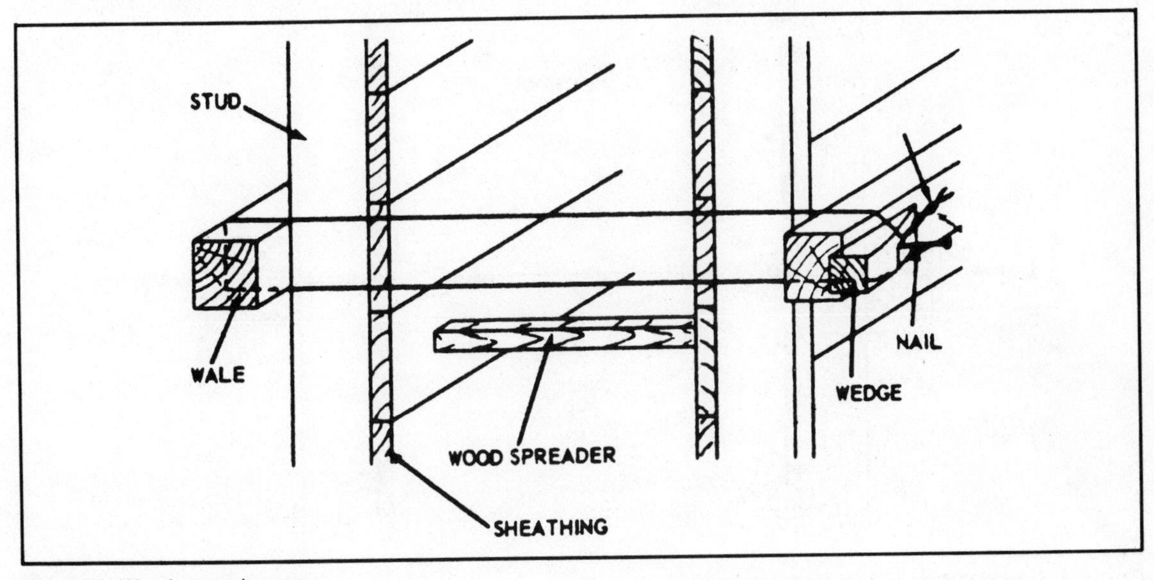

Fig. 4-17. Wood spreader use.

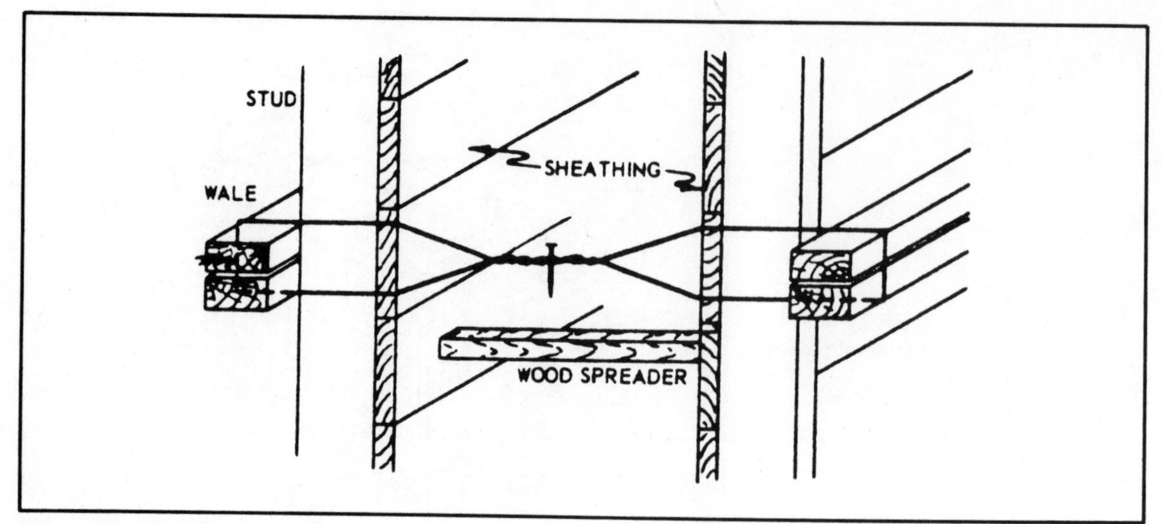

Fig. 4-18. Snap tie for wall forms.

provided hole.) After the concrete has hardened, detach the tie holders to strip the forms. Then use a special wrench to break off the outer sections of rod at their breaking points, located about 1 inch inside the surface of the concrete. Plug the small surface holes that remain with grout if necessary.

Another type of wall form tie is the tie rod shown in Fig. 4-19. The rod in this type consists of three sections: an inner section that is threaded on both ends, and two threaded outer sections. Place the inner section, with the cones set to the thickness of the wall, between the forms. Pass the

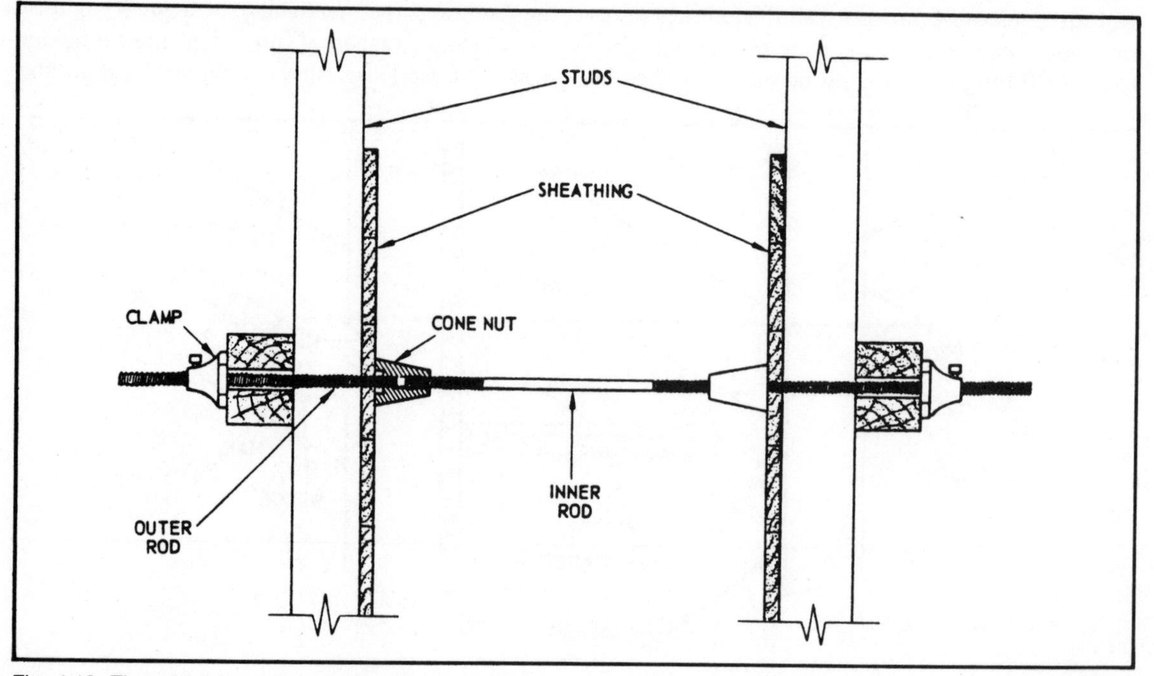

Fig. 4-19. Tie rod joining wall form panels.

outer sections through the wales and sheathing and thread them into the cone nuts. Then thread the clamps up on the outer sections to bring the forms to bear against the cone nuts. After the concrete hardens, loosen the clamps and remove the outer sections of rod by threading them out of the cone nuts.

After you have stripped the forms, remove the cone nuts from the concrete by threading them off the inner sections of rod with a special wrench. You can use grout to plug the cone-shaped surface holes that remain. The inner sections of rod remain in the concrete. You can reuse the outer sections and the cone nuts indefinitely.

Wall forms are usually constructed as separate panels, each made by nailing sheathing to a number of studs. Panels are joined to each other in line as shown in Fig. 4-20. A method of joining panels at a corner is shown in Fig. 4-21.

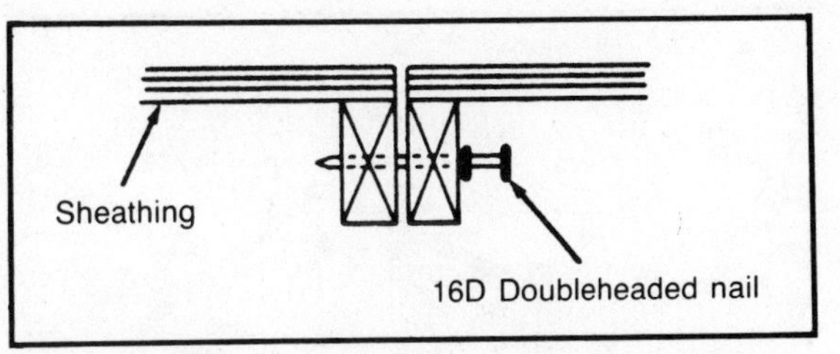

Fig. 4-20. Method of joining wall form panels in a line.

Make up the panels for the form first by nailing the yoke members to the sheathing. On two panels, the yoke members come flush with the edges of the sheathing; on the other two they project beyond the edges. Bore bolt holes in these projections as shown, and insert bolts to back up the wedges that are driven to tighten the yokes.

Column Forms

Figure 4-22 shows a column form. Because the rate of placing in a column form is very high and because the bursting pressure exerted on the form by the concrete increases directly with the rate of placing, you must brace a column form securely with yokes as shown in Fig. 4-22. Because the bursting pressure is greater at the bottom of the form than it is at the top, place the yokes closer together at the bottom than at the top.

Beam Forms

The type of construction used for beam forms depends on whether the sides are to be stripped and the bottom left in place until the concrete has developed enough strength to permit removal of the shoring. Beam and girder forms are shown in Fig. 4-23. Beam forms are subjected to very little bursting pressure, but must be shored up at frequent intervals to prevent sagging under the weight of the fresh concrete.

The bottom of the form has the same width as the beam and is in one piece for the full width. The sides of the form should be 1-inch-thick tongue-and-

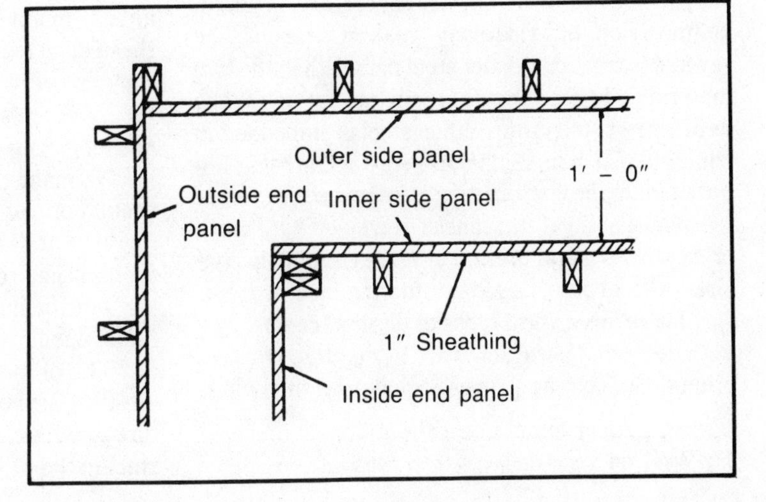

Fig. 4-21. Joining panels at a corner.

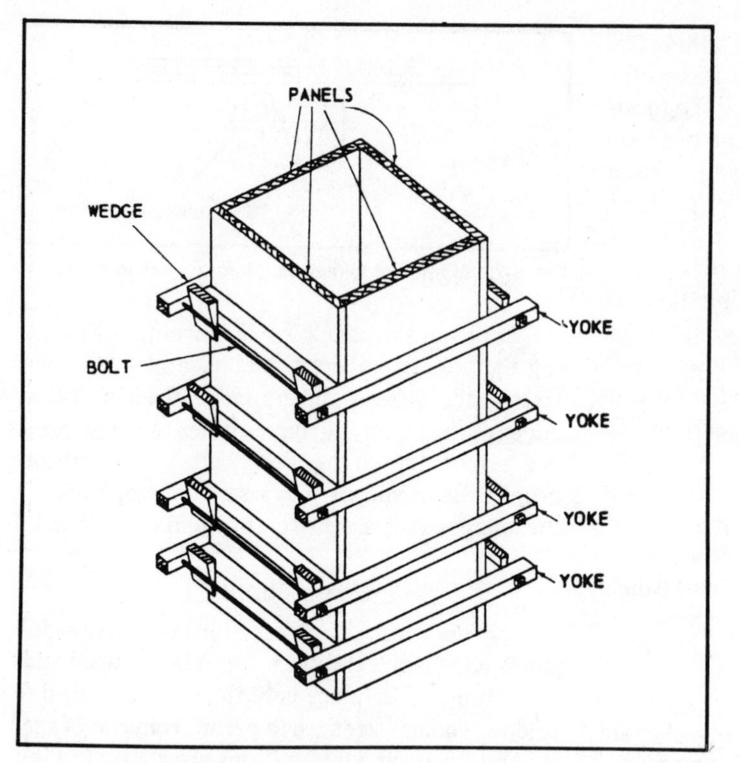

Fig. 4-22. Column form.

groove sheathing and should lap over the bottom as shown. The sheathing is nailed to 2-×-4-inch struts placed on 3-foot centers. A 1-×-4-inch piece is nailed along the struts. These pieces support the joist for the floor panel (Fig. 4-24).

REINFORCED CONCRETE

As you learned in Chapter 3, concrete is strong in compression, but relatively weak in tension. The reverse is true for slender steel bars. When the two materials are used together, one makes up for the deficiency of the other. When steel is embedded in concrete in a manner that assists it in carrying imposed load, the combination is known as *reinforced concrete*. The steel can consist of welded wire mesh or expanded metal mesh, but more commonly consists of steel bars, called reinforcing bars or *rebar*.

There are several types of ties that can be used with deformed bars; some are more effective than others. Six common types include the following:

- ☐ Snap tie or simple tie (Fig. 4-25)
- ☐ Wall tie
- ☐ Saddle tie
- ☐ Saddle tie with twist
- ☐ Double strand single tie
- ☐ Cross tie or figure eight tie

Most builders use the snap tie and the saddle tie.

To make the snap tie or simple tie, simply wrap the wire once around the two crossing bars in a diagonal manner with the two ends on top. Twist these ends together with a pair of sidecutters until they are very tight against the bars. Then cut off the loose ends of the wire. This tie is used mostly on floor slabs.

To make the saddle tie, pass the wires halfway around one of the bars on either side of the crossing bar and bring them squarely or diagonally around the crossing bar. Twist the ends together and cut them off. This tie is used on special locations, such as walls.

The proper location for the reinforcing bars is usually given on drawings and plans. For the structure to withstand the loads it must carry, place the steel in the position given. Secure the bars in posi-

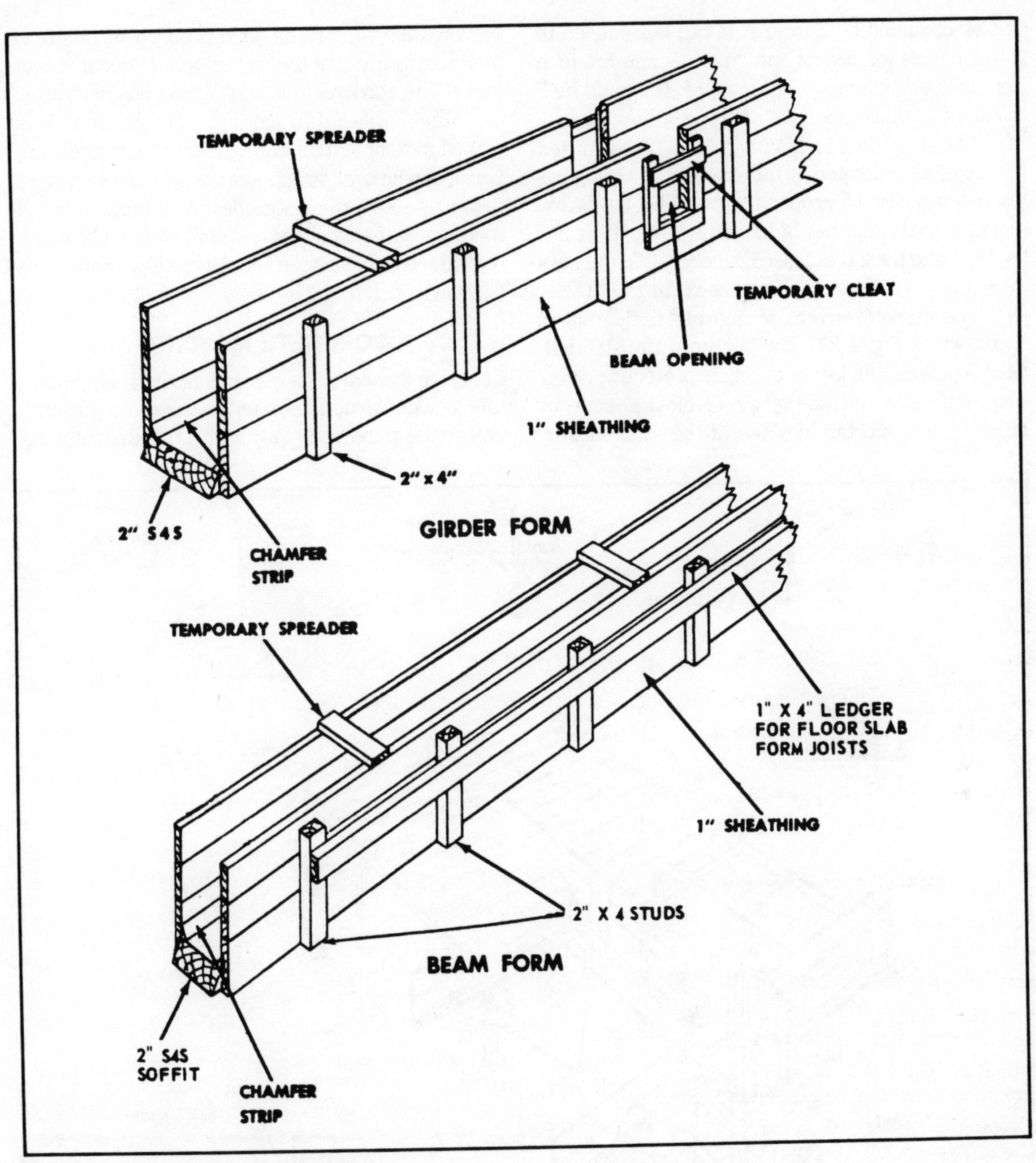

Fig. 4-23. Typical beam and girder forms.

tion in such a way that they will not move when you are placing the concrete. This can be accomplished by using reinforcing bar supports, shown in Figs. 4-26 through 4-28.

Footings and other principal structural members that are against the ground should have at least 3 inches of concrete between steel and ground. If the concrete surface is to be in contact with the ground or exposed to the weather after removal of the forms, the protective covering of concrete over the steel should be 2 inches. It can be reduced to 1 1/2 inches for beams and columns and 3/4 inch for slabs and interior wall surfaces. It should be 2 inches for all exterior wall surfaces, however.

The stress in a tension bar can be transmitted through the concrete and into another adjoining bar by a lap splice of proper length. The lap is expressed as the number of bar diameters. If the bar is #2, make the lap at least 12 inches. Tie the bars together with a snap tie, as shown in Fig. 4-29.

The support for reinforcing steel in floor slabs is shown in Fig. 4-30. The height of the slab bolster is determined by the concrete protective cover required. You can use concrete blocks made of sand-cement mortar in place of the slab bolster. Never use wood blocks for this purpose if there is any possibility that the concrete can become wet and if the construction is of a permanent type.

Steel is placed in footings very much as it is placed in floor slabs. Stones, rather than steel supports, can be used to support the steel at the proper distance above the subgrade. Steel mats in small footings are generally preassembled and placed after the forms have been set. A typical arrangement is shown in Fig. 4-31.

CONCRETE MASONRY

Concrete masonry has become increasingly important as a construction material. Concrete masonry units once were made only with small-diameter ag-

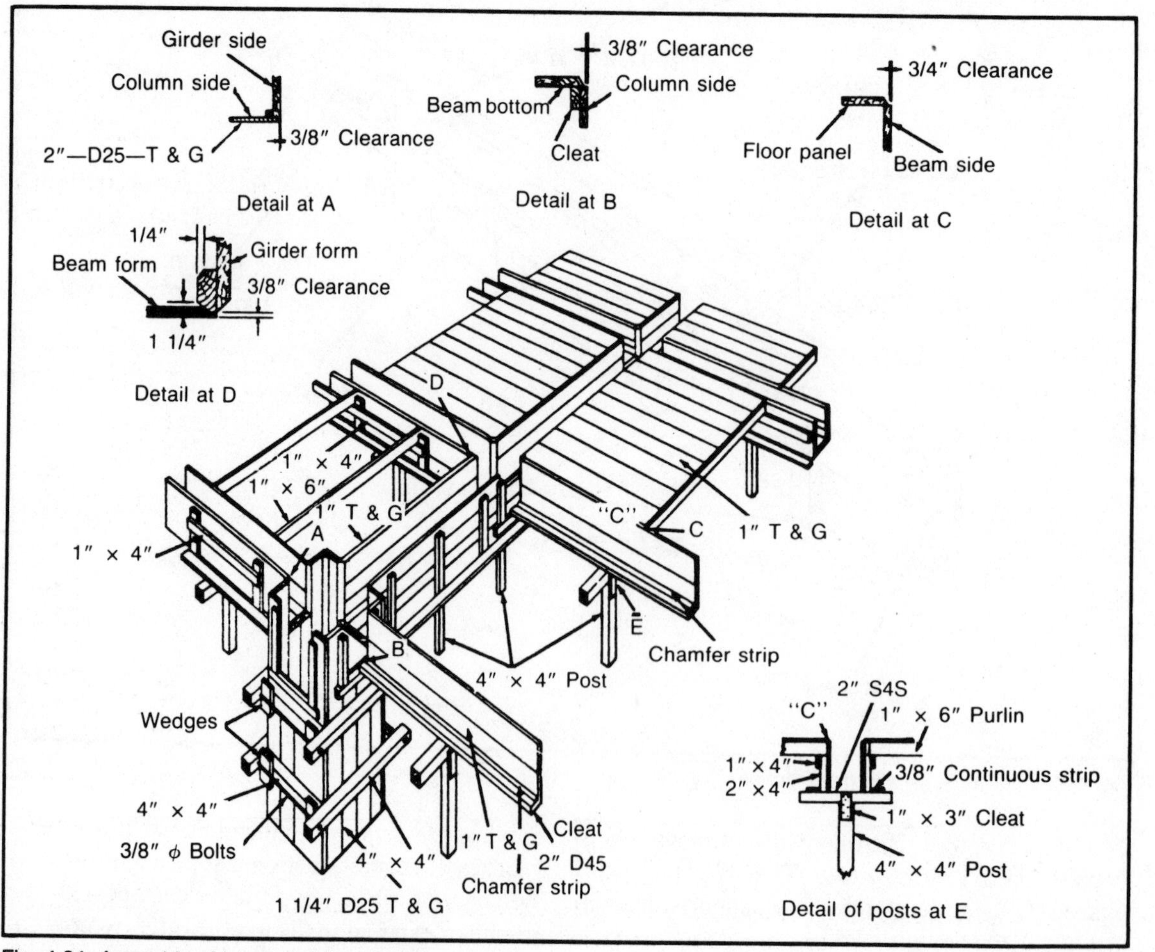

Fig. 4-24. Assembly of beam and floor forms.

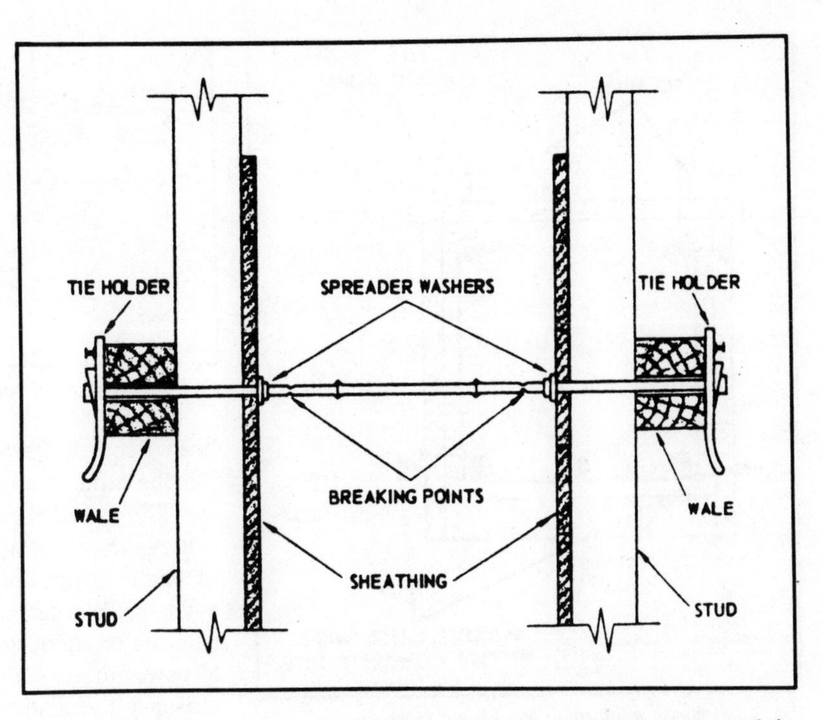

Fig. 4-25. Snap tie.

gregate such as cinder or slag, but now they are made like any other lightweight concrete, except that the maximum diameter of the coarse aggregate is about 5/8 inch.

Concrete units can be manufactured either by machine or by hand. Most builders purchase machine-manufactured units, however. Blocks can be steam-cured to 70 percent of 28-day strength in about 15 hours. They can be ordinary damp-cured in about 7 days.

Concrete Blocks

Concrete building units are made in sizes and shapes to fit different construction needs. Units are

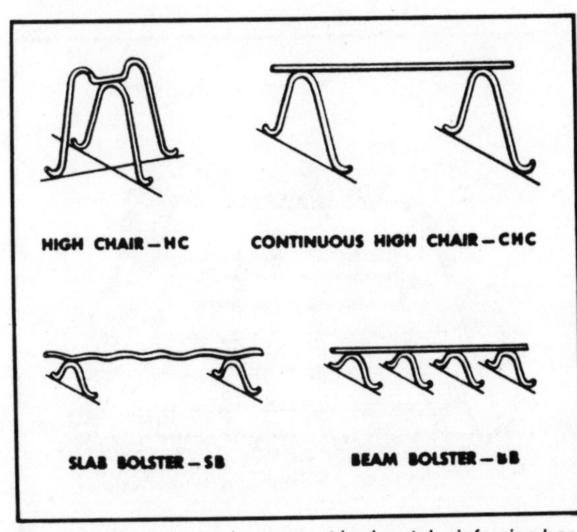

Fig. 4-26. Devices used to support horizontal reinforcing bar.

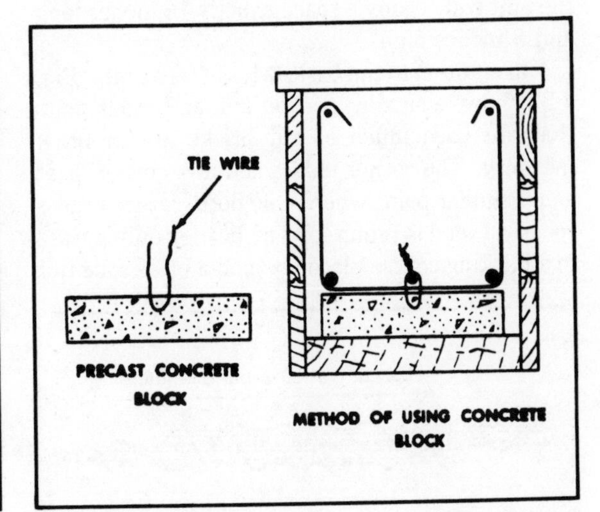

Fig. 4-27. Precast concrete block used for reinforcing steel support.

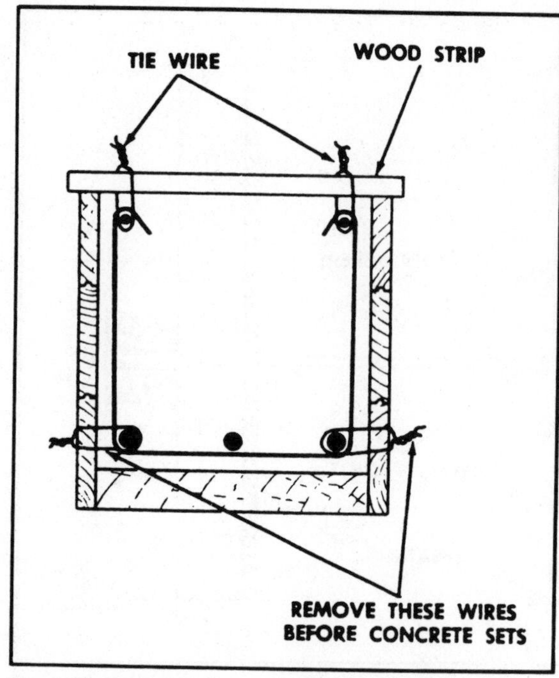

Fig. 4-28. Beam reinforcing steel hung in place.

made in full and half-length sizes (Fig. 4-32).

Concrete unit sizes are usually referred to by their nominal dimensions. A unit measuring 7 5/8 inches wide, 7 5/8 inches high, and 15 5/8 inches long is referred to as an 8-×-8-×-16-inch unit. When it is laid in a wall with 3/8-inch mortar joints, the unit will occupy a space exactly 16 inches long and 8 inches high.

In addition to the basic 8-×-8-×-16 units, Fig. 4-32 shows a smaller portion unit and other units that are used much as cut bricks are in brick masonry. The corner unit is laid at a corner or at some similar point where a smooth, rather than a recessed, end is required. The header unit is used in a backing course placed behind a brick face tier

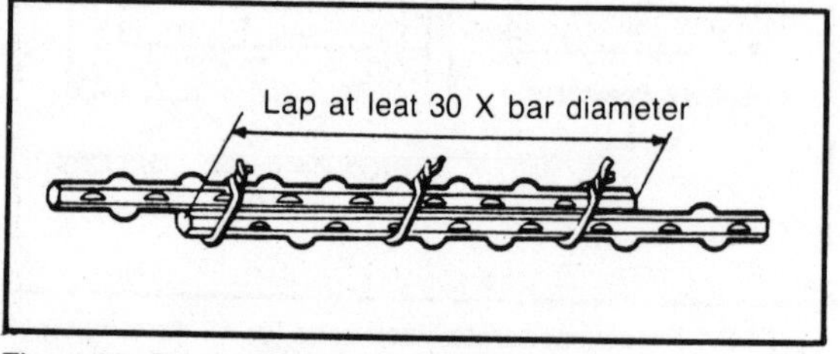

Fig. 4-29. Bars spliced by lapping.

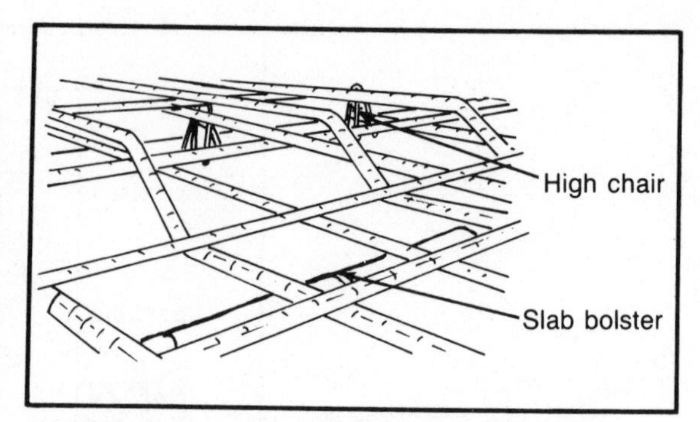

Fig. 4-30. Using reinforcing bar in slab construction.

header course. Part of the block is cut away to admit the brick headers.

The use of the other special units shown are self-evident. In addition to these shapes, a number of smaller shapes for various special purposes are available. You can cut units to the desired size with a bolster or, more conveniently and accurately, with a power-driven masonry saw. Tools used for masonry installation are illustrated in Fig. 4-33.

Mortar Joints

The sides and the recessed ends of a concrete block are called the *shell*. The material that forms the partitions between the cores is called the *web*. Each of the long sides of a block is called a *face shell*, and each of the recessed ends is called an *end shell*. Bed

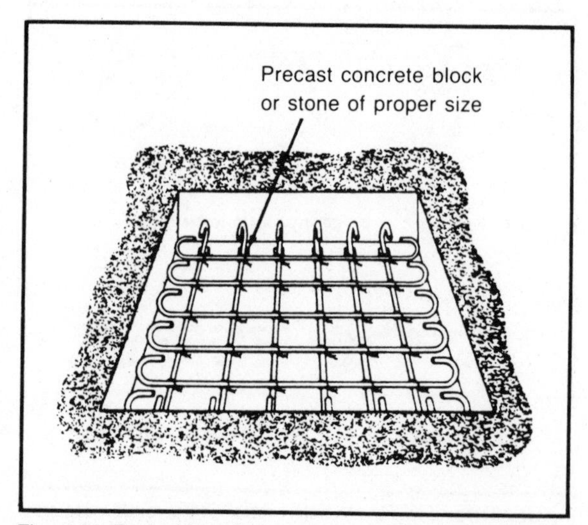

Fig. 4-31. Typical installation of rebar in slab construction.

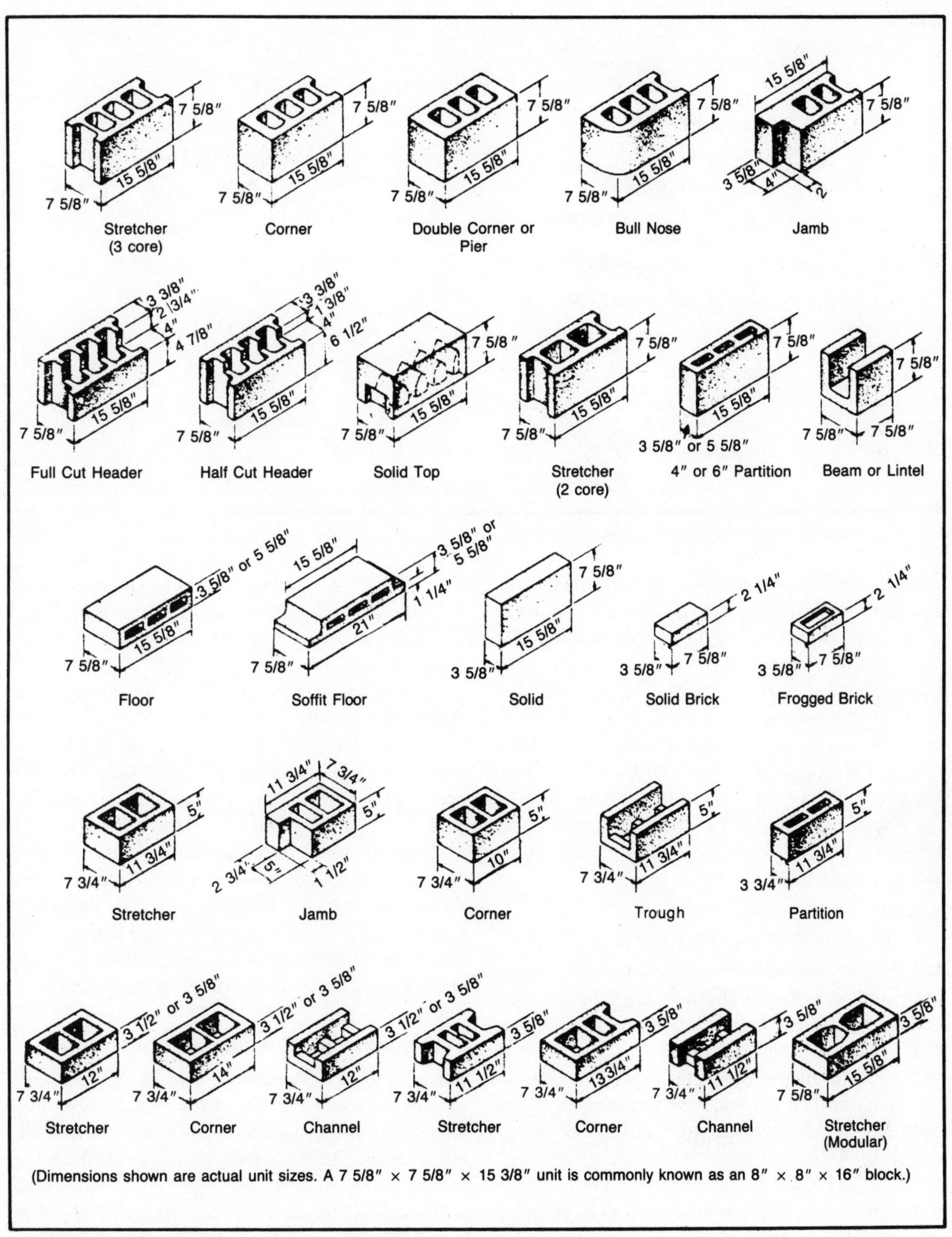

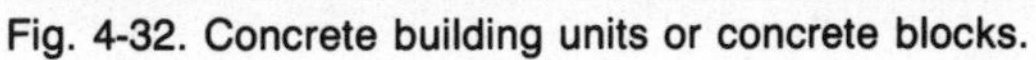
Fig. 4-32. Concrete building units or concrete blocks.

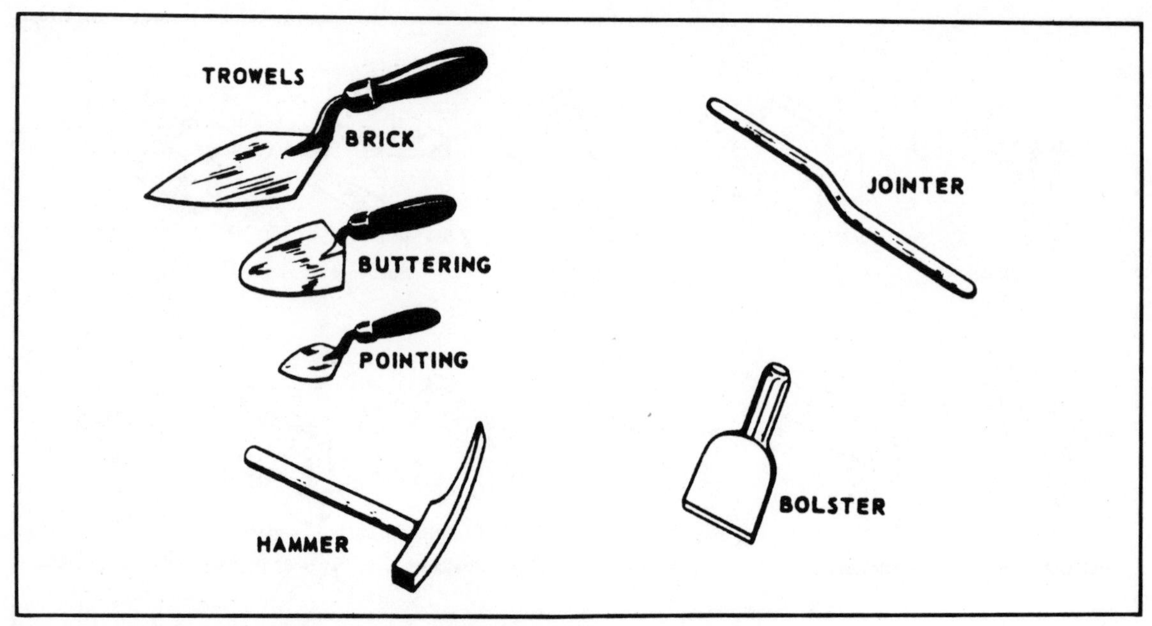

Fig. 4-33. Mason's trowel, bolster, hammer, and jointer.

joints on first courses and bed joints in column construction are mortared by spreading a 1-inch layer of mortar. This is called *full mortar bedding.* For most other bed joints, only the upper edges of the face shells need to be mortared. This is called *face-shell mortar bedding.*

The vertical ends of the face shells, on either side of the end shells, are called the *edges*. Head joints can be mortared by buttering both edges of the block being laid or by buttering one edge on the block being laid and the opposite edge on the block already in place.

Planning the Layout

Lay out concrete masonry foundation walls to make maximum use of full- and half-length units, thus minimizing cutting and fitting of units on the job. Plan the length and height of the foundation wall, and the width and height of openings and wall areas between any doors or windows and corners to use full-size and half-size units, which are usually available (Fig. 4-34).

This procedure assumes that window and door frames are of modular dimensions that fit modular full- and half-size units. In this situation, all horizontal dimensions should be in multiples of nominal full-length masonry units, and both horizontal and vertical dimensions should be designed to be in multiples of nominal full-length masonry units. Both horizontal and vertical dimensions should be in multiples of 8 inches.

Table 4-1 lists nominal lengths of concrete masonry walls by stretchers. Table 4-2 lists nominal lengths of concrete masonry walls by courses, and Table 4-3 by height. When units $8 \times 4 \times 16$ inches are used, the horizontal dimension should be planned in multiples of 8 inches (half-length units) and the vertical dimensions in multiples of 4 inches. If the thickness of the wall is greater or less than the length of a half unit, a special length unit is required at each corner in each course.

Masonry Unit Foundation Construction

Refer to Figs. 4-35 through 4-39. After you have located the corners of the wall, check the layout by stringing out the blocks for the first course without mortar. Use a chalked snap line to mark the footing and align the block accurately. Then spread a full bed of mortar. Furrow it with the trowel to ensure plenty of mortar along the bottom edges of

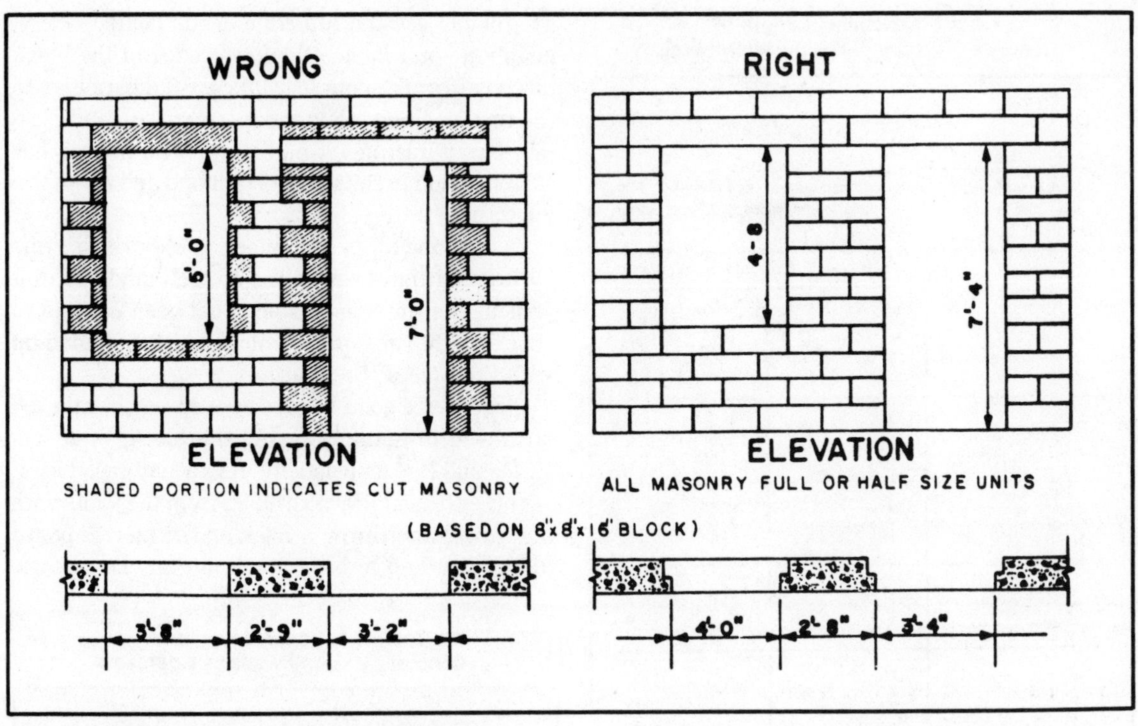

Fig. 4-34. Planning concrete masonry wall openings.

the face shells of the block for the first course.

Lay the corner block first, carefully positioning it. Lay all blocks with the thicker end of the face shell up to provide a larger mortar bedding area. Apply mortar only to the ends of the face shells for vertical joints. Place several blocks on end and apply mortar to the vertical face shells in one operation. Bring each block over its final position and push it downward into the mortar bed and against the previously laid block to obtain a well-filled vertical mortar joint. After you have laid three or four blocks, use a mason's level as a straightedge to ensure correct alignment of the blocks. Carefully check the blocks with the level, bring them to proper grade, and plumb them by tapping with the trowel handle.

Be sure to lay the first course of concrete masonry with great care. Make sure it is properly aligned, leveled, and plumbed to ensure that succeeding courses and the entire wall are straight and true.

After you have laid the first course, apply mortar only to the horizontal face shells of the block (face-shell mortar bedding). You can apply mortar for the vertical joints to the vertical face shells of the block to be placed or to the block previously laid, or both, to ensure well-filled joints.

Build the corners of the wall first, making them four or five courses higher than the center of the wall. As each course is laid at the corner, check it with a level for alignment, for levelness, and plumbness. Check each block carefully with a level or straightedge to make certain that the faces of the block are all in the same plane to ensure true, straight walls.

The use of a *story* or *course pole,* a board with markings 8 inches apart, provides an accurate method of determining the top of the masonry for each course. Joints are 3/8 inch thick. In building the corners, step back each course a half block. Check the horizontal spacing of the block by placing your level diagonally across the corners of the block.

When you are filling in the wall between the

Table 4-1. Nominal Length of Concrete Masonry Walls by Stretchers.

(Actual length of wall is measured from outside edge to outside edge of units and is equal to the nominal length minus 3/8″ (one mortar joint).)

No. of stretchers	Nominal length of concrete masonry walls	
	Units 15 5/8″ long and half units 7 5/8″ long with 3/8″ thick head joints.	Units 11 5/8″ long and half units 5 5/8″ long with 3/8″ thick head joints.
1	1′ 4″	1′ 0″.
1 1/2	2′ 0″	1′ 6″.
2	2′ 8″	2′ 0″.
2 1/2	3′ 4″	2′ 6″.
3	4′ 0″	3′ 0″.
3 1/2	4′ 8″	3′ 6″.
4	5′ 4″	4′ 0″.
4 1/2	6′ 0″	4′ 6″.
5	6′ 8″	5′ 0″.
5 1/2	7′ 4″	5′ 6″.
6	8′ 0″	6′ 0″.
6 1/2	8′ 3″	6′ 6″.
7	9′ 4″	7′ 0″.
7 1/2	10′ 0″	7′ 6″.
8	10′ 8″	8′ 0″.
8 1/2	11′ 4″	8′ 6″.
9	12′ 0″	9′ 0″.
9 1/2	12′ 8″	9′ 6″.
10	13′ 4″	10′ 0″.
10 1/2	14′ 0″	10′ 6″.
11	14′ 8″	11′ 0″.
11 1/2	15′ 4″	11′ 6″.
12	16′ 0″	12′ 0″.
12 1/2	16′ 8″	12′ 6″.
13	17′ 4″	13′ 0″.
13 1/2	18′ 0″	13′ 6″.
14	18′ 8″	14′ 0″.
14 1/2	19′ 4″	14′ 6″.
15	20′ 0″	15′ 0″.
20	26′ 8″	20′ 0″.

corners, stretch a mason's line from corner to corner for each course, and lay the top outside edge of each block to this line. The manner of gripping the block is important. Tip it slightly towards you so you can see the top edge of the course below, enabling you to place the lower edge of the block directly over the course. Make any adjustments to the final position while the mortar is soft and plastic. Any adjustments made after the mortar has stiffened will break the mortar bond and allow water to penetrate.

Level and align each block to the mason's line by tapping lightly with the trowel handle. Limit your use of the mason's level between corners to checking the face of each block to keep it lined up with the face of the wall.

To ensure good bond, do not spread mortar too far ahead of actually laying the block, or it will stiffen and lose its plasticity. As you lay each block, cut off excess mortar extruding from the joints with the trowel and throw it back on the mortar board to be reworked into the fresh mortar. Do not use

Table 4-2. Nominal Height of Concrete Masonry Walls by Courses.

(For concrete masonry units 7 5/8″ and 3 5/8″ in height laid with 3/8″ mortar joints. Height is measured from center to center of mortar joints.)

No. of courses	Nominal height of concrete masonry walls	
	Units 7 5/8″ high and 3/8″ thick bed joint	Units 3 5/8″ high and 3/8″ thick bed joint
1	8″	4″.
2	1′ 4″	8″.
3	2′ 0″	1′ 0″.
4	2′ 8″	1′ 4″.
5	3′ 4″	1′ 8″.
6	4′ 0″	2′ 0″.
7	4′ 8″	2′ 4″.
8	5′ 4″	2′ 8″.
9	6′ 0″	3′ 0″.
10	6′ 8″	3′ 4″.
15	10′ 0″	5′ 0″.
20	13′ 4″	6′ 8″.
25	16′ 8″	8′ 4″
30	20′ 0″	10′ 0″.
35	23′ 4″	11′ 8″.
40	26′ 8″	13′ 4″.
45	30′ 0″	15′ 0″.
50	33′ 4″	16′ 8″.

Table 4-3. Table of Course/Height Relationship.

Courses	Height	Courses	Height	Courses	Height	Courses	Height	Courses	Height
1	0′ 2⅝″	21	4′ 7⅛″	41	8′ 11⅝″	61	13′ 4⅛″	81	17′ 8⅝″
2	0′ 5¼″	22	4′ 9¾″	42	9′ 2¼″	62	13′ 6¾″	82	17′ 11¼″
3	0′ 7⅞″	23	5′ 0⅜″	43	9′ 4⅞″	63	13′ 9⅜″	83	18′ 1⅞″
4	0′ 10½″	24	5′ 3″	44	9′ 7½″	64	14′ 0″	84	18′ 4½″
5	1′ 1⅛″	25	5′ 5⅝″	45	9′ 10⅛″	65	14′ 2⅝″	85	18′ 7⅛″
6	1′ 3¾″	26	5′ 8¼″	46	10′ 0¾″	66	14′ 5¼″	86	18′ 9¾″
7	1′ 6⅜″	27	5′ 10⅞″	47	10′ 3⅜″	67	14′ 7⅞″	87	19′ 0⅜″
8	1′ 9″	28	6′ 1½″	48	10′ 6″	68	14′ 10½″	88	19′ 3″
9	1′ 11⅝″	29	6′ 4⅛″	49	10′ 8⅝″	69	15′ 11⅛″	89	19′ 5⅝″
10	2′ 2¼″	30	6′ 6¾″	50	10′ 11¼″	70	15′ 3¾″	90	19′ 8¼″
11	2′ 4⅞″	31	6′ 9⅜″	51	11′ 1⅞″	71	15′ 6⅜″	91	19′ 10⅞′
12	2′ 7½″	32	7′ 0″	52	11′ 4½″	72	15′ 9″	92	20′ 1½″
13	2′ 10⅛″	33	7′ 2⅝″	53	11′ 7⅛″	73	15′ 11⅝″	93	20′ 4⅛″
14	3′ 0¾″	34	7′ 5¼″	54	11′ 9¾″	74	16′ 2¼″	94	20′ 6¾″
15	3′ 3⅜″	35	7′ 7⅞″	55	12′ 0⅜″	75	16′ 4⅞″	95	20′ 9⅜″
16	3′ 6″	36	7′ 10½″	56	12′ 3″	76	16′ 7½″	96	21′ 0″
17	3′ 8⅝″	37	8′ 1⅛″	57	12′ 5⅝″	77	16′ 10⅛″	97	21′ 2⅝″
18	3′ 11¼″	38	8′ 3¾″	58	12′ 8¼″	78	17′ 0¾″	98	21′ 5¼″
19	4′ 1⅞″	39	8′ 6⅜″	59	12′ 10⅞″	79	17′ 3⅜″	99	21′ 7⅞″
20	4′ 4½″	40	8′ 9″	60	13′ 1½″	80	17′ 6″	100	21′ 10½″

dead mortar that has been picked up from the scaffold or from the floor.

When you are installing the closure block, butter all edges of the opening and all four vertical edges of the closure block with mortar and carefully lower the closure block into place. If any of the mortar falls out leaving an opening joint, remove the block and repeat the procedure.

Weathertight joints and neat appearance of the concrete block foundation walls are dependent on proper tooling. Tool the mortar joints after a section of the wall has been laid and the mortar has become "thumbprint hard." Tooling compacts the

Fig. 4-35. Making a head joint.

Fig. 4-36. Laying a head joint.

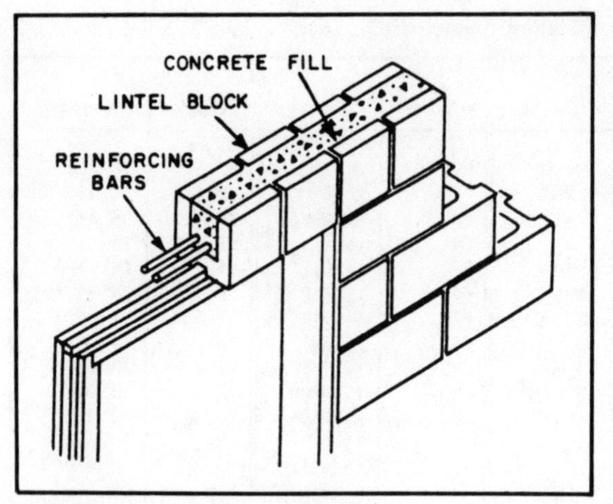

Fig. 4-37. Lintel made from blocks.

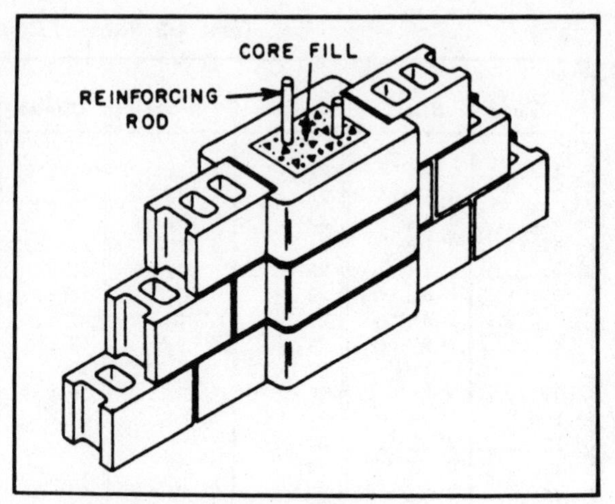

Fig. 4-38. Reinforcing concrete blocks.

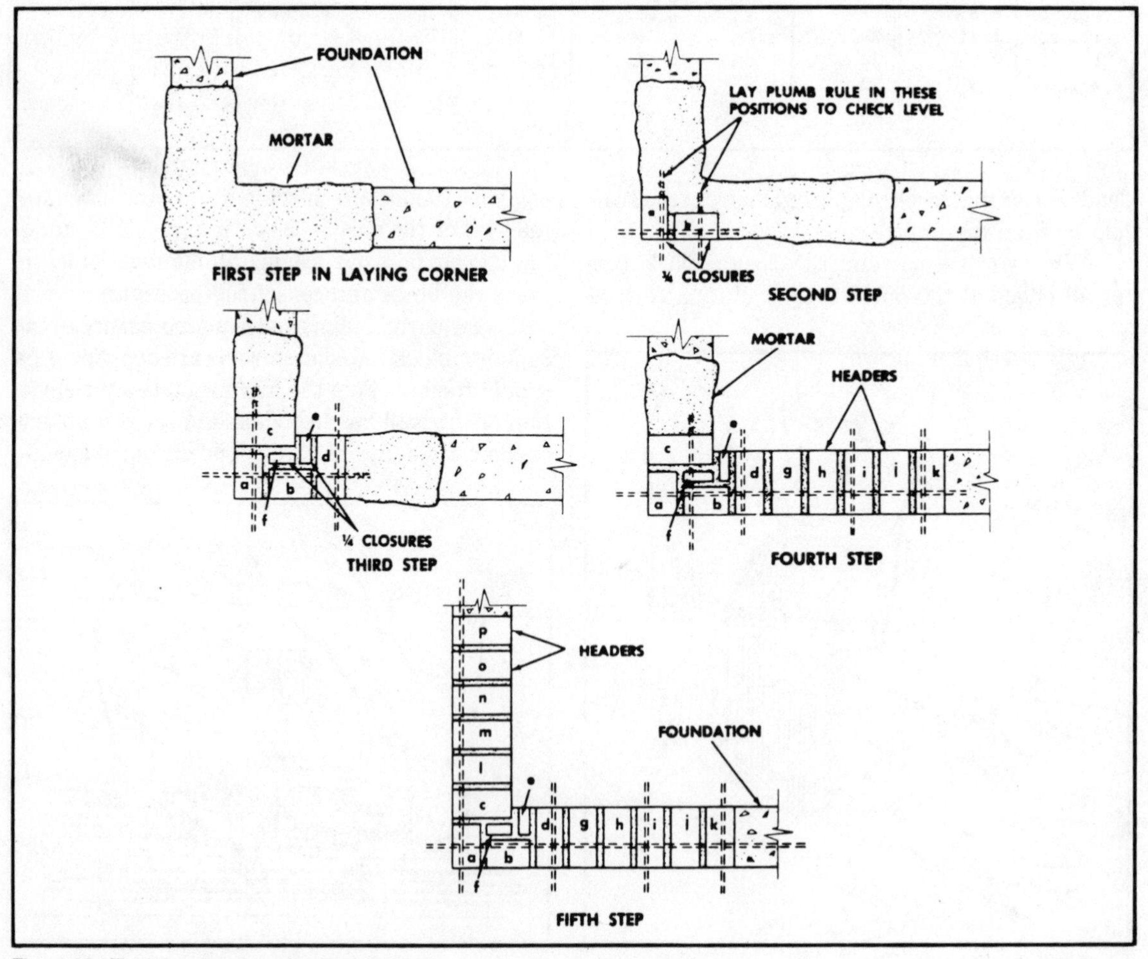

Fig. 4-39. First course of corner lead for wall.

mortar and forces it tightly against the masonry on each side of the joint. Tool all joints either concave or V-shaped. Tool horizontal joints first, then strike the vertical joints with a small S-shaped jointer. Use a trowel to trim mortar burrs remaining after tooling is complete flush with the face of the wall. You also can remove the burrs by rubbing with a burlap bag or soft bristle brush.

BRICK MASONRY

Brick masonry uses units of baked clay or shale of uniform size, small enough to be placed with one hand, and laid in courses with mortar joints to form walls of virtually unlimited length and height. Because of higher costs, brick foundations are not as common as poured concrete or cement block foundations. However, they are often used when both beauty and function are desired in a house foundation.

Bricks are kiln-baked from various clay and shale mixtures. The chemical and physical characteristics of the ingredients vary considerably; these and the kiln temperatures combine to produce a brick in a variety of colors and hardnesses. In some regions, pits are opened and found to yield clay or shale that, when ground and moistened, can be formed and baked into durable brick. In other regions, clays or shales from several pits must be mixed.

The dimensions of a U.S. standard building brick are 2 1/2 × 3 3/4 × 8. The actual dimensions of brick can vary a little because of shrinkage during burning.

Brick Terminology

Frequently, you must cut the brick into various shapes. The more common of these shapes are shown in Fig. 4-40. They are called *half* or *bat, three-quarter closure, quarter closure, king closure, queen closure,* and *split.* They are used to fill in spaces at corners and other places where a full brick will not fit.

The six surfaces of a brick are called the *face*, the *side,* the *end,* and the *beds* (Fig. 4-41).

A finished brick structure contains *face brick*, brick placed on the exposed face of the structure, and *back-up brick,* brick placed behind the face brick. The face brick is often of higher quality than the back-up brick; however, the entire wall can be built of common brick. *Common brick* is brick made from pit-run clay, with no attempt at color control and no special surface treatment like glazing or enameling. Most common brick is red. It is now called *building brick.*

Although any surface brick is a face brick as distinguished from a back-up brick, the term *face brick* is also used to distinguish high-quality brick from brick that is of common or lower quality. Applying this criterion, face brick is more uniform in color than common brick, and might be available in a variety of colors as well. It might be specifically finished on the surface and has a better sur-

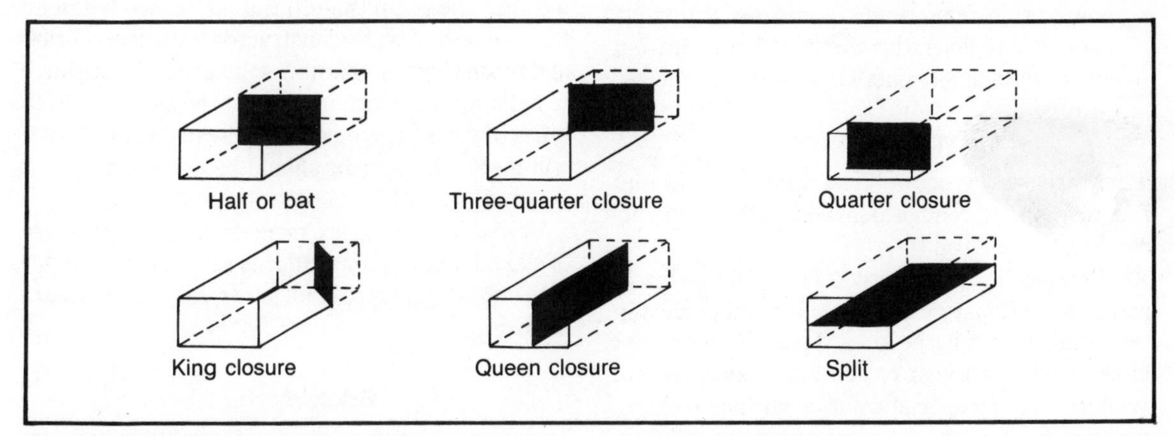

Fig. 4-40. Common shapes of cut brick.

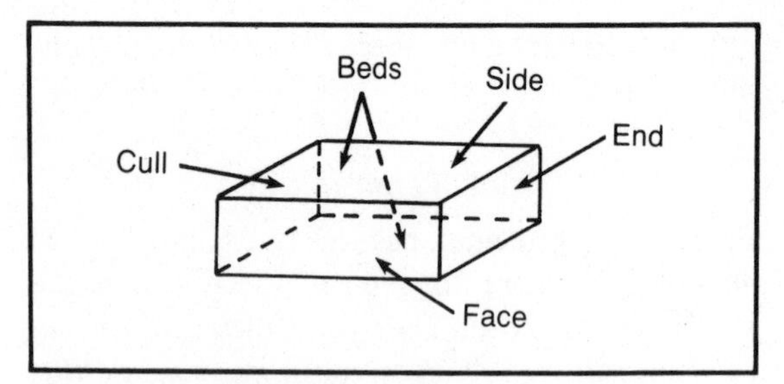

Fig. 4-41. Names of brick surfaces.

face appearance than common brick. It also might be more durable as a result of the use of select clay and other materials or as a result of special manufacturing methods.

Back-up brick might consist of brick that is inferior in quality even to common brick. Brick that has been underburned or overburned, or brick made with inferior clay or by inferior methods, is often used for back-up brick.

Still another type of classification divides brick into grades based upon probable climatic conditions to which it is to be exposed, as follows:

- ☐ *GRADE SW* is brick designed to withstand exposure to temperatures below freezing in a moist climate like that of the northern regions of the United States.
- ☐ *GRADE MW* is brick designed to withstand exposure to temperatures below freezing in a drier climate than Grade SW.
- ☐ *GRADE NW* is brick primarily intended for interior or back-up brick. It can be used exposed, however, in regions where no frost action occurs, or in regions where frost action occurs but where the annual rainfall is less than 15 inches.

Types of Brick

There are many types of brick. Some are different in formation and composition, while others vary according to their use.

Building brick, formerly called common brick, is made of ordinary clays or shales and burned in the usual manner in the kilns. These bricks do not have special scorings or markings and are not produced in any special color or surface texture. Building brick is also known as *hard* and *kiln-run brick*. It is generally used for the backing courses in solid or cavity brick walls. The harder and more durable kinds are preferred for this purpose.

Face brick is used in the exposed face of a wall and is of a higher quality unit than back-up brick. It has better durability and appearance. The most common colors of face brick are various shades of brown, red, gray, yellow, and white.

When bricks are overburned in the kilns, they are called *clinker brick*. This type of brick is usually hard and durable and might be irregular in shape. Rough-hard corresponds to the clinker classification.

The dry-press process is used to make *press brick,* which has regular smooth faces, sharp edges, and perfectly square corners. Ordinarily, all press brick is used as face brick.

Glazed brick has one surface of each brick glazed in white or another color. The ceramic glazing consists of mineral ingredients that fuse together in a glasslike coating during burning. This type of brick is particularly suited for walls or partitions in hospitals, dairies, laboratories, or other buildings where cleanliness and ease of cleaning is necessary.

Fire brick is made of a special type of fire clay that will withstand the high temperatures of fireplaces, boilers, and similar usages without cracking or decomposing. Fire brick is generally larger than regular structural brick.

Cored bricks are bricks made with two rows of five holes extending through their beds to reduce weight. There is no significant difference between the strength of walls constructed with cored brick and those constructed with solid brick. Resistance to moisture penetration is about the same for both types of walls. The most easily available brick that will meet requirement should be used, whether it is cored or solid.

Sand-lime bricks are made from a lean mixture of slaked lime and fine silicous sand molded under mechanical pressure and hardened under steam pressure.

Brick Mortar

Mortar is used to bond the brick together and, unless properly mixed and applied, will be the weakest part of brick masonry. Both the strength and resistance to rain penetration of brick masonry walls are dependent to a great degree on the strength of the bond. Water in the mortar is essential to the development of bond, and if the mortar contains insufficient water, the bond will be weak and spotty. When brick foundation walls leak, it is usually through the mortar joints. Irregularities in the dimension and shape of bricks are corrected by the mortar joint.

Mortar should be plastic enough to work with a trowel. The properties of mortar depend largely upon the type of sand used in it. Clean, sharp sand produces excellent mortar. Too much sand in mortar causes it to segregate, drop off the trowel, and weather poorly.

The selection of mortar for brick construction depends on the use requirements of the structure. For example, the recommended mortar for use in laying up interior nonload-bearing partitions would not be satisfactory for foundation walls. In many cases, the builder relies upon a fixed portion of cement, lime, and sand to provide a satisfactory mortar. The following types of mortar are proportioned on a volume basis:

- ☐ *Type M* is 1 part portland cement, 1/4 part hydrated lime or lime putty, and 3 parts sand, *or* 1 part portland cement, 1 part type II masonry cement, and 6 parts sand. This mortar is suitable for general use and is recommended specifically for masonry below grade and in contact with earth, such as foundations, retaining walls, and walks.
- ☐ *Type S* mortar is 1 part portland cement, 1/2 part hydrated lime or lime putty, and 4 1/2 parts sand, *or* 1/2 part portland cement, 1 part type II masonry cement, and 4 1/2 parts sand. This mortar is suitable for general use and is recommended where high resistance to lateral forces is required.
- ☐ *Type N* is 1 part portland cement, 1 part hydrated lime or lime putty, and 6 parts sand, *or* 1 part type II masonry cement, and 3 parts sand. This mortar is suitable for general use in exposed masonry above grade and is recommended specifically for exterior walls subjected to severe exposure.
- ☐ *Type O* mortar is 1 part portland cement, 2 parts hydrated lime or lime putty, and 9 parts sand, *or* 1 part type I or type II masonry cement and 3 parts sand. This mortar is recommended for load-bearing walls of solid units where the compressive stress doesn't exceed 100 pounds per square inch and where the masonry will not be subjected to freezing and thawing in the presence of excessive moisture.

MASONRY TERMS

Specific terms are used to describe the various positions of masonry units and mortar joints in a wall. Knowing them will help you understand brick foundation construction.

course—One of the continuous horizontal layers or rows of masonry that, bonded together, form the masonry structure (Fig. 4-42).

whythe—A continuous, vertical, 4-inch or greater section or thickness of masonry, such as the thickness of masonry separating the flues in a chimney (Fig. 4-43).

stretcher—A masonry unit laid flat with its longest dimension parallel to the face of the wall (Fig. 4-44).

header—A masonry unit laid flat with its longest dimension perpendicular to the face of the wall; generally used to tie two whythes of masonry together (Fig. 4-45).

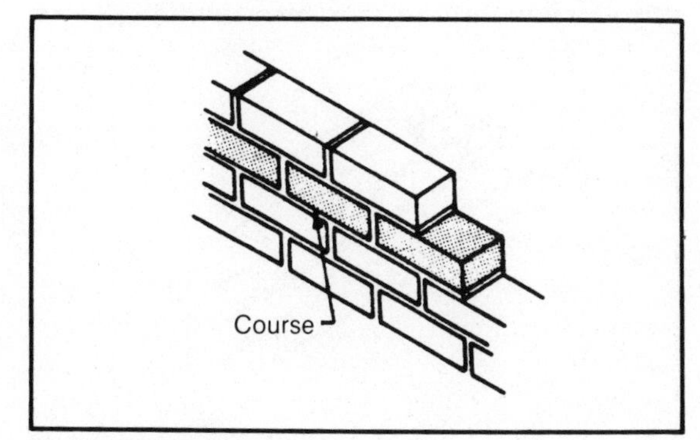

Fig. 4-42. Course.

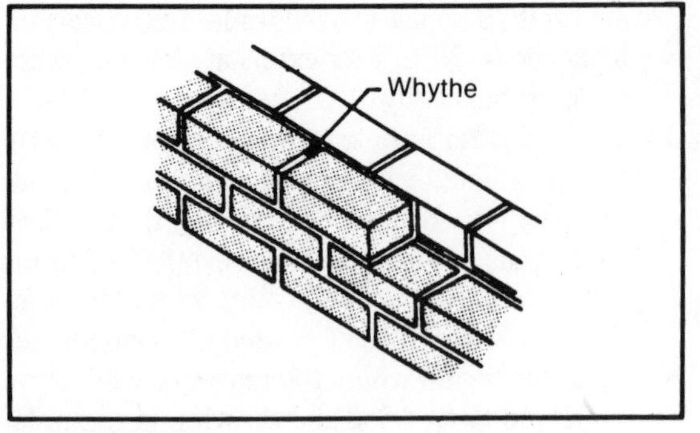

Fig. 4-43. Whythe.

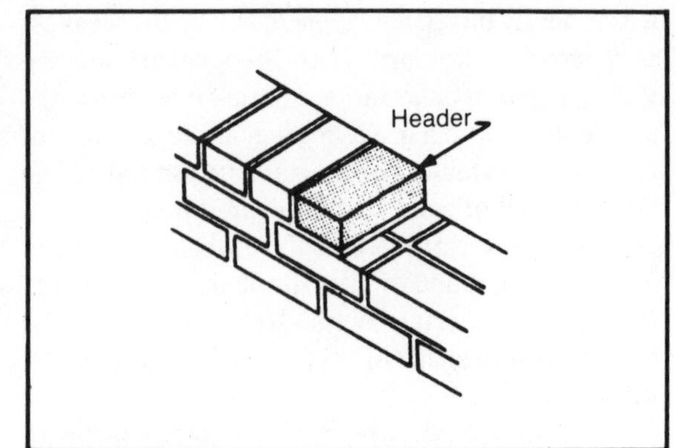

Fig. 4-45. Header.

rowlock—A brick laid on its edge (face).

bull-stretcher—A rowlock brick laid with its longest dimension parallel to the face of the wall (Fig. 4-46).

bull-header—A rowlock brick laid with its longest dimension perpendicular to the face of the wall.

soldier—A brick laid on its end so that its longest dimension is parallel to the vertical axis of the face of the wall (Fig. 4-47).

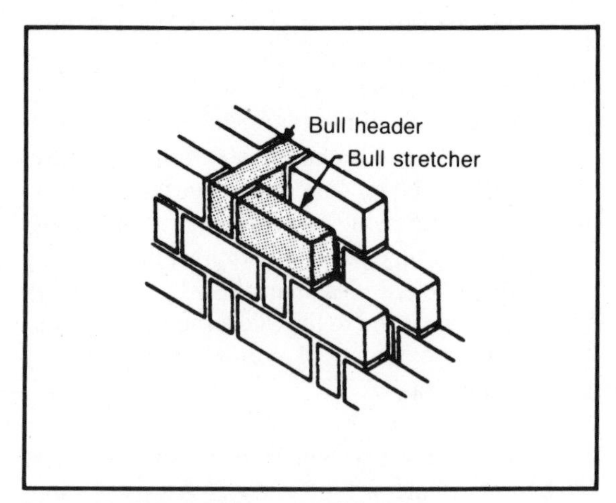

Fig. 4-46. Bull stretcher and bull header.

LAYING BRICK FOUNDATIONS

Good bricklaying procedure depends on efficiency and good workmanship (Fig. 4-48). Efficiency involves doing the work with the fewest possible motions. After learning the fundamentals, every do-it-yourselfer develops his own methods for achieving maximum efficiency.

There are three different meanings to *bond* when used in reference to masonry. *Structural bond*

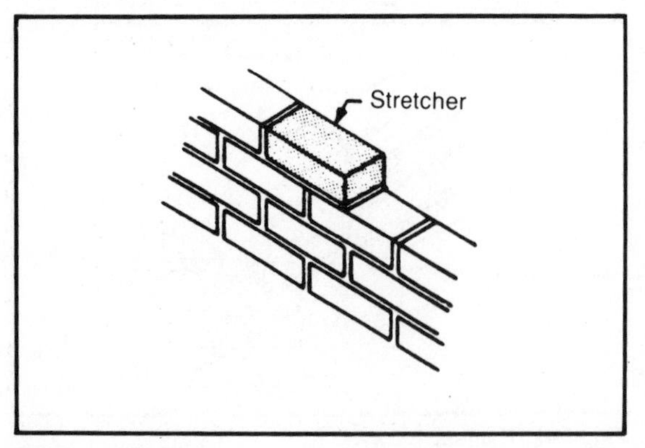

Fig. 4-44. Stretcher.

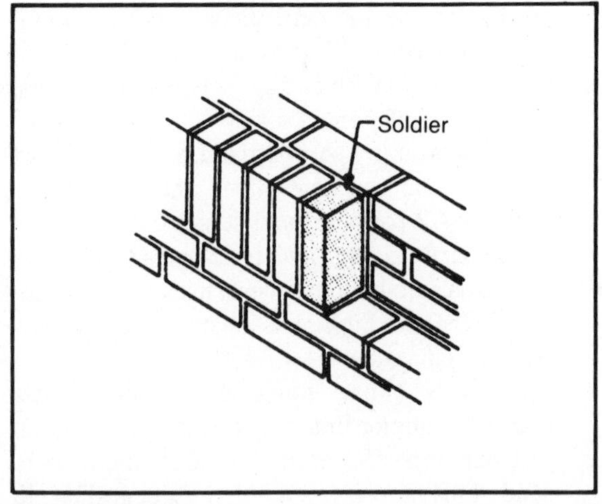

Fig. 4-47. Soldier.

Fig. 4-48. Good bricklaying is good workmanship and efficiency.

is the method by which individual masonry units are interlocked or tied together to cause the entire assembly to act as a single structural unit. Structural bonding of brick foundations and other walls can be accomplished in three ways: first, by overlapping (interlocking) masonry units; second, by using metal ties imbedded in connecting joints; and third, by the adhesion of grout to adjacent whythes of masonry.

Mortar bond is the adhesion of the joint mortar to the masonry units or to the reinforcing steel.

Finally, *pattern bond* is the pattern formed by the masonry units and the mortar joints on the face of a wall. The pattern can result from the type of structural bond used or can be purely a decorative one in no way related to the structural bond. There are five basic pattern bonds in common use today (Fig. 4-49): running bond, common or American bond, Flemish bond, English bond, and block or stack bond.

Running bond, the simplest of the basic pattern bonds, consists of all stretchers. Because there are no headers in this bond, metal ties are usually used. Running bond is used largely in cavity wall construction and veneered walls of brick and often in

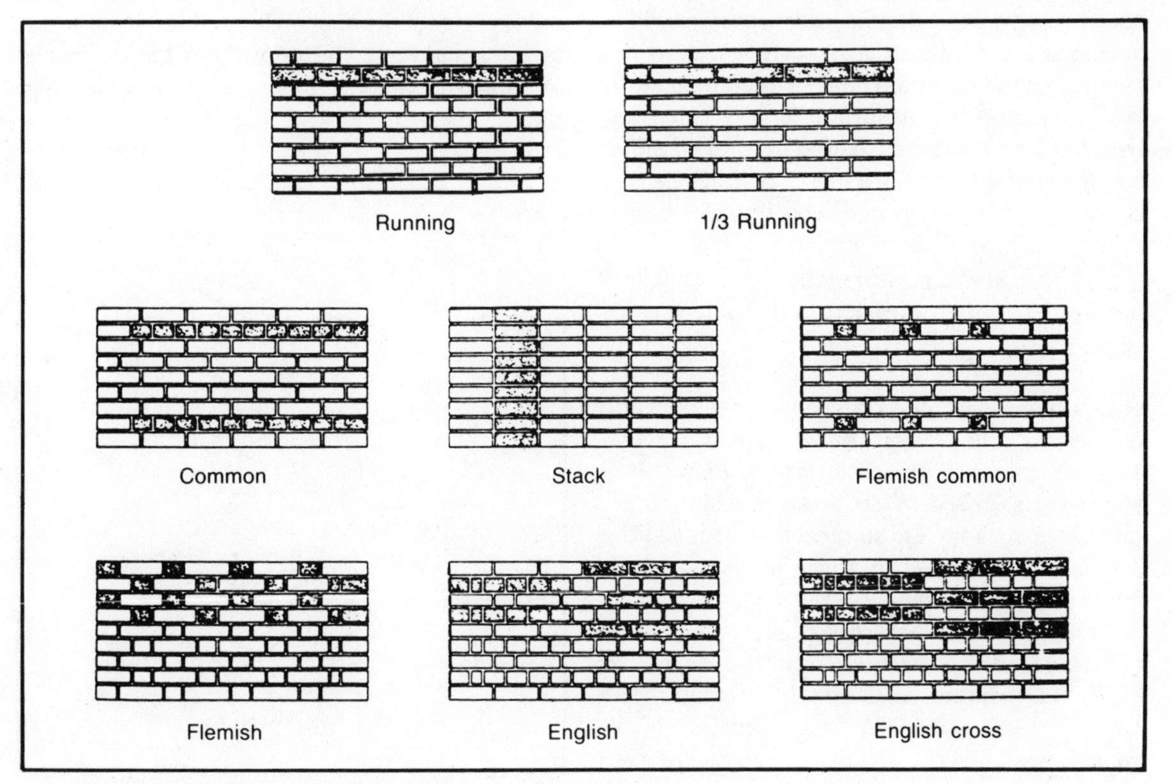

Fig. 4-49. Types of brick and masonry bond.

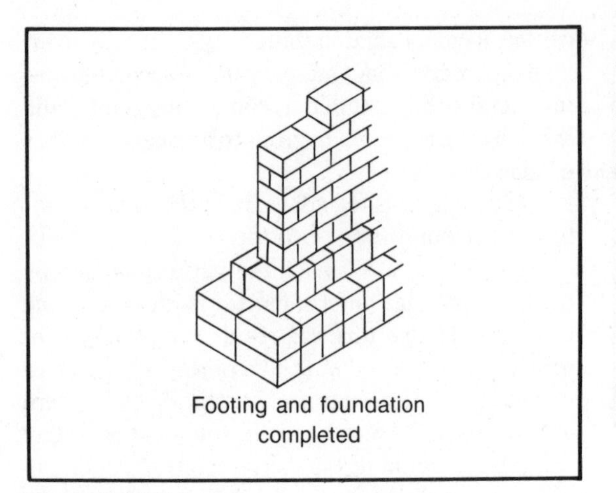

Fig. 4-50. Wall footings.

facing tile walls where the bonding can be accomplished with extra stretcher tile.

Common or *American bond* is a variation of running bond with a course of full-length headers at regular intervals. These headers provide structural bonding, as well as pattern. Header courses usually appear at every fifth, sixth, or seventh course, depending on the structural bonding requirement. In laying out any bond pattern, it is very important that you start the corners correctly. For common bond, a three-quarter brick must start each header course at the corner. Common bond can be varied by using a Flemish header course.

Flemish bond is made up of alternate stretchers and headers, with the headers in alternate courses centered over the stretchers in the intervening courses. Where the headers are not used for the structural bonding, they can be obtained by using half brick, called *blind headers.* There are two methods used in starting the corners. Figure 4-49 shows the so-called *Dutch corner,* in which three-quarter brick is used to start each course, and the *English corner,* in which 2-inch or quarter-brick closures must be used.

English bond is composed of alternate courses of headers and stretchers. The headers are centered on the stretchers and joints between stretchers. The vertical (head) joints between stretchers in all courses line up vertically. Blind headers are used in courses that are not structural bonding courses.

Block or *stack bond* is purely a pattern bond. There is no overlapping of the units; all vertical joints are aligned. Usually this pattern is bonded to the backing with rigid steel ties, but when 8-inch thick stretcher units are available, they can be used. In large wall areas and in load-bearing construction, it is advisable to reinforce the wall with steel pencil rods placed in the horizontal mortar joints. The vertical alignment requires dimensionally accurate units, or carefully prematched units, for each vertical joint alignment. Variety in pattern can be achieved by numerous combinations and modifications of the basic patterns shown.

English cross or *Dutch bond* is a variation of English bond and differs only in that vertical joints between the stretchers in alternate courses do not line up vertically. These joints center on stretchers themselves in courses above and below.

Brick Footings

A footing is required under a wall when the bearing capacity of the supporting soil is not sufficient to withstand the wall load without further means of redistribution. The footing must be wider than the thickness of the wall (Fig. 4-50).

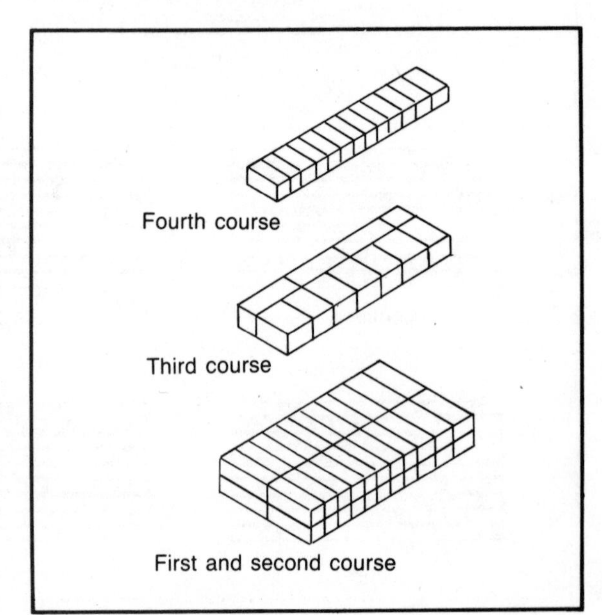

Fig. 4-51. Brick courses.

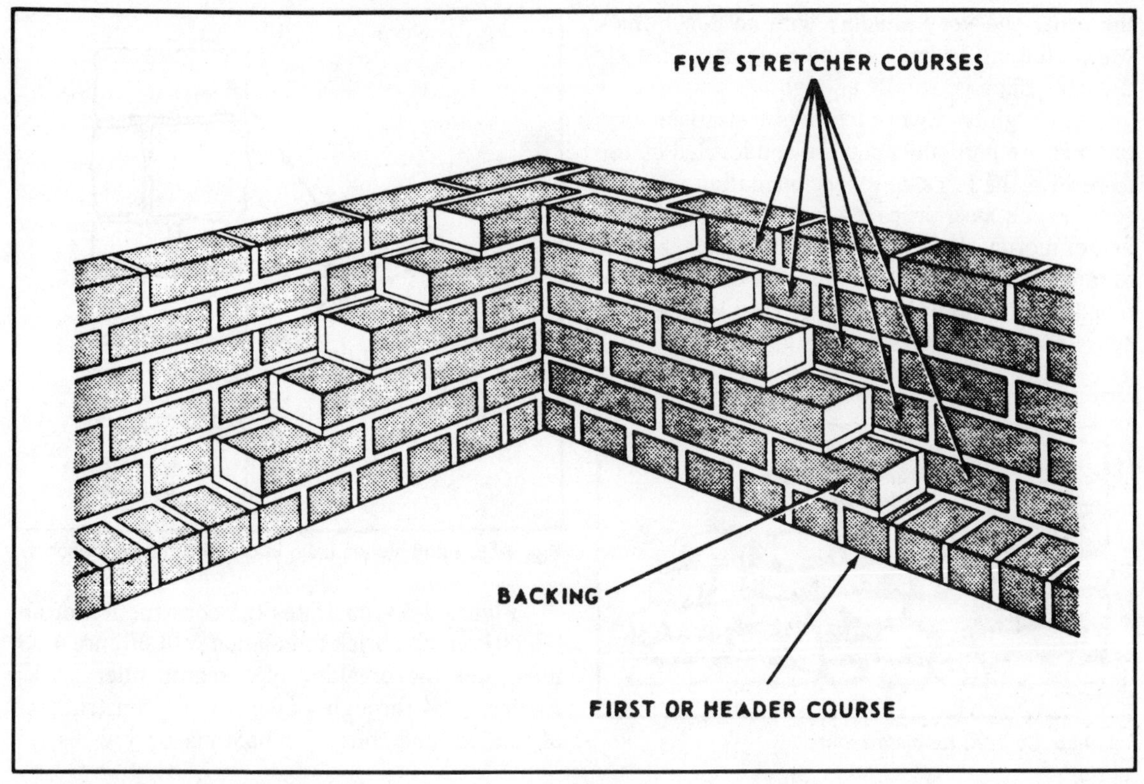

Fig. 4-52. Backing brick at the corner.

The required footing width and thickness for walls of considerable height or for walls that are to carry a heavy load should be determined using the guidelines given under *Concrete Footings*. Every footing should be below the frost line to prevent heaving and settlement of the foundation. For

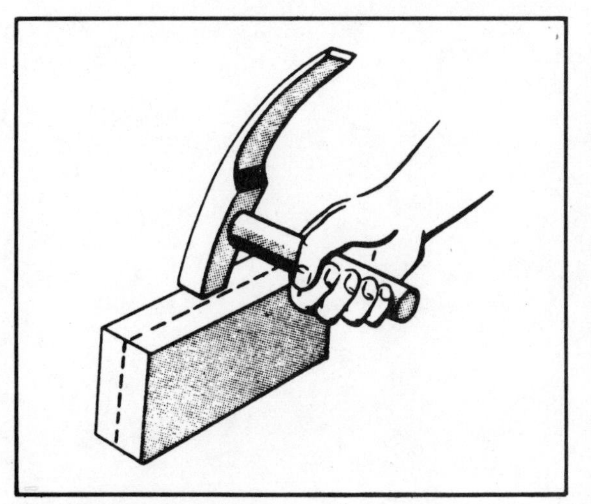

Fig. 4-53. Splitting a brick.

Fig. 4-54. Random rubble masonry.

the usual one-story building with an 8-inch-thick wall, a footing 16 inches wide and approximately 8 inches thick is usually enough.

Although brickwork footings are satisfactory, footings are normally concrete and leveled on top to receive the brick or stone foundation wall. As soon as you have prepared the subgrade, place a bed of mortar about 1 inch thick on the subgrade to tape up all irregularities. Lay the first course of foundation on this bed or mortar. Then lay the other courses on this first course (Fig. 4-51).

Fig. 4-55. Coursed rubble masonry.

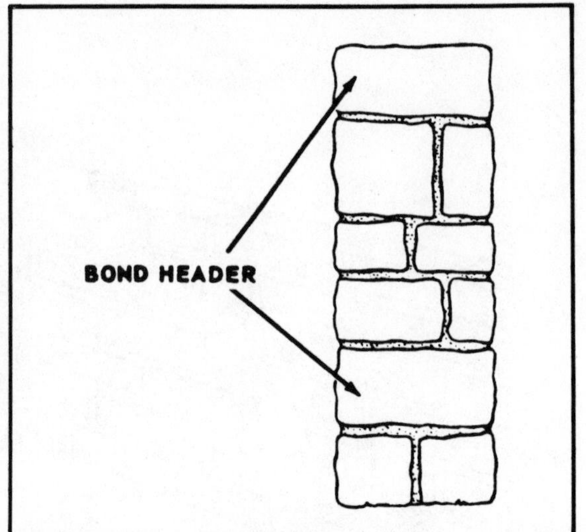

Fig. 4-56. Bond stones used in coursed rubble masonry.

Figure 4-52 illustrates the construction of an 8-inch common brick foundation wall. Figure 4-53 illustrates the breaking of a scored filler brick. Figures 4-54 through 4-56 show the construction of random and coursed rubble masonry walls.

Chapter 5

Concrete Floor Slabs

The number of new one-story houses with full basements has declined in recent years, particularly in the warmer areas of the United States. The decline is in part a result of lower construction costs of houses without basements and an apparent decrease in need for the basement space.

The primary function of a basement in the past was to provide space for a central heating plant and for the storage and handling of bulk fuel and ashes. It also housed laundry and utilities. With the wide use of liquid and gas fuels, however, the need for fuel and ash storage space has greatly diminished. Because space can be compactly provided on the ground-floor level for the heating plant, laundry, and utilities, the need for a basement often disappears.

PROS AND CONS OF SLAB FLOORS

One common type of floor construction for basementless houses is a concrete slab over a suitable foundation (Fig. 5-1). Sloping ground or low areas are usually not ideal for slab-on-ground construction because structural and drainage problems would add to costs. In split-level houses, a portion of the foundation is often designed for a grade slab. In such use, the slope of the lot is taken into account, and the objectionable features of a sloping ground become an advantage.

The finish flooring for concrete floor slabs on the ground was initially asphalt tile laid in mastic directly on the slab. These concrete floors didn't prove satisfactory in a number of instances, and considerable prejudice has grown toward this method of construction. The common complaints have been that the floors are cold and uncomfortable and that condensation sometimes collects on the floor, near the walls in cold weather, and elsewhere during warm, humid weather. Some of the undesirable features of concrete floors on the ground apply to both warm and cold climates; others, only to cold climates.

Improvements in construction methods based on experience and research have materially reduced the common faults of the slab floor. Consequently,

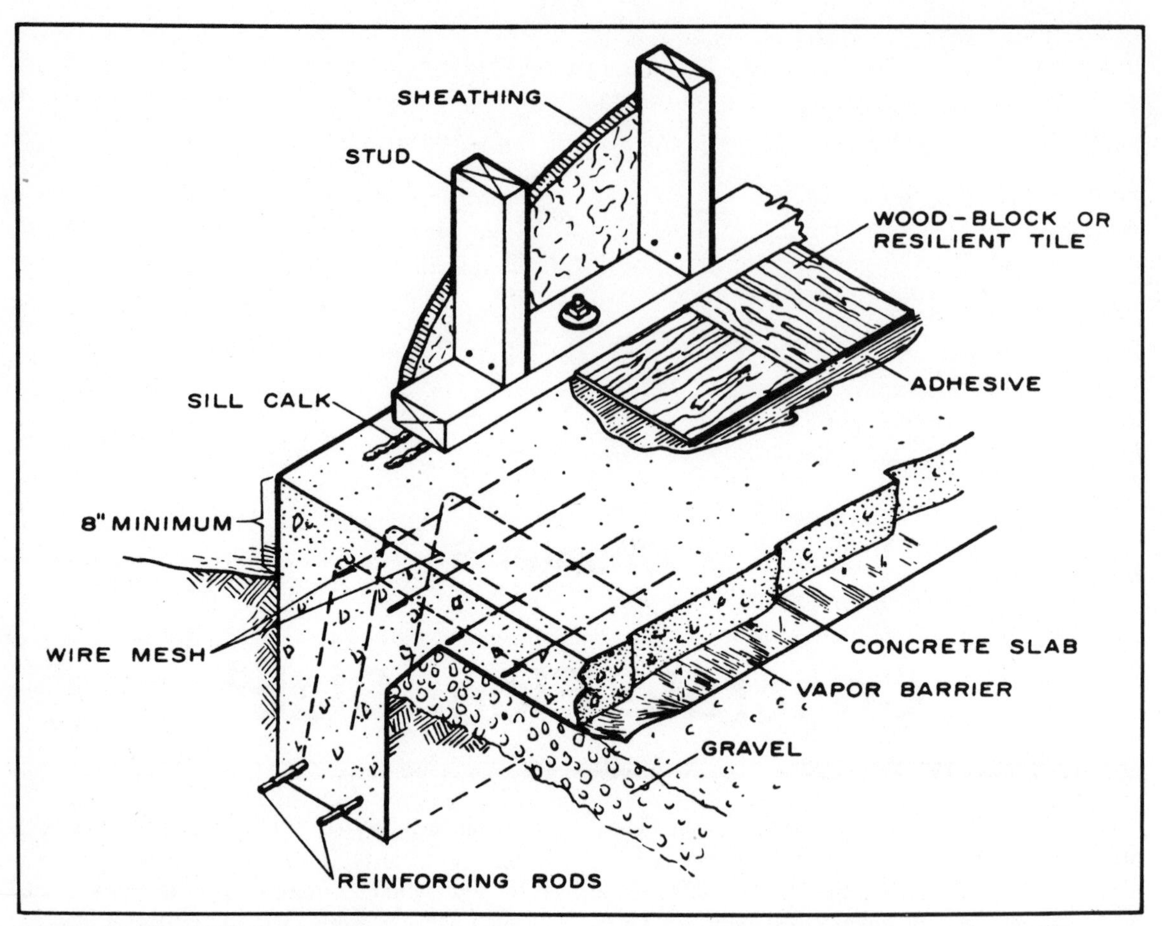

Fig. 5-1. Combined slab and foundation.

though, these improvements have increased the cost of slab-on-grade construction.

Floors are cold principally because of heat loss through the floor and the foundation walls, with most loss occurring around the exterior walls. Suitable insulation around the perimeter of the house helps reduce the heat loss. Radiant floor-heating systems are effective in preventing cold floors and floor condensation problems. Peripheral warm-air heating ducts are also effective in this respect. Vapor barriers over a gravel fill under the floor slab prevent soil moisture from rising through the slab.

BASIC SLAB FLOOR REQUIREMENTS

Construction of concrete floor slabs should meet certain basic requirements to provide a satisfactory floor. They are:

- ☐ Establish the finish floor level high enough above the natural ground level so that the finish grade around the house can be sloped away for good drainage. Locate the top of the slab no less than 8 inches above the ground, and the siding no less than 6 inches.
- ☐ Remove topsoil and install sewer and water lines. Then cover the lines with 4 to 6 inches of gravel or crushed rock well tamped in place.
- ☐ Under the concrete slab, use a vapor barrier consisting of a heavy plastic film, such as 6-mil polyethylene, asphalt laminated duplex sheet,

or 45-pound or heavier roofing, with a minimum of 1/2-perm rating. Lap joints at least 4 inches and seal them. The barrier should be strong enough to resist puncturing during placing of the concrete.

- ☐ Install a permanent, waterproof, nonabsorptive type of rigid insulation around the perimeter of the wall. Insulation can extend down on the inside of the wall vertically or under the slab edge horizontally.
- ☐ Reinforce the slab with 6- × -6-inch No. 10 wire mesh or other effective reinforcement. The concrete slab should be at least 4 inches thick and should conform to guidelines in Chapter 3. A monolithic slab is preferred in termite areas.
- ☐ After leveling and screeding, float the surface with wood or metal floats while the concrete is still plastic. If you need a smooth, dense surface for the installation of wood or resilient tile adhesives, trowel the surface. These procedures will be covered later in this chapter.

Combined Slab and Foundation

The combined slab and foundation, sometimes referred to as the *thickened-edge slab,* is useful in warm climates where frost penetration is not a problem and where soil conditions are especially favorable. It consists of a shallow, perimeter-reinforced footing poured integrally with the slab over a vapor barrier (Fig. 5-2). The bottom of the footing should appear at least 1 foot below the natural grade line and be supported on solid, unfilled, and well-drained ground.

Independent Concrete Slab and Foundation Walls

Where ground freezes to any appreciable depth during winter, the walls of a house must be supported by foundations or piers that extend below the frost line to solid bearing on unfilled soil. In such construction, the concrete slab and foundation wall are usually separate. Three typical systems shown in Figs. 5-3 through 5-5 are suitable for such conditions.

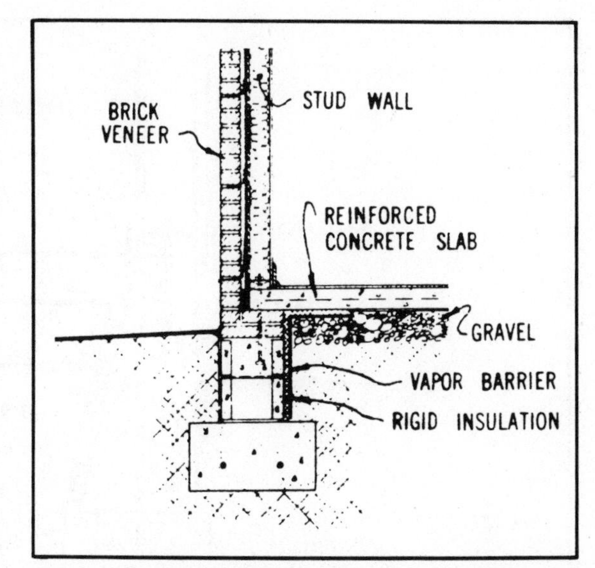

Fig. 5-2. Reinforced concrete slab with vapor barrier.

VAPOR BARRIERS

The most desirable properties in a vapor barrier to be used under a concrete slab are good vapor-transmission rating (less than 0.5 perm), resistance to damage by moisture and rot, and ability to withstand normal usage during pouring operations. Such properties are included in the following types of materials:

- ☐ 55-pound roll roofing or heavy asphalt laminated duplex barriers
- ☐ Heavy plastic film, such as 6-mil or heavier polyethylene, or similar plastic film laminated to a duplex-treated paper
- ☐ Three layers of roofing felt mopped with hot asphalt
- ☐ Heavy asphalt-impregnated and vapor-resistant rigid sheet material with sealed joints

INSULATION REQUIREMENTS

The use of perimeter insulation for slabs is necessary to prevent heat loss and cold floors during the heating season, except in warm climates. Various types of home insulation are shown in Fig. 5-6.

The thickness of the insulation depends on the requirement of the climate and on the materials

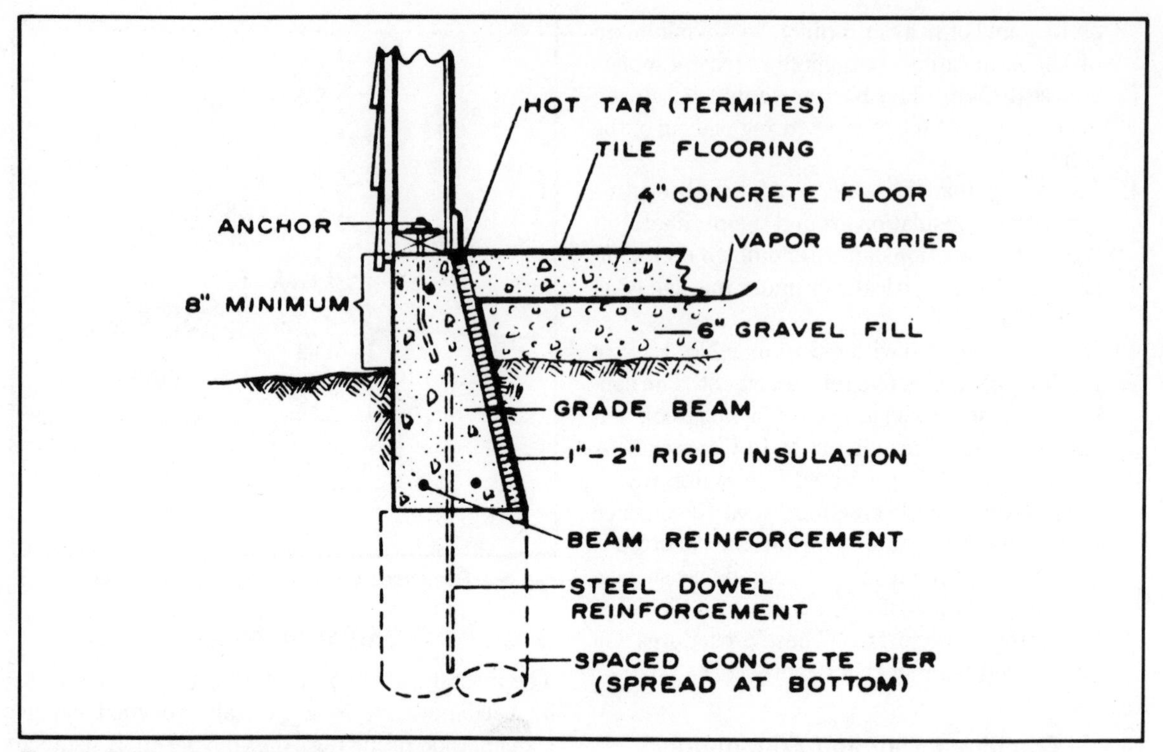

Fig. 5-3. Reinforced grade beam for concrete slab.

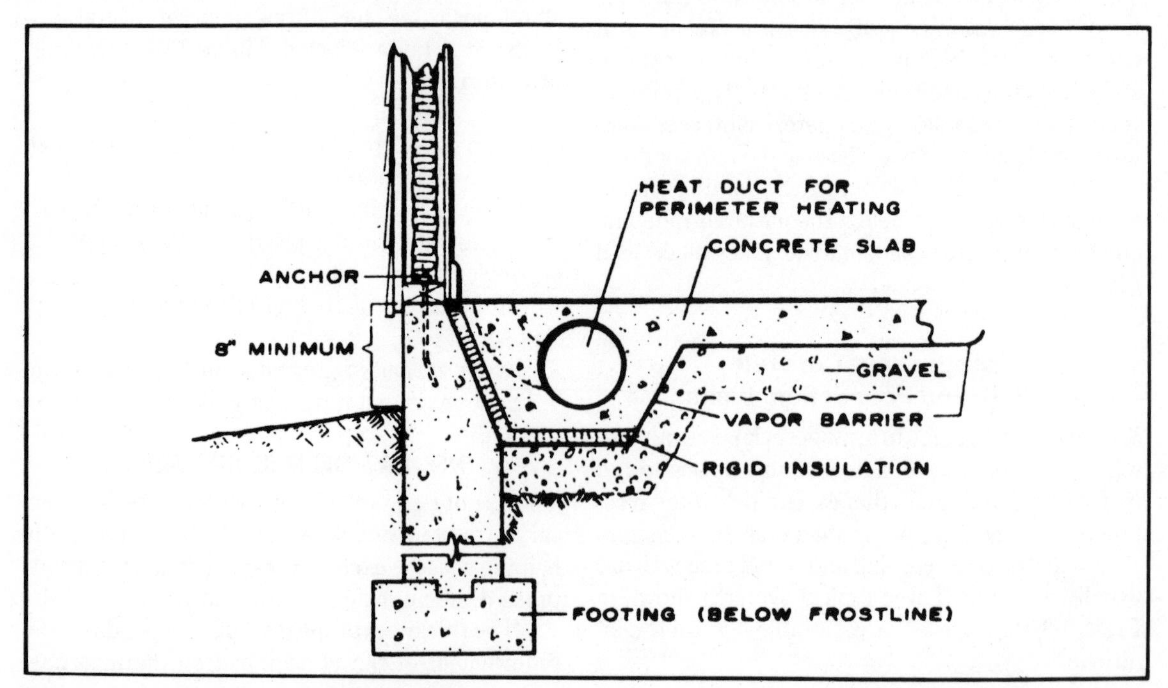

Fig. 5-4. Full foundation wall for cold climates.

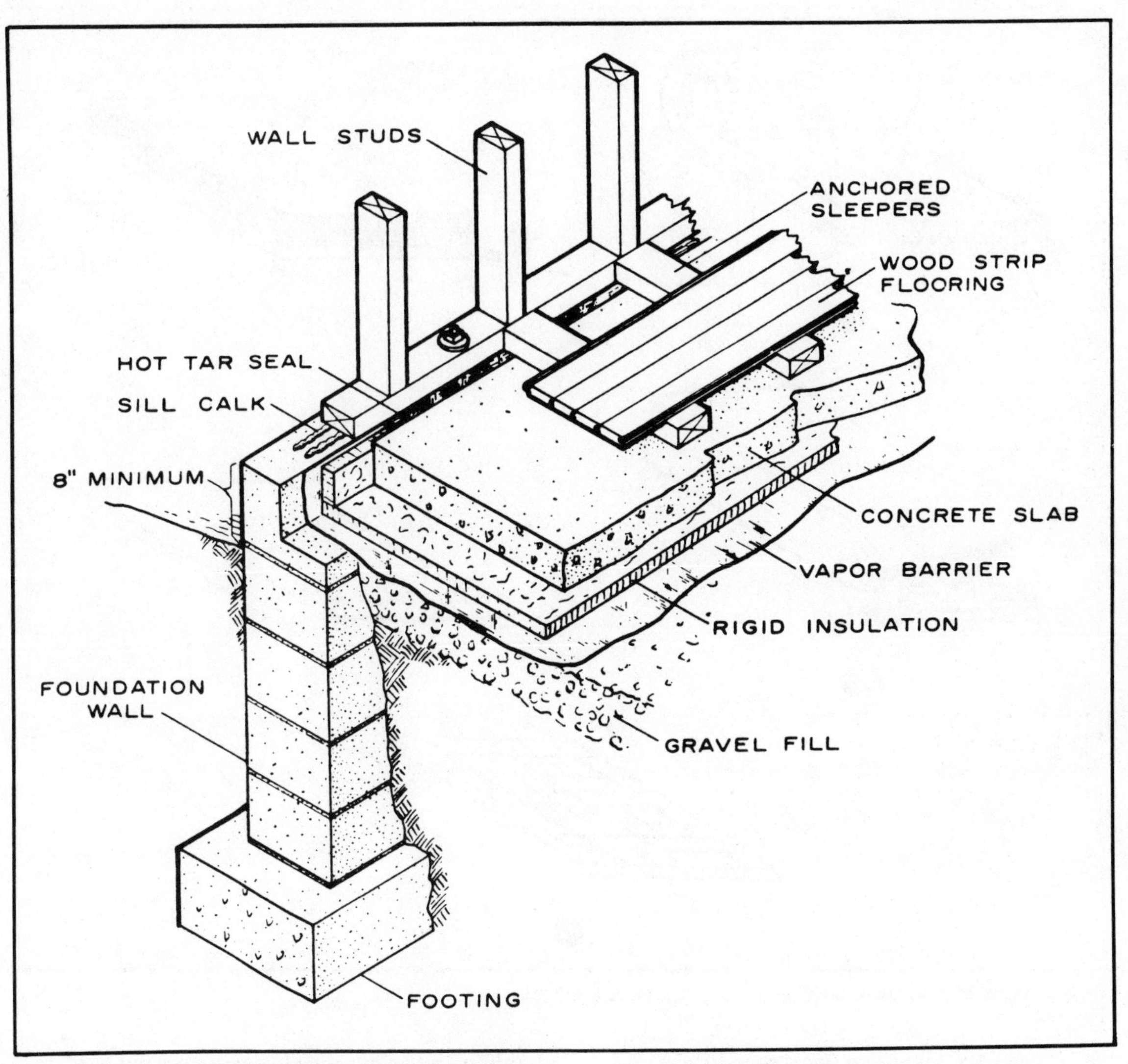

Fig. 5-5. Independent concrete floor slab and wall.

used. Some insulations have more than twice the insulating value of others. The resistance per inch of thickness (R factor), as well as the heating design temperature, should govern the amount required. Perhaps two good general rules to follow are:

- For average winter low temperatures of 0 degrees F and higher (moderate climates), the total R should be about 2.0 and the depth of the insulation or the width under the slab should not be less than 1 foot.
- For average winter low temperatures of −20 degrees F and lower (cold climates), the total R should be about 3.0 without floor heating, and the depth or width of insulation should not be less than 2 feet.

Refer to Table 5-1.

Insulation Types

The properties desired in insulation for floor slabs are high resistance to heat transmission, permanent

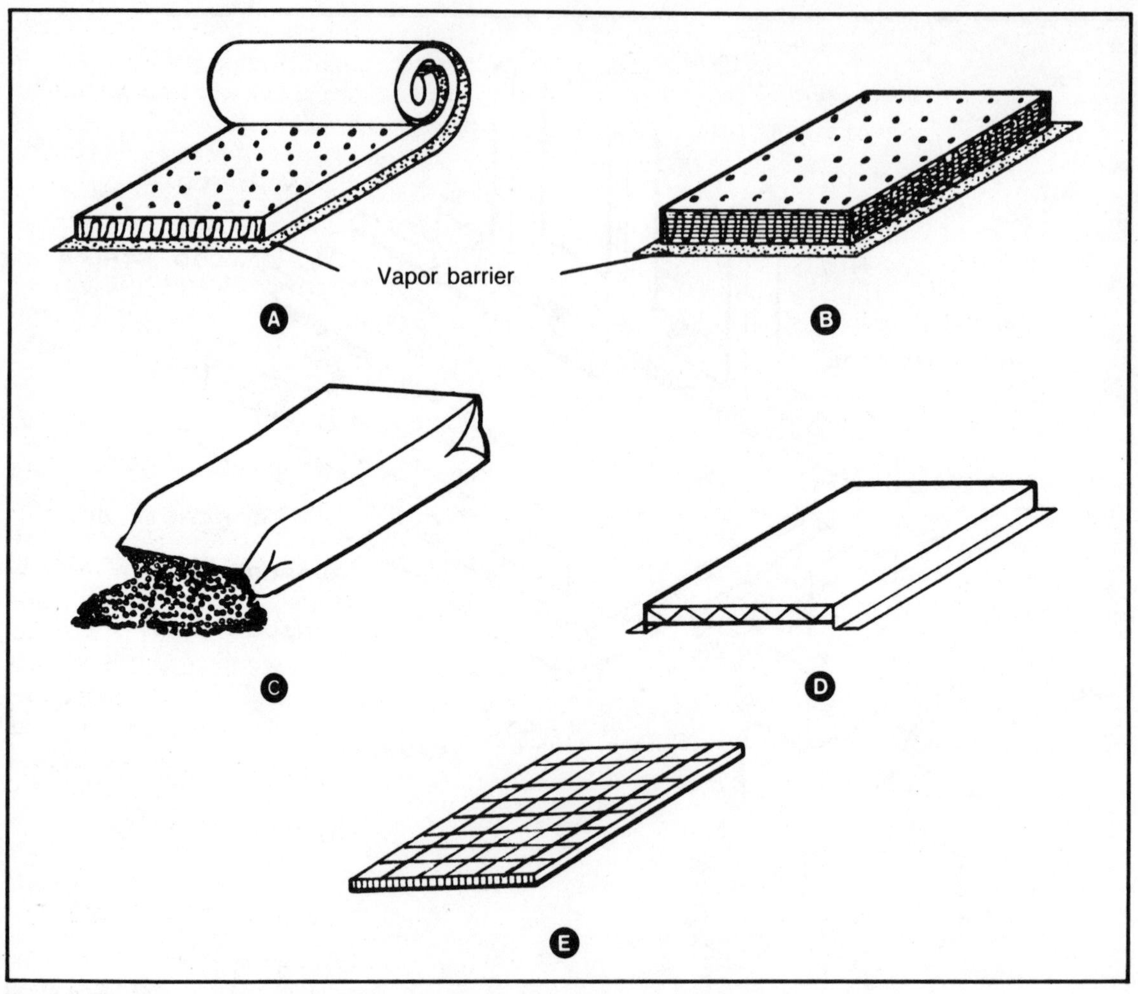

Fig. 5-6. Types of insulation: blanket (A), batt (B), fill (C), reflective (D), rigid (E).

Table 5-1. Insulation Resistance Values.

Low temper-atures °*F.*	Depth insulation extends below grade *Ft.*	Resistance (R) factor	
		No floor heating	Floor heating
−20	2	3.0	4.0
−10	1½	2.5	3.5
0	1	2.0	3.0
+10	1	2.0	3.0
+20	1	2.0	3.0

durability when exposed to dampness and frost, and high resistance to crushing as a result of floor loads, weight of slab, or expansion forces. The slab also should be immune to fungus and insect attack, and should not absorb or retain moisture. Examples of materials considered to have these properties follow.

Cellular-Glass Insulation Board. Available in slabs 2, 3, 4, and 5 inches thick. The R factor is 1.8 to 2.2 per inch of thickness. Crushing strength is approximately 150 pounds per square inch. It is easily cut and worked. The surface can *spall* (chip or crumble) away if subjected to moisture and freezing. It should be dipped in roofing pitch or asphalt for protection. Insulation should be located above or inside the vapor barrier for protection from moisture. This type of insulation has been replaced to a large extent by the newer foamed plastics, such as polystyrene and polyurethane.

Glass Fibers with Plastic Binder. Coated or uncoated; available in thickness of 3/4, 1, 1 1/2, and 2 inches. The R factor is 3.3 to 3.9 per inch of thickness. Crushing strength is about 12 pounds per square inch. Water penetration into coated board is slow and inconsequential unless the board is exposed to a constant head of water, in which case this water can disintegrate the binder. Use a coating board or apply coal-tar pitch or asphalt to the uncoated board. Coat all edges. Follow the manufacturer's instructions for cutting. Placement of the insulation inside the vapor barrier will afford some protection.

Foamed Plastic (polystyrene, polyurethane, and others) Insulation, in sheet for, usually available in thicknesses of 1/2, 1, 1 1/2, and 2 inches. At normal temperatures, the R factor varies from 3.7 for polystyrenes to over 6.0 for polyurethane with a 1-inch thickness. These materials generally have low water-vapor transmission rates. Some are low in crushing strength and perhaps are best in a vertical position and not under the slab, where crushing could occur.

Insulating Concrete. Expanded mica aggregate, 1 part cement to 6 parts aggregate, thickness used as required. R factor is about 1.1 per inch of thickness. Crushing strength is adequate. It can take up moisture when subjected to dampness, and consequently, its use should be limited to locations where there will be no contact with moisture from any source.

Concrete Made with Lightweight Aggregate. Examples: expanded slag, burned clay, or pumice, using 1 part cement to 4 parts aggregate; thickness used as required. R factor is about 0.40 per inch of thickness. Crushing strength is high. This lightweight aggregate also can be used for foundation walls in place of stone or gravel aggregate.

Under service conditions, there are two sources of moisture that might affect insulating material: vapor from inside the house, and moisture from soil. Vapor barriers and coatings can retard, but not entirely prevent, the penetration of moisture into the insulation. Dampness can reduce the crushing strength of insulation, which, in turn, can permit the edge of the slab to settle. Compression of the insulation, moreover, reduces its efficiency. Insulating materials should perform satisfactorily in any position if they do not change dimensions and if they are kept dry.

TERMITE PROTECTION

In areas where termites are a problem, certain precautions are necessary for concrete slab floors on the ground. Leave a countersink-type opening 1 inch wide and 1 inch deep around plumbing pipes where they pass through the slab. Fill the opening with hot tar when the pipe is in place. If you are using insulation between the slab and the foundation wall, keep the insulation 1 inch below the top of the slab, and fill the space with hot tar. Figure 5-7 illustrates how a metal shield is installed at a porch slab to protect wood from termites.

FINISH FLOORS OVER CONCRETE SLABS

A natural concrete surface is sometimes used for the finish floor, but generally is not considered wholly satisfactory. Special dressings are required to prevent dusting. Moreover, such floors tend to feel cold.

Asphalt or vinyl tile laid in mastic is comparatively economical and easy to clean, but it also feels cold. You can use wood tile in various forms, as well as wood parquet flooring. You also can install tongued-and-grooved wood strip flooring 25/32 inch thick but you should use it over pressure-treated wood sleepers anchored to the slab. For existing concrete floors, use a vaporproof coating before you install the treated sleepers.

Refer to *Tile Floors: Installing, Maintaining and Repairing* (TAB #1998) and *Hardwood Floors: Installing, Maintaining and Repairing* (TAB #1928)

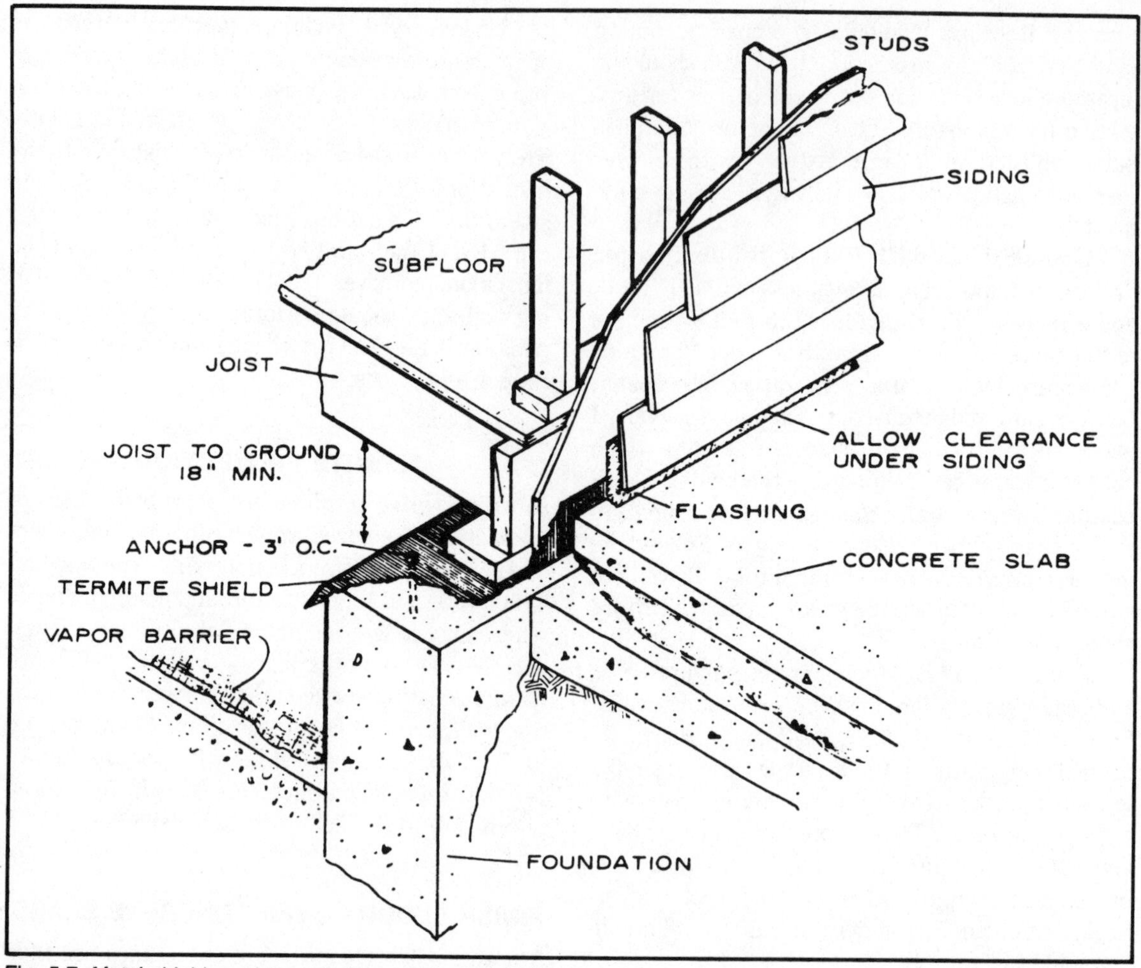

Fig. 5-7. Metal shield used to protect wood at porch slab.

for additional information on installing floors over concrete slabs and other surfaces.

FINISHING CONCRETE SLABS

Figure 5-8 illustrates how forms are set for concrete slabs. The concrete finishing process can be performed in many ways, depending on the effect desired. Some concrete slabs require only screeding to proper contour and elevation; others must be broomed, floated, or troweled finished, as specified.

Screeding

The first step in finishing a slab is screeding. The chief purpose of screeding is to level the surface of the slab by striking off the excess concrete. The strike-off board rides on the edges of the side forms or on wood or metal strips (screeds) set up for this purpose. Workers on each side of the board give a sawing motion while moving it along the slab.

The vibrating screed is being used more often for constructing floor and deck slabs. It can be rented by the do-it-yourselfer at most home-equipment rental stores. The screed, incorporating the use of vibration, permits the use of stronger and more economical low-slump concrete because it strikes off this relatively dry material smoothly and quickly. The advantage of vibration are twofold: greater density and stronger concrete.

Not only do vibratory finishing screeds gives a better finish and reduce maintenance, but they also save considerable time as a result of the speed at which they operate. Then, too, screeds are much less fatiguing to operate than hand strike-offs.

A vibratory finishing screed usually consists of a beam or beams with a gasoline engine or an electric motor and vibrating mechanism that is mounted in the center of the beam. Most screeds are very heavy and equipped with wheels and a raised device to help roll it back for a second pass. There are lightweight screeds, however, that are not equipped with wheels. They are easily lifted by two men and set back for the second pass if required. The vibration is normally transmitted through the length of the beam directly to the concrete.

Screeds are pulled by either ropes or pipe handles with a worker at each end. The speed at which it is pulled is directly related to the slump of the concrete; the less the slump, the faster the speed. The finishing screed, having no transverse (crosswise) movement of the beam, is merely drawn directly forward riding on the forms or rails. Whether the screed is operated by an electric motor or gas engine, a method is provided to quickly start or stop the screed's vibration. It is important to prevent overvibration when the screed is standing still.

Place the concrete from 15 to 20 feet ahead of the strikeboard and shovel it as close as possible to its final resting place. Then put the screed into operation and pull it along by two workers.

It's very important that sufficient concrete is kept in front of the screed. Should the concrete be below the level of the screed beam, voids or bare spots will appear on the concrete surface as the screed passes over the slab. If this occurs, throw a shovelful or so of concrete onto the bare spot, lift the screed, and move it back for a second pass over the slab. After the vibratory finishing screed has passed over the slab, the surface is then ready for broom or burlap finishing.

If possible, lay out the concrete slab specifically for using a vibratory finishing screed. Lay out forms in lanes of equal widths so that the same length of screed can be used on all lanes or slabs. Plan that any vertical columns will be next to the forms so that you can easily lift the screed or maneuver it around the column.

Vibratory finishing screeds make it possible to start troweling floor slabs sooner since drier mixes can be used that set up more quickly.

Floating

If you need a smoother surface than the one obtained by screeding, work the surface sparingly with a wood or metal float or finishing machine. A wood float is shown in Fig. 5-9.

This process should take place shortly after screeding and while the concrete is still plastic and

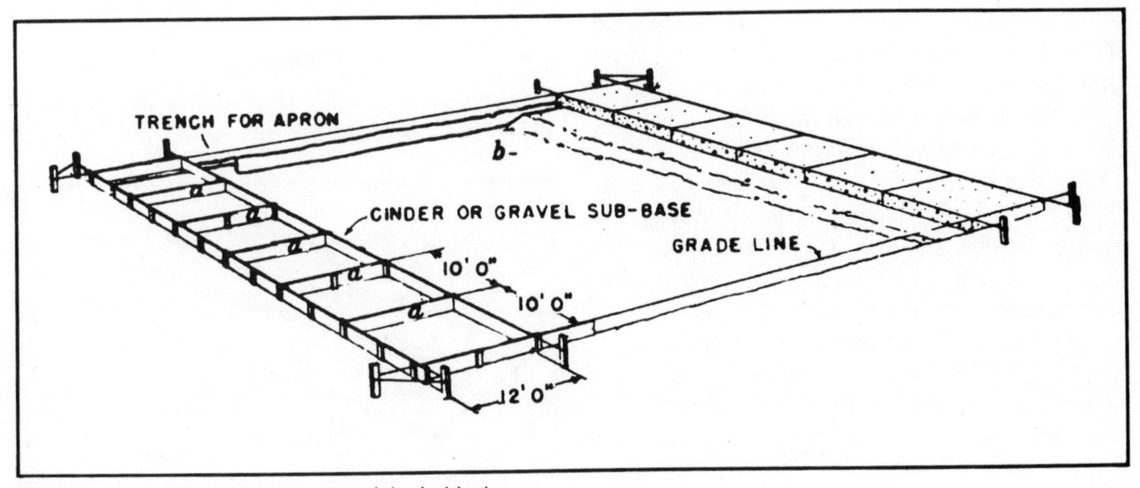

Fig. 5-8. Forms for casting concrete slabs in blocks.

Fig. 5-9. Wood float.

workable. High spots are eliminated, low spots filled in, and enough mortar is brought (floated) to the surface to produce the desired finish. You must not overwork the concrete while it is still plastic to avoid bringing an excess of water and mortar to the surface. This fine material will form a thick, weak layer that will scale or wear off under usage. When you desire a coarse texture as the final finish, float the surface a second time after it has partially hardened.

In slab construction, long-handled wood floats are used. The steel float is used the same way as the wood float, but it gives the finish a much smoother surface. Steel floating should begin when the water sheen disappears from the concrete surface to avoid cracking and dusting the finished concrete. Cement or water should not be used to aid in finishing the surface.

Troweling

If you desire a dense, smooth surface, you must follow the floating procedure with steel troweling. Trowel some time after the moisture film or sheen disappears from the floated surface and the concrete has hardened enough to prevent fine material and water from being worked to the surface. Delay this step as long as possible. Excessive troweling too early tends to produce crazing and lack of durability; too long a delay in troweling results in a surface too hard to finish properly. The usual tendency is to start to trowel too soon.

Troweling should leave the surface smooth, even, and free of marks and ripples. Spreading dry cement on a wet surface to take up excessive water is not good practice where a wear-resistant and durable surface is required.

Avoid wet spots if possible. When they do occur, do not resume finishing operations until the water has been absorbed, evaporated, or mopped up.

You can obtain a surface that is fine-textured but not slippery by troweling lightly over the surface with a circular motion immediately after the first regular troweling. In this process, keep the trowel flat on the surface of the concrete.

When you require a hard steel-troweled finish, follow the first regular troweling with a second troweling after the concrete has become hard enough so that no mortar adheres to the trowel and a ringing sound is produced as the trowel passes over the surface. During this final troweling, tilt the trowel slightly and exert heavy pressure to thoroughly compact the surface.

Hair cracks usually result from a concentration of water and fine material at the surface, which comes from overworking the concrete during finishing operations. Such cracking is aggravated by too rapid drying or cooling. You can usually close checks that develop before troweling by pounding the concrete with a hand float. A steel trowel is shown in Fig. 5-10.

The mechanical (troweler) finishing machine is used tc good advantage on flat slabs with stiff consistency. The concrete must be set enough to support the weight of the machine and the operator. Machine finishing is faster than by hand where the machine will fit in with the type of construction, such as concrete slabs and patios.

Brooming

You can produce a nonskid surface by brooming the concrete before it has thoroughly hardened and

Fig. 5-10. Steel trowel.

after floating has been done. For some floors and sidewalks where severe scoring is not desirable, you can produce the broomed finish with a hairbrush after you have troweled the surface to a smooth finish. Where rough scoring is required, use a stiff broom made of steel wire or coarse fiber. Brooming should be done in such a way that the direction of the scoring is at right angles to the direction of the traffic.

SACK-RUBBED FINISH

A sack-rubbed finish is sometimes necessary when the appearance of formed concrete falls considerably below expectations. This treatment is performed after all required patching and correction of major imperfections have been completed.

Wet the surfaces thoroughly. Start sack rubbing while they are still damp. The mortar used consists of 1 part cement, 2 parts by volume of sand passing a No. 16 screen, and enough water so that the mortar consistency is like thick cream. Blend the cement with white cement to obtain a color that matches the surrounding concrete surface.

Rub the mortar thoroughly over the area with clean burlap or a sponge rubber flat so that it fills all pits. While the mortar in the pits is still plastic, rub the surface over with a dry mix of the same proportions and materials as the mortar. This step removes all of the excess plastic material and places enough dry material in the pits to stiffen and solidify the mortar so that the fillings will be flush with the surface. No material should remain on the surface except that within the pits. Then cure the surface.

CURING CONCRETE

Concrete hardens as a result of the hydration of the cement by the water. Freshly placed concrete contains more than enough water to hydrate the cement completely, but if the concrete is not protected against drying out, the water content, especially at and near the surface, will drop below that required for complete hydration.

The procedure called *curing* is designed to prevent surface evaporation of the water during the period between the beginning and final set. Concrete takes a beginning set in about 1 hour; a final set takes about 7 days.

Curing is accomplished by keeping the concrete surface continuously moist. Depending upon the type of structure, you can cure by spraying or ponding; by covering with continually moistened earth, sand, burlap, or straw; or by covering with a water-retaining membrane.

Make sure that concrete made with ordinary cement is kept moist for a minimum of 7 days. It should be protected from the direct sunlight for at least the first 3 days of the curing period. Wet burlap is excellent for this purpose. Wood forms left in place also furnish good protection against the sunlight, but you should loosen them at the time when they might safely be removed. Use water to flood the space between the forms and the concrete at frequent intervals.

Curing by *ponding* is usually confined to large slabs. An earth dike is built around the area to be cured, and the space inside the dike is filled with water.

Coverings can be either burlap or cotton mats. Burlap covers consist of two or more layers of burlap having a combined weight of 14 ounces or more per square yard when dry. Burlap should either be new or have been used only for curing concrete.

Cotton mats and burlap strips should have a length, after shrinkage, at least 1 foot greater than necessary to cover the entire width and edges of the slab. Let the mats overlap each other at least 6 inches. Wet the mats thoroughly before placing, and keep them continuously wet and in direct contact with the concrete edges and surface for the remaining curing time.

Hay and straw absorb moisture readily and retain it well. The minimum depth of layer should be at least 6 inches. Whatever wet method of curing is used, the entire slab from edge to edge must be kept wet during the entire curing period.

Curing compounds are available that you can apply using hand-operated pressure sprayers after the forms are all removed. Apply a second coat at right angles to the first coat. The compound should form a uniform, continuous, cohesive film that will not check, crack, or peel, and be free from pinholes and other imperfections.

Make sure that the concrete is properly cured at the joints, but that no curing compound enters the joints that are to be sealed with joint-sealing compound. Tightly seal the top of the joint opening and the joint groove at exposed edges as soon as you have completed any joint-sawing operations. After you have applied the seal, spray the concrete around the joint with curing compound.

REMOVING FORMS

Whenever possible, leave forms in place for the entire curing period of about 7 days. If you are reusing forms, however, strip them for reuse as soon as possible. In any event, do not strip forms until the concrete has hardened enough to hold its own weight and any other weight it might be carrying. The surface must be hard enough to remain uninjured and unmarked when reasonable care is used in stripping the forms. Under ordinary circumstances, you can remove floor slab forms after 10 days, wall forms after 1 day, and column forms after 3 days.

After you have removed the forms, you should inspect the concrete for surface defects. These defects include rock pockets, inferior quality, ridges at form joints, bulges, bolt holes, and form-stripping damage. Sometimes repairs might not be necessary, but if they are, they should be done within 24 hours of stripping the forms.

You can repair defects in various ways. We will discuss each of these ways.

You can repair *ridges* and *bulges* by careful chipping followed by rubbing with a grinding stone.

Defective areas such as *honeycomb* must be chipped out to solid concrete. Cut the edges as straight as possible at right angles to the surface or slightly undercut to provide a key at the edge of the patch. If a shallow layer of mortar were directly placed on top of the honeycomb concrete, moisture would form in the voids and subsequent weathering would cause the mortar to spall off. Fill shallow patches with mortar placed in layers not more than 1/2 inch thick. Scratch-finish each layer to match the surrounding concrete by floating, rubbing, or tooling. On formed surfaces, scratch-finish by pressing the form material against the patch while the mortar is still in place.

Fill *deep patches* with concrete held in place by forms. Reinforce and dowel these patches to the hardened concrete (Fig. 5-11).

Patches usually appear darker than the surrounding concrete. Use some white cement in the mortar or concrete used for patching if appearance is important. Try a trial mix to determine the proportion of white and gray cements to use. Before you place mortar or concrete in patches, keep the surrounding concrete wet for several hours. Mix a grout of cement and water to the consistency of paint and brush it into the surfaces to which the new material is to be bonded. Start curing as soon as possible to avoid early drying. Damp burlap, tarpaulines, and membrane curing compounds are useful for this purpose.

Fill *bolt holes* with grout carefully packed into place in small amounts. Mix the grout as dry as possible, with just enough water so that it will be tightly compacted when forced into place. Fill tie rod holes extending through the concrete with grout with a pressure gun similar to an automatic grease gun.

Rock pockets and other defective concrete should be completely chipped out. Give the chipped-out hole sharp edges and shape it so that the grout patch can be keyed in place (Fig. 5-12). Make sure

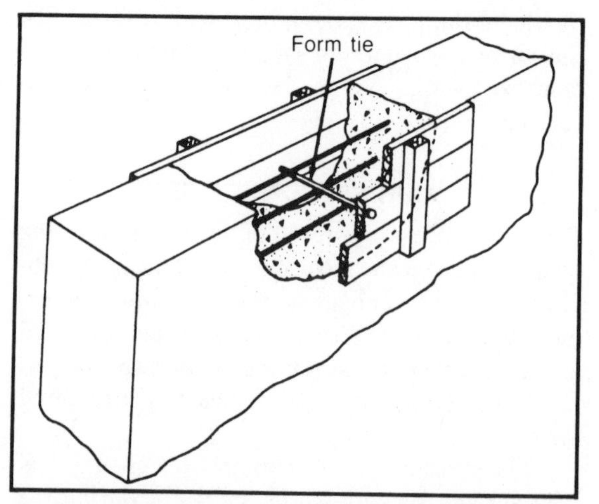

Fig. 5-11. Repairing a large volume of concrete.

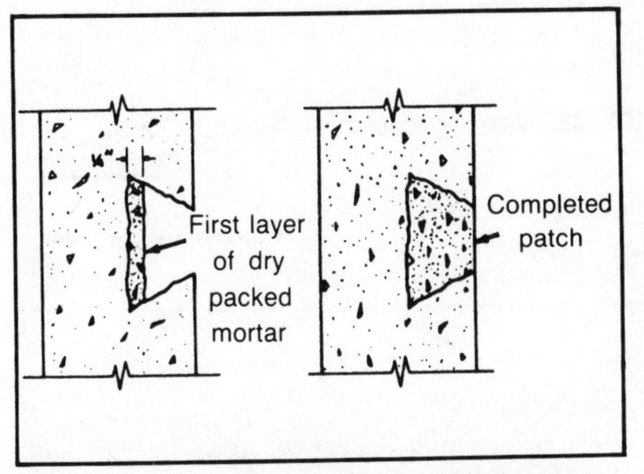

Fig. 5-12. Repairing concrete with dry-packed mortar.

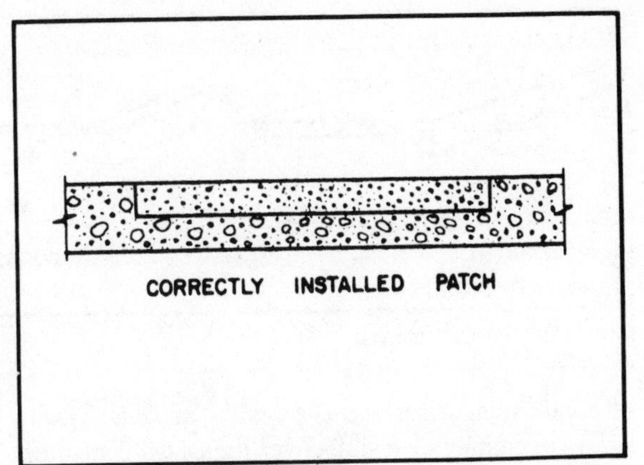

Fig. 5-14. Chipped area should be at least 1 inch deep.

the surfaces of all holes that are to be patched are kept moist for several hours before you apply the grout. Place the grout in these holes in layers not over 1/4 inch thick and compact well. Allow the grout to set as long as possible before using to reduce the amount of shrinkage and make a better patch. Rough-scratch each layer to improve the bond with the succeeding layer and smooth the last layer to match the adjacent surface. Where you have used absorptive form lining, match the patch to the rest of the surface by pressing a piece of the form lining against the fresh patch.

Feathered edges around a patch (Fig. 5-13) will break down. Make the chipped area at least 1 inch deep with the edges at right angles to the surface (Fig. 5-14). The correct method of screeding a patch is shown in Fig. 5-15. Project the new concrete slightly beyond the surface of the old concrete. Allow it to stiffen and then trowel and finish it to match the adjoining surfaces.

PRECAST CONCRETE

Precasting is the fabrication of a structural member at a place other than its final position of use. It can be done anywhere, although this procedure is best adapted to a factory or yard. The most commonly used precast slabs or panels for floor and roof decks are the channel and double-T types (Fig. 5-16).

The channel slab varies in size with a depth of 9 to 12 inches, a width of 2 to 5 feet, and a thickness of 1 to 2 inches. If desired or needed, you can extend the legs of the channels across the ends; if used in combination with the top slab, you can stiffen the legs with occasional cross-ribs.

The double-T slab varies in size from 4 to 6 feet in width and 9 to 16 inches in depth, and has been used in spans as long as 50 feet. When the top slab size ranges from 1 1/2 to 2 inches in thickness, you should reinforce it with wire mesh.

You can use a concrete floor slab with a

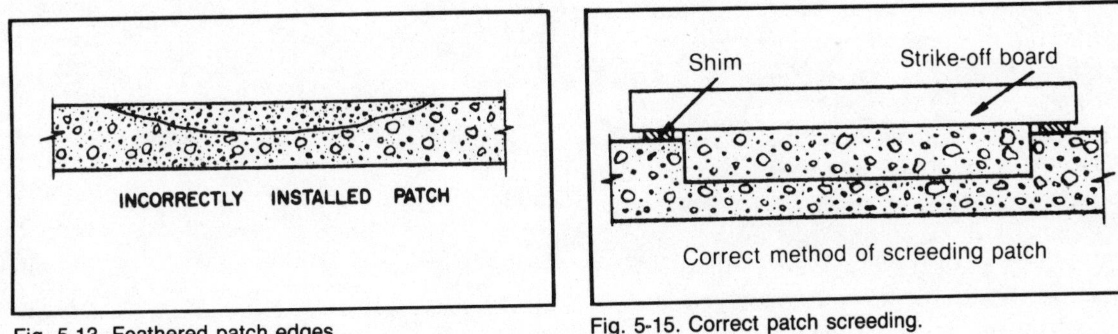

Fig. 5-13. Feathered patch edges.

Fig. 5-15. Correct patch screeding.

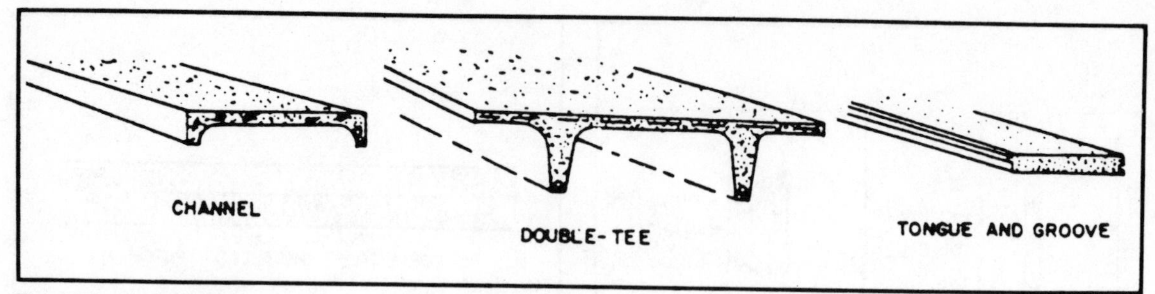

Fig. 5-16. Precast planks.

smooth, regular surface as a casting surface. When you are casting on a slab cover the casting surface with some form of liquid or sheet material to prevent bonding between the surface and the wall panel. You can finish the upper surface of the panel in the same way as regular concrete—by troweling, floating, or brooming.

Chapter 6

Wood Foundations

Many homeowners and builders are discovering a unique type of foundation. The all-weather wood foundation is essentially a below-grade, wood-frame wall built of pressure-preservative-treated framing lumber and plywood.

The all-weather wood foundation is an engineered foundation system that is appropriate for both crawl-space and basement foundations. It can support single-story and multistory buildings, can work on a level or a sloping lot, and can be prefabricated in a shop or constructed at the building site.

The system has been accepted by all nationally recognized model building codes, but not all local codes. It is also recognized by federal agencies, such as Farmers Home Administration and the Federal Housing Administration, as well as lending and mortgage insurance institutions and warranty and fire insurance organizations.

Wood foundations have been used for over two decades. They now support many thousands of buildings around the country.

The all-weather wood foundation has several advantages noted by builders who have used the system. Carpentry crews install the system; no masonry crews or skills are needed. Fast installation is typical; the foundation can be installed by a small crew in less than a day. Cost is usually competitive or lower than a masonry foundation. Wood foundation walls need not be as thick as typical masonry walls. Insulating is much easier in the wood-frame system than with masonry. Lower heat loss is typical with the wood foundation. For example, heat loss from an uninsulated concrete foundation (18 inches exposed above grade) can be as much as three times more than an insulated wood foundation. Finally, interior finish materials are easy to apply, and plumbing and wiring are simplified.

BASIC COMPONENTS

Wood, of course, comes from trees (Figs. 6-1 through 6-3). Lumber for foundation framing must be of a species for which strength values have been assigned. It must be structurally graded material and be grade-marked by an approved inspection

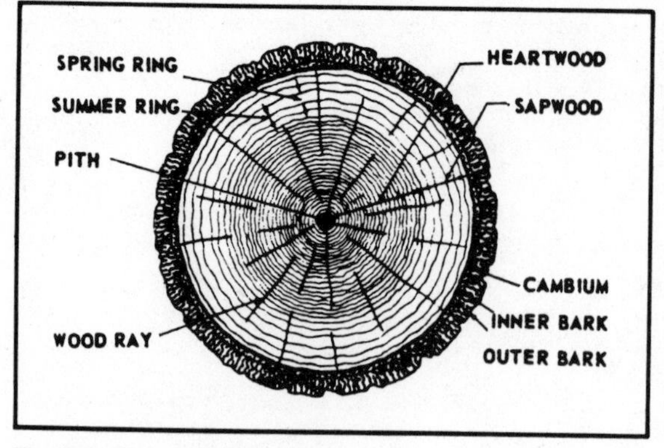

Fig. 6-1. Cross section of tree.

agency. The wood used for framing lumber in this system must accept the pressure-preservative treatment specified.

Plywood must be an exterior type or interior type bonded with an exterior (waterproof) glue. It must be produced according to "U.S. Product Standard PS1 for Construction and Industrial Plywood" and grade-marked by an approved agency.

All lumber used in a wood foundation must be pressure-treated with appropriate waterborne chemicals to resist decay and insect attack. Approved preservatives are ammoniacal copper arsenate (ACA) and three forms of chromated copper arsenate (CCA-A, CCA-B, and CCA-C). Required preservative salt retention is 0.60 pound per cubic foot of wood. After the material is impregnated, it must be dried to a moisture content of 18 percent or less for plywood and 19 percent or less for lumber. All treated lumber and plywood should bear

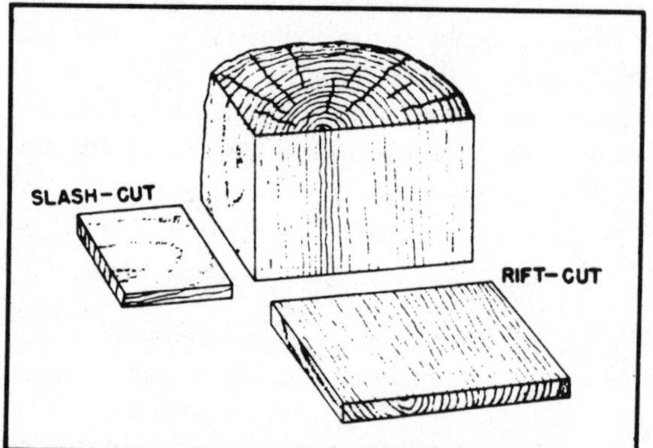

Fig. 6-2. Two types of wood cuts.

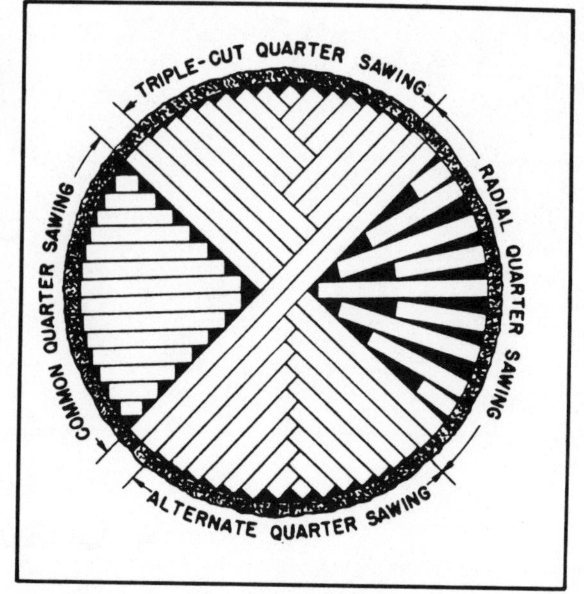

Fig. 6-3. Rift cutting: four methods of quartersawing.

the grade mark of the American Wood Preservers Bureau (AWPB-FDN) to certify conformance with the penetration and retention requirements for foundation material (Fig. 6-4). A typical lumber grade mark is shown in Fig. 6-5. Figure 6-6 shows a typical treated plywood grade mark.

Fasteners used to assemble the lumber and plywood into framed walls must be corrosion resistant. Stainless steel nails or staples (Type 304 or 316) should be used below grade for attaching the plywood to the lumber. Lumber-to-lumber fasteners

Fig. 6-4. Typical treated wood stamp.

below grade also should be stainless steel, but can be hot-dipped or hot-tumbled galvanized nails.

Washed gravel (3/4 inch or less), crushed stone (1/2 inch or less), or coarse sand (1/16 inch or greater) is used for fill under the footings and basement slab. Gravel is preferred. Any of the materials used must be clean and free of all silt, clay, and organic material.

SYSTEM DESIGN

Different soils have been classified according to allowable bearing loads, drainage characteristics, frost-heave potential, and volume-change potential. Structural requirements relating to resistance to lateral pressure of the soil and vertical loads from the building have been developed and tabled. Lumber and plywood sizes have been determined for different loading situations. These tables and requirements are discussed later in this chapter.

A good drainage system has been developed and is an integral part of the all-weather wood foundation system. Proper placement and depth of gravel around and under the foundation and basement slab are key elements. The gravel provides an unobstructed path for water to flow away from the foundation or into a sump.

CONSTRUCTION AND INSTALLATION

Perform the excavation in the usual manner to provide the required depth for a crawl space, partial basement, or full basement. Footings are required under both perimeter and interior loading walls. Gravel footings below the frost line are usually installed, but you can use concrete footings on gravel.

The treated foundation wall panels rest on a treated footing plate. Attach the footing plate to the bottom wall before installation. Any treated lumber cut during installation must be field-treated with

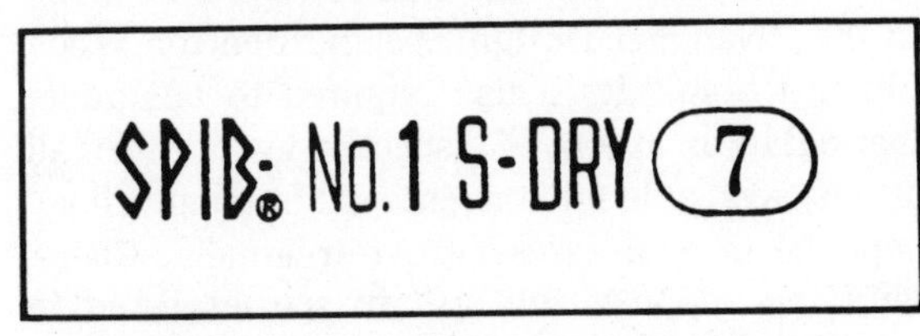

Fig. 6-5. Typical lumber grade mark.

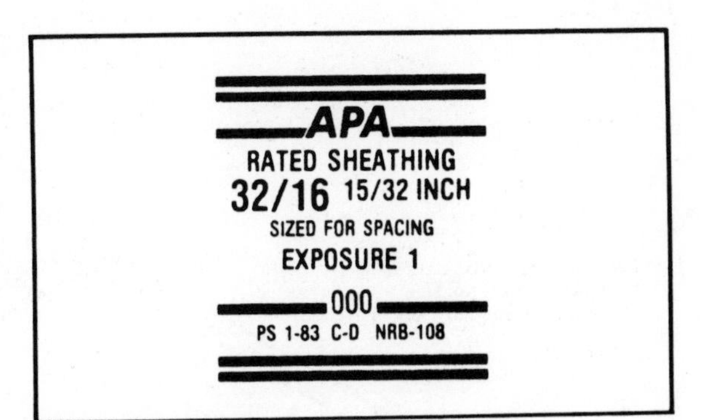

Fig. 6-6. Typical treated plywood grade mark.

the same preservative (minimum concentration of 3 percent in solution).

After you have installed the treated wall panels square and plumb, install an untreated wood second top plate. Then attach the first-story floor framing and subfloor to stabilize the top of the foundation. Then install the concrete slab floor or a special wood floor in a basement foundation to resist lateral forces at the basement of the foundation wall. For best results, use 6 inches of gravel under the basement slab.

A vital part of this system is moisture control. For basement foundations, you must carefully seal all panel joints with a water-durable, high-performance caulking compound. Use a 6-millimeter thick polyethylene plastic to cover the below-grade portion of the foundation walls. Spot-bond the polyethylene with an appropriate adhesive, lap the vertical joints 6 inches, and seal with an adhesive. Protect the plastic film at the top with a treaded wood strip. Install a vapor barrier under the basement floor as well.

Both the first-story floor and basement slab must be in place (concrete slab cured) before you backfill against the basement foundation. In crawl-space construction, brace the walls before backfilling. Use gravel for the lower half of the backfill to enhance drainage. Always slope the ground away from the building. Use gutters, downspouts, and splash blocks to direct the roof water away from the structure.

WOOD FOUNDATION SYSTEM

Figure 6-7 gives a closer look at how a permanent

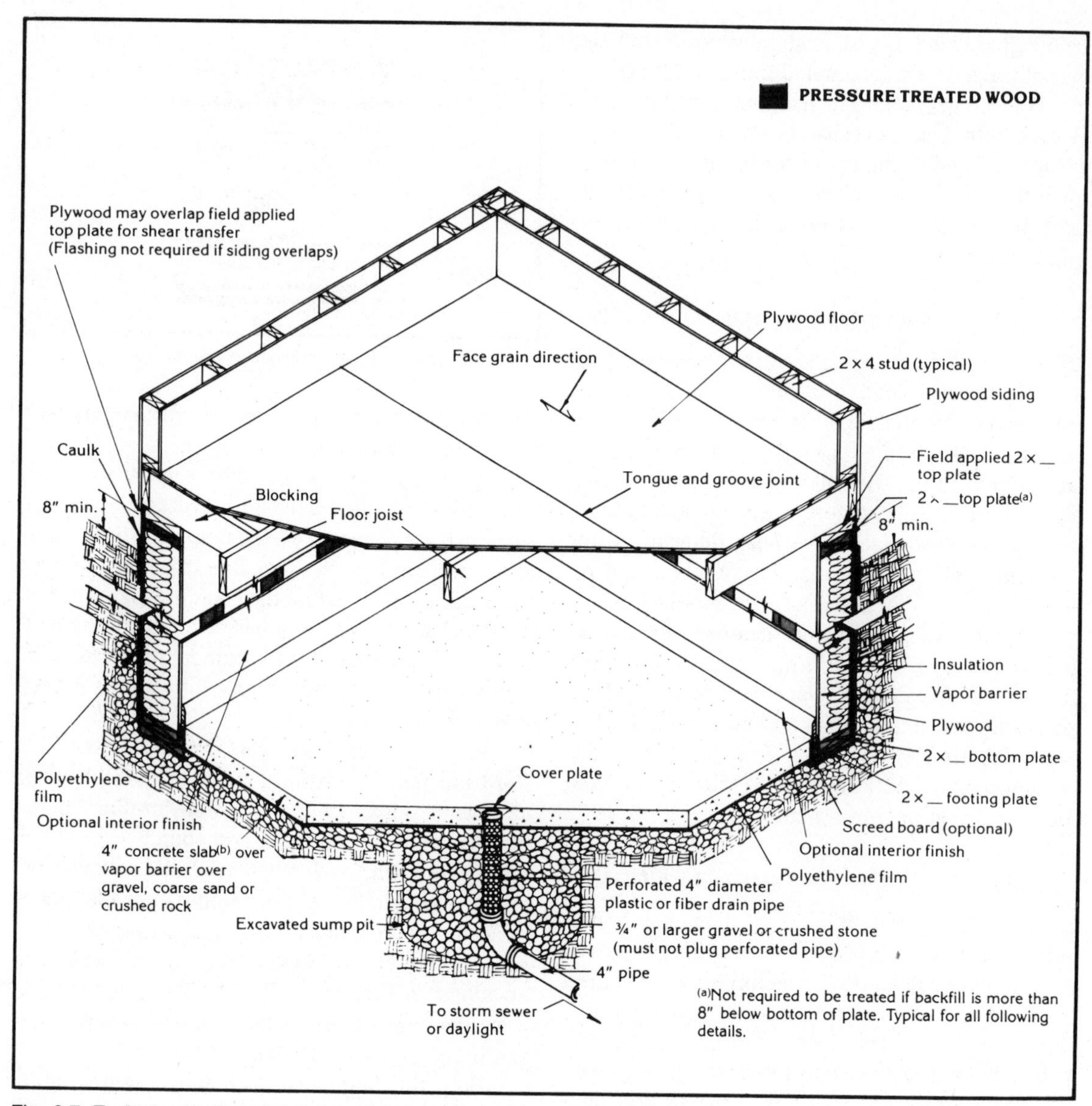

Fig. 6-7. Typical wood foundation (courtesy American Plywood Association).

wood foundation is planned and constructed. A permanent wood foundation is basically a below-grade stud wall built of pressure-preservative-treated lumber and American Plywood Association (APA) trademarked plywood. Plywood must be exterior type or interior type bonded with exterior glue (Exposure 1) and produced according to "U.S. Product Standard PS 1-83 for Construction and Industrial Plywood."

Lumber for foundation framing must be of a species for which allowable unit stresses are given in the "National Design Specification for Wood Construction." It is also required to be grade-marked by an approved inspection agency. Not all species available for conventional framing will accept the pressure-preservative treatment. Check with the supplier and specify species listed in "AWPB-FDN Standard."

All plywood and lumber used for foundation construction must be pressure-treated in accordance with the treating and drying requirements specified in the American Wood Preservers Bureau's "*AWPB-FDN Standard*." It must also meet the penetration and retention requirements of that Standard.

The treatment consists of impregnating certain preservative salts into the wood, as discussed previously. Required preservative salt retention is 0.60 pound per cubic foot of wood, 50 percent higher than codes require for normal ground contact application.

After impregnation, you must dry the material to a moisture content of 18 percent or less for plywood and 19 percent or less for lumber. The preservative treatment darkens the wood, but it does not have an odor and is not sticky to the touch.

The only exceptions to the treatment requirements are members more than 8 inches above the ground. These might be, for example, window and door headers, upper plates, or the upper course of plywood in a basement wall with low backfill.

If possible, cut the pieces to size before treatment. Field-treat lumber that is cut or drilled after treatment by repeated brushing, dipping, or soaking until the wood absorbs no more preservative. Use one of the following preservatives for field treatment of cut surfaces:

- ☐ ACA, or CCA Types A, B, or C (as used in the original treatment) with a minimum concentration of 3 percent in solution

  or

- ☐ Copper napthenate containing a minimum of 2 percent metallic copper in solution

Studs cut after pressure treatment need not be field-treated if the cut ends are at least 8 inches above the backfill. Footing plates can be extended past the corner of the foundation to minimize end cutting and field treating.

Fasteners used in wood foundations must be corrosion resistant. Specific fastener schedules are shown in Tables 6-1 through 6-5.

BUILDING WOOD FOUNDATIONS

A permanent wood foundation is basically conventional frame construction with a few exceptions, such as the offset footer and bottom plates. It's easy to design and build for most common buildings. The wood foundation must be designed to withstand backfill and vertical loading, but it doesn't have to be engineered for most single-family houses.

The recommendations in this chapter have been developed by experts in the wood industry and the National Association of Home Builders Re-

Table 6-1. Plywood Fastener Schedule for PWF (courtesy American Plywood Association).

Fastener	Equal Backfill (Max. 2-ft. Differential)		Unequal Backfill (Max. 4-ft. differential)			
			Long Walls[3]		Short Walls	
	Panel Edges	Intermediate	Panel Edges	Intermediate	Panel Edges	Intermediate
16 ga. × 1½" Staple	4"	8"	4"	8"	4" Blocking[4] required	8"
8d Common Nail	6"	12"	6"	12"	6" Blocking[4] required	12"
					6"	6"

[1]In crawl space construction, provide a fastener within 1½" of the bottom of each stud.

[2]In above-grade areas, hot-dipped or hot-tumbled galvanized steel, silicon bronze or copper nails may also be used.

[3]Schedule applies when backfill on one long wall is higher than on the opposite one. When backfill is higher on one short wall than on the opposite one, use the schedule from the "Short wall" columns for the long walls, but for a length at least equal to one-half the length of the short wall. Then the "Long wall" schedule applies to the rest of the long walls, and all of the short walls.

[4]All plywood panel edges shall be fastened to blocking or framing of 2-inch nominal lumber.

Table 6-2. General Nailing Schedule (courtesy American Plywood Association).

Joint description	Minimum[1] nail size	Number or spacing
Bottom plate to footing plate—Face nail	10d	12" o.c.
Bottom plate to stud—End nail—2" plate	16d	2
—1" plate	8d	2
Top plate to stud—End nail minimum (See Table 3)	16d	2
Upper top plate to top plate—Face nail minimum (See Table 3) (No overlap of plywood)	10d	8" o.c.
Header joist to upper top plate—Toe nail minimum (See Table 4)	8d	16" o.c.
Joist to upper top plate—Toe nail minimum (See Table 4)	8d	3
	10d	2
End joist to plate (joists parallel to wall)—Toe nail minimum (See Table 4)	8d	4" o.c.
Plywood flooring to blocking at end walls (See Table 5)	—	—
Window header support studs to window sill—End nail minimum	16d	2
Window sill to studs under—End nail minimum (See Table 3)	16d	2
Window header to stud—End nail	16d	4
Knee wall top plate to studs—End nail	16d[2]	2
Knee wall bottom plate to studs—End nail	8d[2]	2
Knee wall top plate to foundation wall—Toe nail	16d[2]	1 per stud
Knee wall stud over 5' long to foundation wall stud—Toe nail at mid-height of stud	16d[2]	2 per stud
Knee wall bottom plate to footing plate—Face nail	8d[2]	2 per stud space
Window, door or beam pocket header support stud to stud—Face nail	10d	24" o.c.
Corner posts—stud to stud—Face nail	16d	16" o.c.

[1]Heavy loads may require more or larger fasteners or framing anchors. All lumber-to-lumber fasteners below grade may be hot-dipped or hot-tumbled galvanized, except as noted.
[2]Stainless-steel nails required.

search Foundation. The details are applicable for basic installation on a variety of soils.

In larger or more unusual buildings, engineering might be necessary because of differing structural requirements. If so, refer to the *All-Weather Wood Foundation: Design, Fabrication, Installation Manual* available from the National Forest Products Association. It contains complete engineering design data and is the source of some of the tabular material and many of the details given here.

SOIL PREPARATION AND DRAINAGE

Proper excavation and site drainage are important in keeping any type of foundation—masonry, concrete, or wood—dry and trouble-free.

A good drainage system has been developed for the wood foundation to keep the basement dry under most conditions. It is an integral part of the

Table 6-3. Minimum Nailing Schedules: Top Plate to Stud and Plate to Plate Connections (courtesy American Plywood Association).

Height of fill (inch)	Treated lumber species	End-nail treated top plate to treated studs		Face-nail untreated top plate to treated top plate			
				No overlap of plywood		¾" plywood overlap	
		Nail size[2]	Number per joint	Nail size[2]	Spacing (inch)	Nail size[2]	Spacing (inch)
24	All	16d	2	10d	8	10d	16
48	All	16d	2	10d	8	10d	16
72	A, B	16d[3]	3	10d	6	10d	8
	C, D	16d[3]	4	10d	4[4]	10d	4[4]
86	A, B	20d[3]	3	10d	3[4]	10d	4[4]
	C, D	20d[3]	4	10d	2[4]	10d	3[4]

[1]Based on 30 pcf equivalent-fluid density soil pressure and dry lumber.
[2]Hot-dipped, hot-tumbled or stainless-steel common wire nails.
[3]Alternatively, may use "U" type framing anchor or hanger with nails and steel plate meeting requirements of DFI 2.4 and having a minimum load capacity (live plus dead load, normal duration) of 340 pounds in species combination "B"
[4]Alternatively, two nails 2½ inches apart across the grain at twice the spacing indicated may be used.

Table 6-4. Minimum Nailing Schedules: Floor Joists to Wall Connections (courtesy American Plywood Association).

Height of fill (inch)	Joist spacing (inch)	Joists perpendicular to wall: Toe-nail[2] header joist to plate — Nail size[5]	Spacing (inch)	Toe-nail[2] each joist to plate — Nail size[5]	No. per joist	Framing anchor[4] each joist to plate
48 or less	16	8d	16	8d	3	none
		10d	16	10d	2	none
	24	8d	8	8d	3	none
		10d	8	10d	2	none
72	16	8d	8	8d	3	none
		10d	8	10d	2	none
		8d	16	none	none	1
	24	10d	8	10d	3	none
		8d	16	none	none	1
86	16	8d	8	none	none	1
	24	8d	4	none	none	1

Height of fill (inch)	Blocking[3] between joists, spacing (inch)	Joists parallel to wall: Toe-nail[2] end joist to plate — Nail size[5]	Spacing (inch)	Toe-nail[2] blocking to plate — Nail size[5]	No. per block	Framing anchor[4] each block to plate
48 or less	No blocking	8d	4	none	none	none
72	48	8d	4	8d	3	none
		10d	4	10d	2	none
		10d	6	10d	4	none
		8d	6	none	none	1
86	24	8d	4	none	none	1

[1]Based on 30 pcf equivalent-fluid density soil pressure and dry lumber. Untreated top plate not less than species combination "D" from Table 10, or species Group III from National Design Specification.

[2]Toe-nails driven at angle of approximately 30° with the piece and started approximately one-third the length of the nail from the end or edge of the piece.

[3]See Table 5 for additional spacing requirements for blocking, and for subfloor to blocking nailing schedule.

[4]Framing anchors shall have a minimum load capacity (live load plus dead load, normal duration) of 320 pounds per joist. If plate or joist is species combination "C" or "D", then rated load capacity of anchors when installed in species combination "B" shall be not less than 395 pounds per joist.

[5]Common wire steel nails.

Table 6-5. Minimum Nailing Schedules: Subfloor to End Wall Blocking[2,3] (courtesy American Plywood Association).

Height of fill (inch)	Species of joist and blocking lumber[4]: "A" or "B" — Block spacing (inch)	Nails per block[1]: 6d	8d	"C" or "D" — Block spacing (inch)	Nails per block[1]: 6d	8d
60	48	4	3	48	6	4
72	48	8	6	48	11	8
	24	3	2	24	5	3
86	24	7	5	24	9	7

[1]Common wire nails. Nails shall be spaced 2 inches on center or more; where block length requires, nails may be in two rows.

[2]See Table 4 for additional requirements for block spacing and nailing.

[3]Based on 30 pcf equivalent-fluid density soil pressure and dry lumber.

[4]See Table 10 for minimum properties of lumber species combinations. Other species in Group III, National Design Specification, Table 8-1A, require the same nailing as species grade combination "D".

wood foundation system and should be incorporated in the installation.

The gravel is a key element in the drainage system. It provides an unobstructed path for the water to flow away from the foundation or into a sump in a full-basement house. The gravel thus prevents a buildup of pressure against the foundation and helps avoid leaks.

Soils are classified by their makeup and how well they drain. Table 6-6 lists the type of soils commonly found and describes the important properties. Soil classifications for most areas are listed in the standard series of soil surveys published by the U.S. Department of Agriculture's Soil Conservation Service.

The permanent wood foundation can be built in Group I, II, or III soils. Group IV soils are unsatisfactory for wood foundations unless special precautions are taken.

If you are excavating to a depth of 4 to 6 feet in Group I soils, and if the ground slopes away from the building at a grade of 1/2 inch per foot for 6 or more feet and drainage is provided for surface water, you might not need a polyethylene film around the foundation. Even in Group I soils, however, it is still a good idea to use the polyethylene for full basements.

For full basements in Group I soils, a sump is recommended. It should drain to daylight or into a storm sewer.

In Group II soils, grade slope should be 1/2 inch per foot. The backfill should be free of organic material, voids, or chunks of clay. It should be compacted and no more permeable than the surrounding soil.

For full basements in Group II soils, a moisture barrier is essential. A sump should be included in full-basement construction.

Group III soils are satisfactory for wood foundations under the same conditions as Group II soils. For full basements, granular fill under the slab must be at least 6 inches deep as opposed to the 4-inch minimum for Group I and II soils.

Before excavation, it's wise to obtain a plot plan. This is a quick reference to the location of plumbing, sewer, gas, and electrical locations.

Site clearing and excavation begin with staking out the building and utility location. Staking helps avoid overexcavating or underexcavating—costly but easily avoided mistakes. Batter boards also help locate areas to be trenched for footings, plumbing, and utilities and foundation drainage (Fig. 6-8). Separate topsoil and organic material from excavated earth to be used for backfill.

Excavation must be wide and deep enough into the undisturbed soil so the footings will be centered under the foundation walls. Properly installed footings will not settle and distort the foundation wall. Footings are required under both perimeter and interior load-bearing walls.

The gravel footing distributes the load from the foundation walls and functions as a structural element in all cases. It is also an integral part of the drainage system. To prevent frost heave, it is essential that the bottom of the gravel footing be below the frost line. If a concrete footing is used, it should be on gravel to maintain continuity of the drainage system, or drains through the concrete must be provided.

After all of the utility trenches are dug and leveled, you can line them with fine gravel or sand before you set pipes and conduit in place. Then fill the trenches the rest of the way with gravel, coarse sand, or crushed rock.

CONSTRUCTING THE FOUNDATION WALL

You can build the permanent wood foundation in place, fabricate it on site, or, perhaps easiest, deliver it to the site prefabricated. No matter what method is used, it is very important to use high-quality materials of the proper size and to assemble them correctly.

The wood foundation is easily adaptable to almost any design just like conventional above-grade stud walls. It can fit any floor plan and will fit on a level lot or into a hillside.

Frame a typical wall panel (Fig. 6-9) with 2x studs. The panel can be any convenient width. Height will depend on whether it is for a crawl-space or full-basement foundation. You can orient the plywood with the face grain horizontal or ver-

Table 6-6. Types of Soils and Their Design Properties (courtesy American Plywood Association).

Soil group	Unified soil classification system symbol	Soil description	Allowable bearing in pounds per square foot with medium compaction or stiffness[1]	Drainage Characteristics[2]	Frost heave potential	Volume change potential expansion[3]
Group I Excellent	GW	Well-graded gravels, gravel-sand mixtures, little or no fines.	8000	Good	Low	Low
	GP	Poorly graded gravels or gravel-sand mixtures, little or no fines.	8000	Good	Low	Low
	SW	Well-graded sands, gravelly sands, little or no fines.	6000	Good	Low	Low
	SP	Poorly graded sands or gravelly sands, little or no fines.	5000	Good	Low	Low
	GM	Silty gravels, gravel-sand-silt mixtures.	4000	Good	Medium	Low
	SM	Silty sand, sand-silt mixtures.	4000	Good	Medium	Low
Group II Fair to Good	GC	Clayey gravels, gravel-sand-clay mixtures.	4000	Medium	Medium	Low
	SC	Clayey sands, sand-clay mixture.	4000	Medium	Medium	Low
	ML	Inorganic silts and very fine sands, rock flour, silty or clayey fine sands or clayey silts with slight plasticity.	2000	Medium	High	Low
	CL	Inorganic clays of low to medium plasticity, gravelly clays, sandy clays, silty clays, lean clays.	2000	Medium	Medium	Medium[4]
Group III Poor	CH	Inorganic clays of high plasticity, fat clays.	2000	Poor	Medium	High[4]
	MH	Inorganic silts, micaceous or diatomaceous fine sandy or silty soils, elastic silts.	2000	Poor	High	High
Group IV Unsatisfactory	OL	Organic silts and organic silty clays of low plasticity.	400	Poor	Medium	Medium
	OH	Organic clays of medium to high plasticity, organic silts.	—0—	Unsatisfactory	Medium	High
	Pt	Peat and other highly organic soils.	—0—	Unsatisfactory	Medium	High

[1]Allowable bearing value may be increased 25 percent for very compact, coarse grained gravelly or sandy soils or very stiff fine-grained clayey or silty soils. Allowable bearing value shall be decreased 25 percent for loose, coarse-grained gravelly or sandy soils, or soft, fine-grained clayey or silty soils.

[2]The percolation rate for good drainage is over 4 inches per hour, medium drainage is 2 to 4 inches per hour, and poor is less than 2 inches per hour.

[3]For expansive soils, contact local soils engineer for verification of design assumptions.

[4]Dangerous expansion might occur if these soil types are dry but subject to future wetting.

tical. Blocking between studs is not required, except as noted in the illustrations or where end walls serve as highly loaded shear walls.

Attach the treated footing plate to the bottom wall plate before foundation installation. This method helps make certain that the footing plates are properly aligned and offsets the joints in the plates from those of the wall section. It also prevents loose gravel from lodging between the footing plate and the bottom wall plate.

Frame corners in the same manner as for panelized wood frame construction. Set back the end stud of the corner panel to the stud depth plus the plywood thickness from the edge of the plywood sheathing. Construct wall sections so all vertical joints between plywood sheathing are backed by a stud.

In full basements, use caulking compound to seal the full length of all plywood joints. Apply the compound before you install the adjacent panels.

Squareness of construction is extremely important for any foundation. Check the framed panels for squareness by measuring diagonals. Diagonal measurement should not differ by more than 1/8 inch for a 4- × -8-foot unit (proportionately more or less for other sizes).

Nailed connections are also important in the permanent wood foundation, particularly at the top plate. The top plate provides the support necessary to withstand the soil pressure.

After you have installed the foundation on the prepared site, install an untreated second top plate with staggered joints. Next install the floor framing. You must have the entire floor system in place, including blocking and subflooring, before backfilling. In basement construction, you also must cure

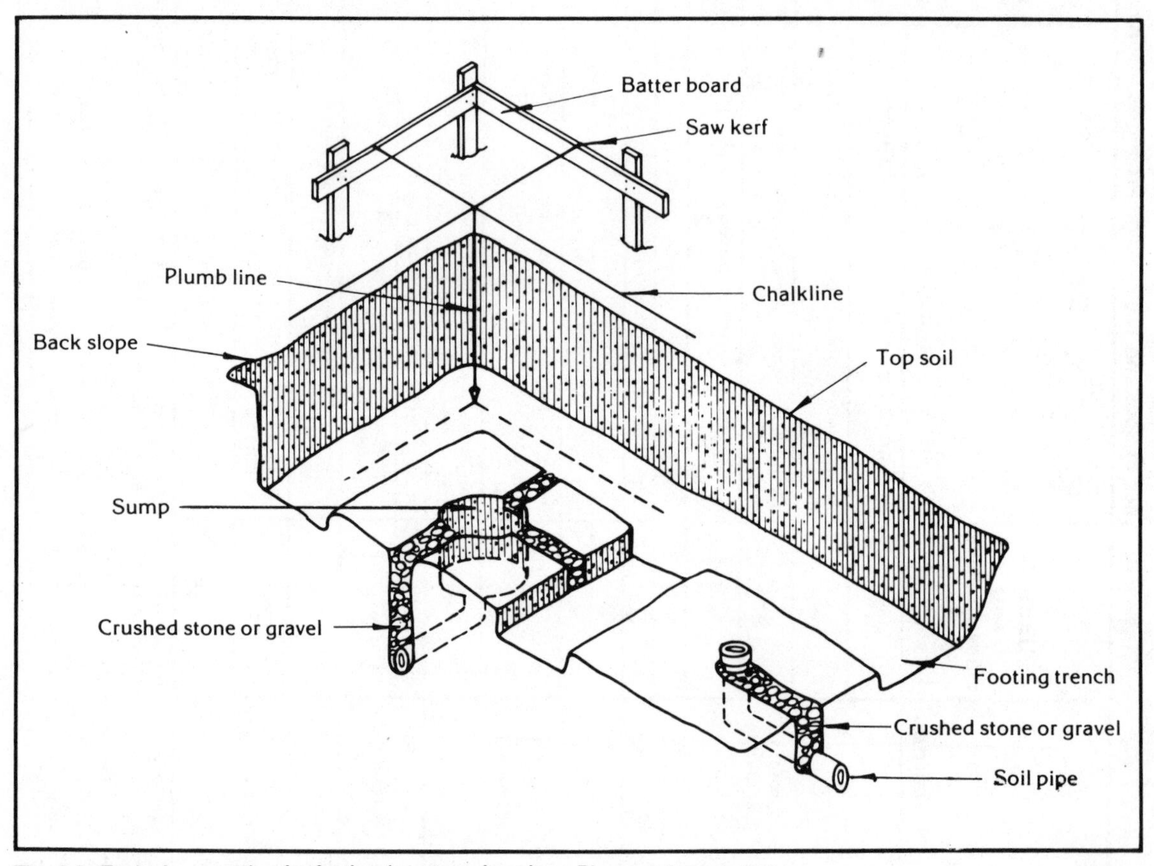

Fig. 6-8. Typical excavation for footing (courtesy American Plywood Association).

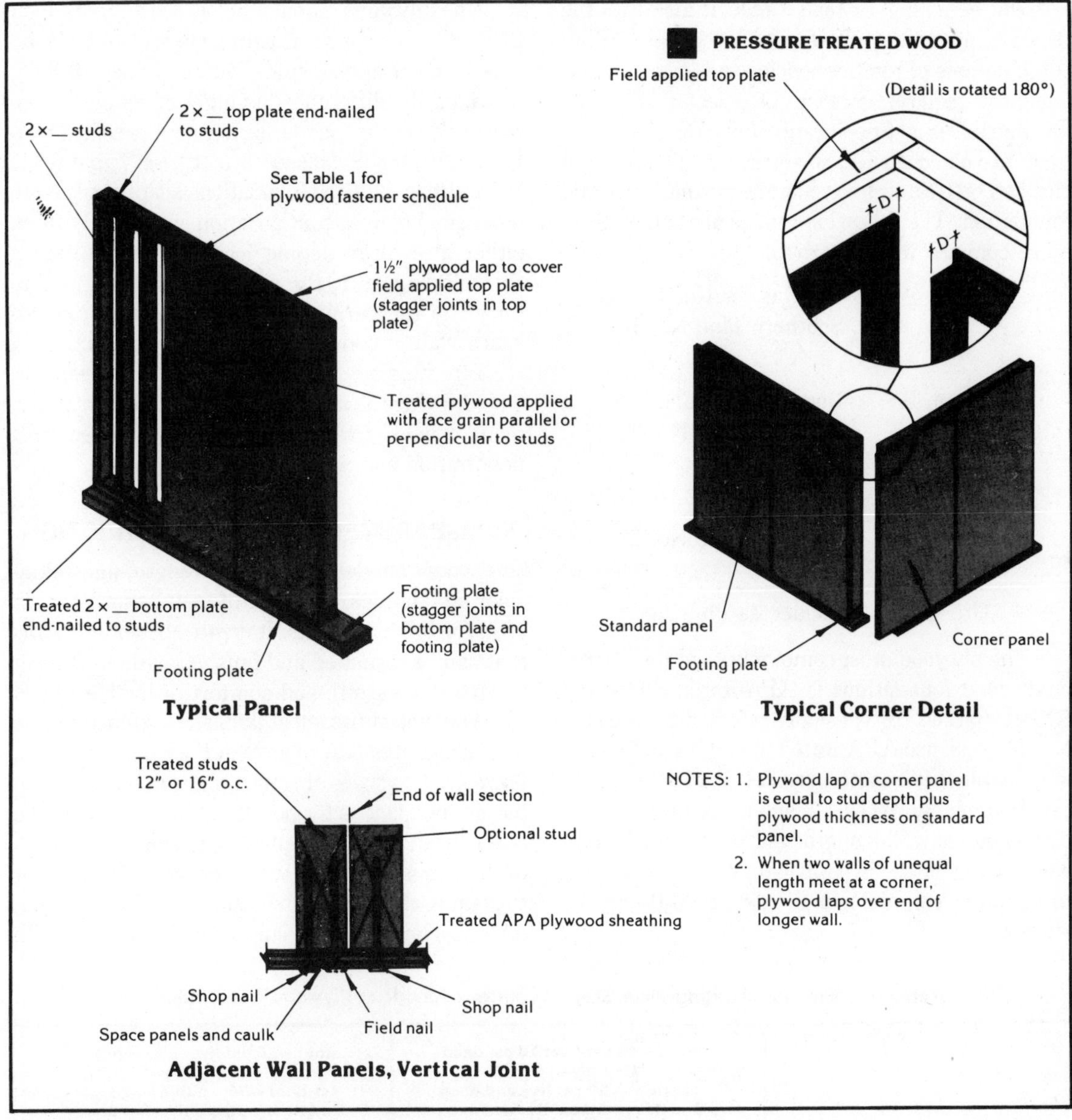

Fig. 6-9. Panel construction (courtesy American Plywood Association).

the concrete slab before backfilling. In crawl-space construction, brace the wall as required before backfill.

CRAWL-SPACE CONSTRUCTION

Footing plate size is determined by the size of the building. Gravel footings are twice the width of the plate, and the depth should be at least three-quarters of the width of the footer plate. Footings for residential construction are about 4 to 7 inches deep and 11 to 19 inches wide.

The size of the framing members and footing plate for one- and two-story homes can be selected from Tables 6-7 and 6-8. Sizes listed in Table 6-8 assume that the fill against the crawl-space foundation will not exceed 4 feet and that the wall itself will be no higher than 6 feet. It also lists the required stresses for the framing species. Many combinations of lumber species and grades meeting these requirements can be selected from the "National Design Specification for Wood Construction." As noted earlier, however, not all of the species can be treated for use in the permanent wood foundation. The following grades are some of the more common in each group.

Group A	Douglas Fir No. 1 Southern Pine No. 1
Group B	Douglas Fir No. 2 Southern Pine No. 2
Group C	Hem-Fir No. 2 Northern Pine No. 2
Group D	Ponderosa Pine No. 2

The plywood most commonly used for permanent wood foundations is APA-Rated Sheathing EXP 1 marked PS 1, usually referred to as CDX. You also can use APA-Rated Sheathing EXT and, if appearance is a factor, use A-C EXT, B-C EXT, C-C Plugged, or MDO. All should be group 1 species. You can treat ungrooved textured 303 plywood siding for use in permanent wood foundations to match or complement the siding on the upper stories.

The plywood should be at least 15/32-inch CDX, if the face grain is parallel to the studs. If the face grain is across studs, you can use 3/8 inch, provided the difference in height of fill on the opposite side of the foundation does not exceed 3 feet. If the difference exceeds 3 feet, use Table 6-11.

Permanent wood foundations adapt easily to almost any building configuration and handily fit on either a level or sloping lot. The design details shown in Figs. 6-10 through 6-14 illustrate the simplicity of the permanent wood foundation crawl-space wall in conventional house construction.

Intermediate center posts are easily installed. The drawings also illustrate the installation of permanent wood foundations in areas of deep frost penetration and with slab-on-grade floors.

FULL-BASEMENT WALL CONSTRUCTION

The wood foundation's great strength, natural insulating qualities, and ease of installation make it ideal for full basements. It creates basement walls that can be insulated and finished without furring to give the warmth and comfort of livable space.

The type of basement drainage system depends on soil conditions. As shown in Figs. 6-15 and 6-16, there are two types of sumps that can be used with basements. One is for poorly drained soils and the other for medium to well-drained soils. The depth of the sump should allow for the gravel bed under the sump, the height of the sump, and the underslab fill. The top of the sump should be level with

Table 6-7. Minimum Footing Plate Size[1,2] (courtesy American Plywood Association).

	Roof—40 psf live; 10 psf dead Ceiling—10 psf 1st floor—50 psf live and dead 2nd floor—50 psf live and dead		Roof—30 psf live; 10 psf dead Ceiling—10 psf 1st floor—50 psf live and dead 2nd floor—50 psf live and dead	
House width (feet)	2 stories	1 story	2 stories	1 story
32	2 × 10	2 × 8	2 × 10[3]	2 × 8
28	2 × 10	2 × 8	2 × 8	2 × 6
24	2 × 8	2 × 6	2 × 8	2 × 6

[1] Footing plate shall be not less than species and grade combination "D" from Table 8.

[2] Where width of footing plate is 4 inches (nominal) or more wider than that of stud and bottom plate, use 23/32 inch thick continuous treated plywood strips with face grain perpendicular to footing: minimum grade APA Rated Sheathing 48/24 EXP 1 marked PS 1. Use plywood of same width as footing and fasten to footing with two 6d galvanized nails spaced 16 inches.

[3] This combination of house width and height may have 2 x 8 footing plate when second floor design load is 40 psf live and dead load.

Table 6-8. Minimum Structural Requirements for Crawl-Space Wall Framing (courtesy American Plywood Association).

Apply to installations with outside fill height not exceeding 4 feet and wall height not exceeding 6 feet. Roof supported on exterior walls. Floors supported on interior and exterior bearing walls.[1,2] 30 lbs. per cu. ft. equivalent fluid density soil pressure—2000 lbs. per sq. ft. allowable soil bearing pressure.

		Uniform load conditions					
		Roof—40 psf live; 10 psf dead Ceiling—10 psf 1st floor—50 psf live and dead 2nd floor—50 psf live and dead			Roof—30 psf live; 10 psf dead Ceiling—10 psf 1st floor—50 psf live and dead 2nd floor—50 psf live and dead		
Construction	House width (feet)	Lumber species and grade[3]	Stud and plate size (nominal)	Stud spacing (inches)	Lumber species and grade[3]	Stud and plate size (nominal)	Stud spacing (inches)
2 Stories	32 or less	B	2×6	16	B	2×6	16
					D	2×6	12
	28 or less	D	2×6	12	D	2×6	12
	24 or less	D	2×6	12	C	2×6	16
1 Story	32 or less	B	2×4	12	A	2×4	16
		B	2×6	16	B	2×4	12
		D	2×6	12	D	2×6	16
	28 or less	A	2×4	16	B	2×4	12
		D	2×6	16	D	2×6	16
	24 or less	D	2×6	16	C	2×4	12

[1] Studs and plates in interior bearing walls supporting floor loads only must be of lumber species and grade "D" or higher. Studs shall be 2 inches by 4 inches at 16 inches on center where supporting one floor and 2 inches by 6 inches at 16 inches on center where supporting two floors. Footing plate shall be 2 inches wider than studs.

[2] If brick veneer is used, see page 23 for knee wall requirements.

[3] Species, species groups and grades having the following minimum (surfaced dry or surfaced green) properties as provided in the National Design Specification:

		A	B	C	D
F_b (repetitive member) psi	2×6	1,700	1,400	1,100	975
	2×4	1,950	1,650	1,300	1,150
F_c psi	2×6	1,250	1,000	825	700
	2×4	1,250	975	775	675
$F_{c\perp}$ psi		385	385	245	235
F_v psi		90	90	70	70
E psi		1,700,000	1,600,000	1,300,000	1,100,000
Typical lumber grades		Douglas Fir No. 1 Southern Pine No. 1	Douglas fir No. 2 Southern Pine No. 2	Hem fir No. 2 Northern Pine No. 2	Ponderosa Pine No. 2

Where indicated (*), length of end splits or checks at lower end of studs not to exceed width of piece.

the floor fill's surface in medium- to well-drained soils. A pump will be needed if the sump cannot be drained by gravity to daylight or to sewer.

Moisture Control

A vital part of the permanent wood foundation is moisture control. The most important factors in moisture control are the use of downspouts and splash blocks to direct the water away from the building and to slope the ground away.

Caulking the panel joints and applying polyethylene over the foundation—two steps not required in crawl-space construction—are important in providing watershed in full-basement construction. Seal the plywood joints their full lengths with a high-performance caulking compound, such as those listed in Table 6-9.

Use a 6-mil polyethylene sheet to cover the below-grade portions of the foundation walls. This sheet directs moisture to the gravel fill and footing so it can drain without causing pressure against the wall.

You also can install the polyethylene any time before backfilling. Backfill after the first floor is in

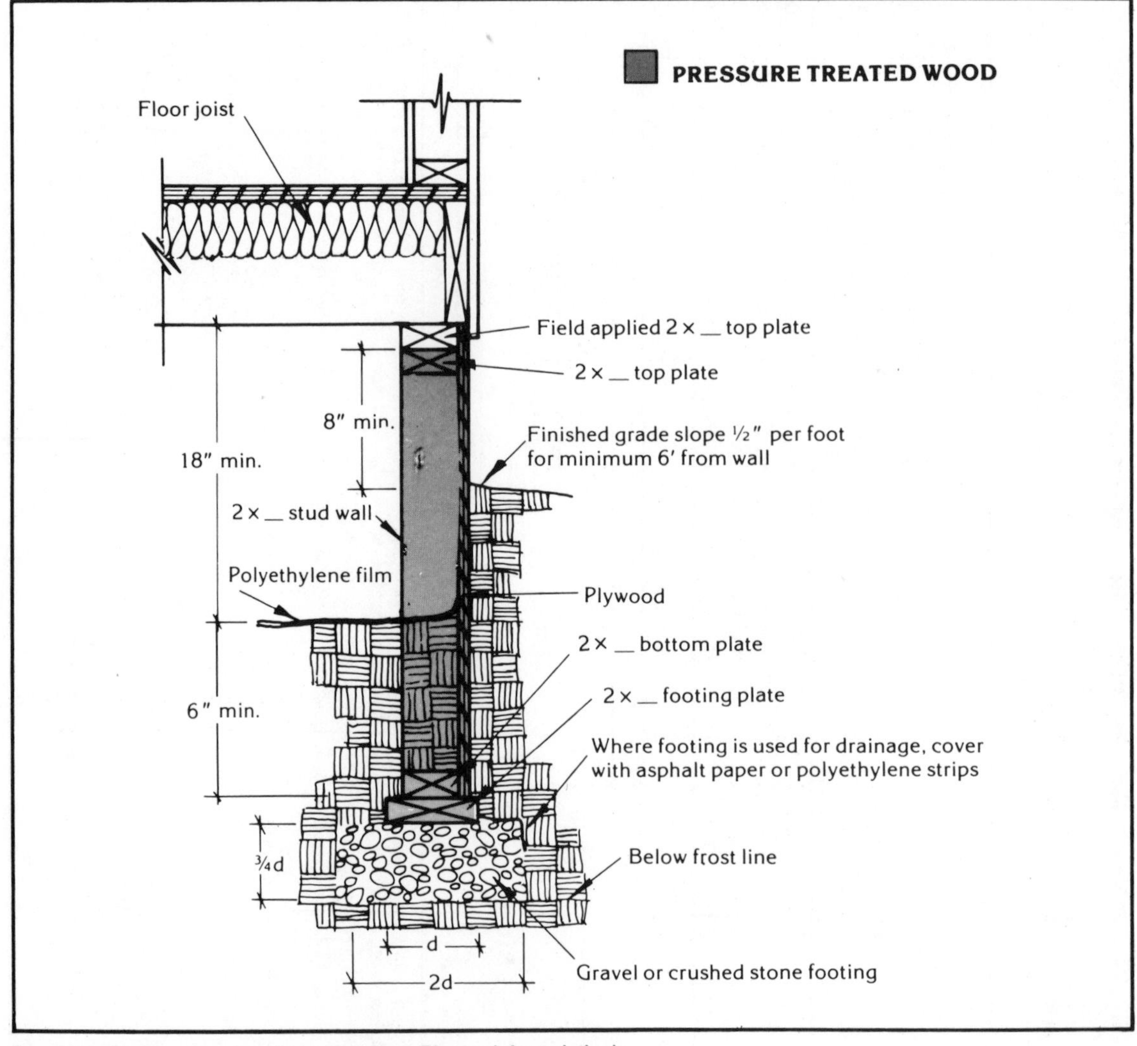

Fig. 6-10. Crawl space (courtesy American Plywood Association).

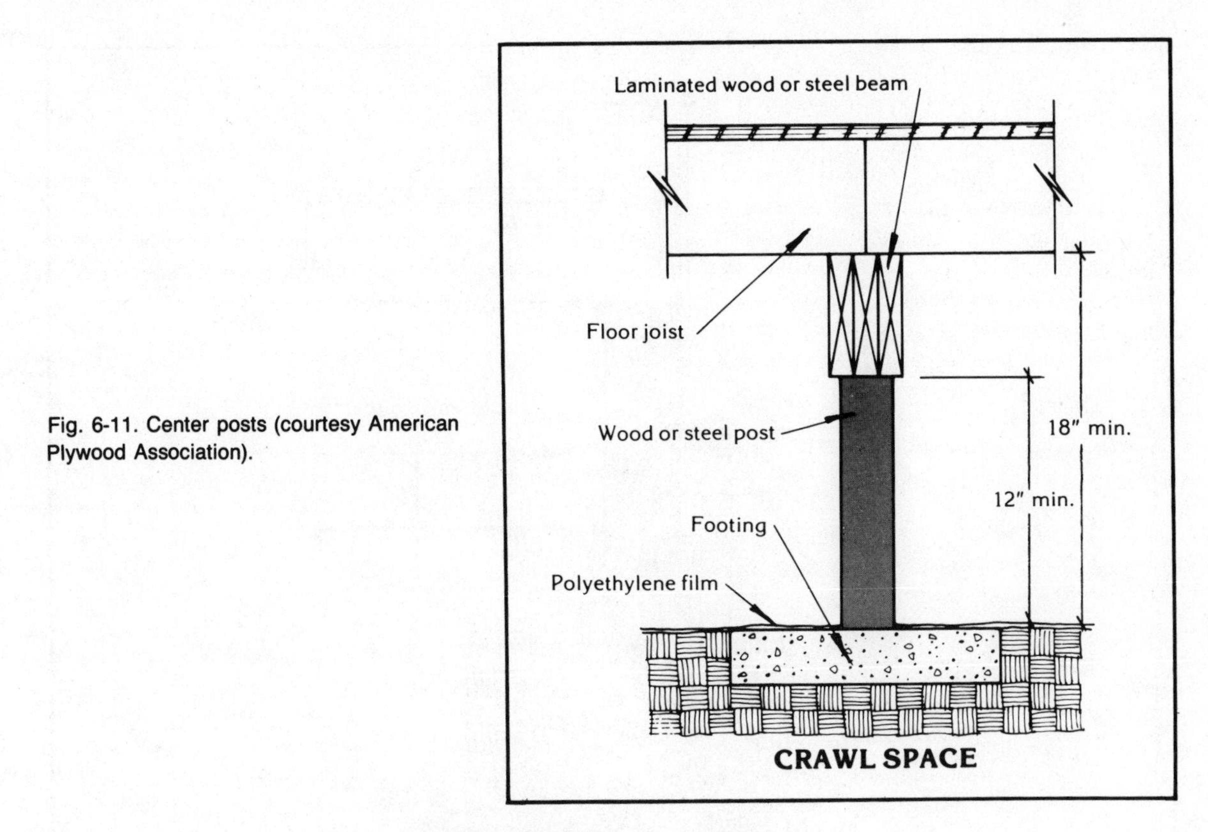

Fig. 6-11. Center posts (courtesy American Plywood Association).

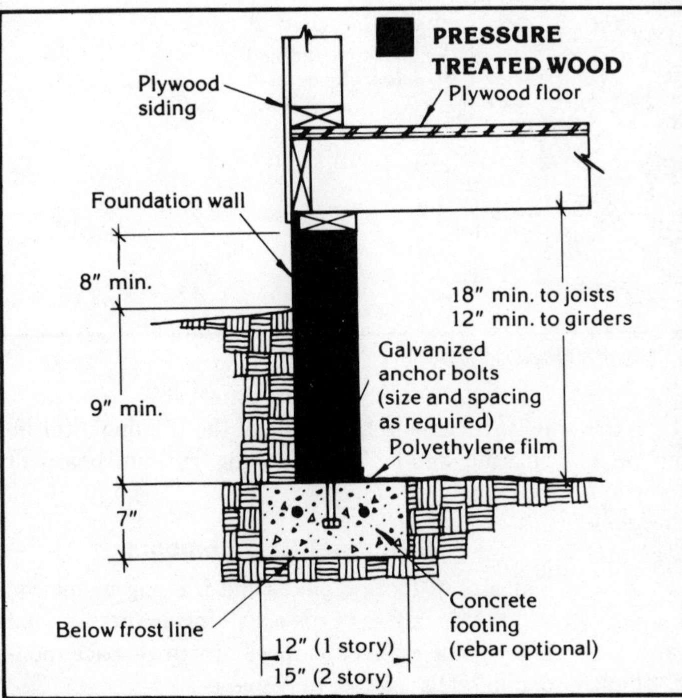

Fig. 6-12. Crawl space PWF on concrete footing (courtesy American Plywood Association).

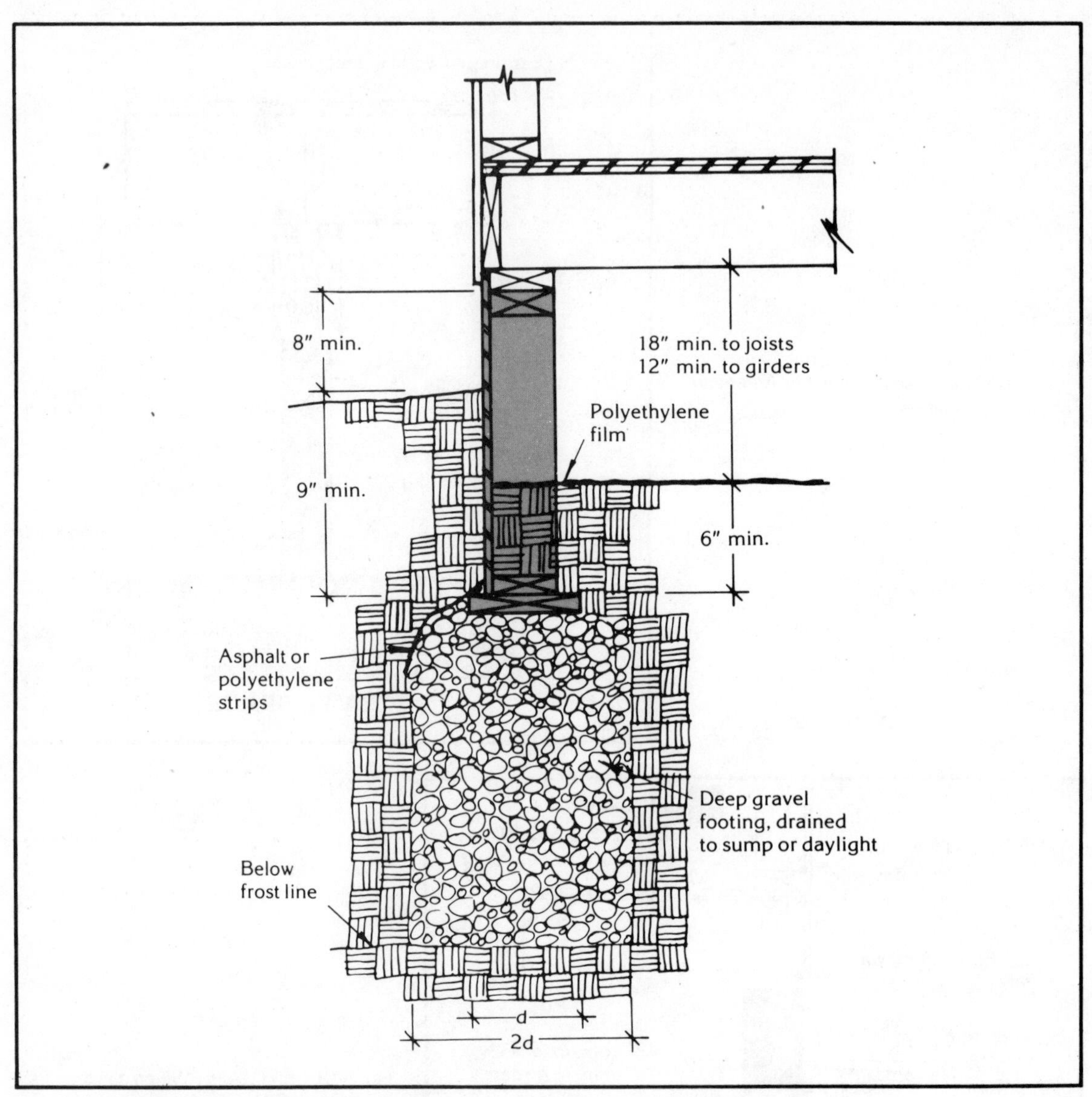

Fig. 6-13. Deep gravel footing for PWF (courtesy American Plywood Association).

place and the basement slab has been poured or the wall suitably braced. Spot-bond the polyethylene to the plywood with an adhesive, lap the vertical joints in the sheet 6 inches, and seal with the adhesive. Let the film hang freely down the wall and protect it at the top with a treated plywood nailing strip at least 12 inches in nominal width.

After basement walls are insulated and finished, add a vapor barrier on the interior. Staple polyethylene to the inside of the framing after insulating and before installing gypsum board or paneling.

Structural Requirements

The size of footing plates and framing members, and the thickness of the sheathing are determined in the same manner as those for crawl-space foundation walls. (Refer to Table 6-7.)

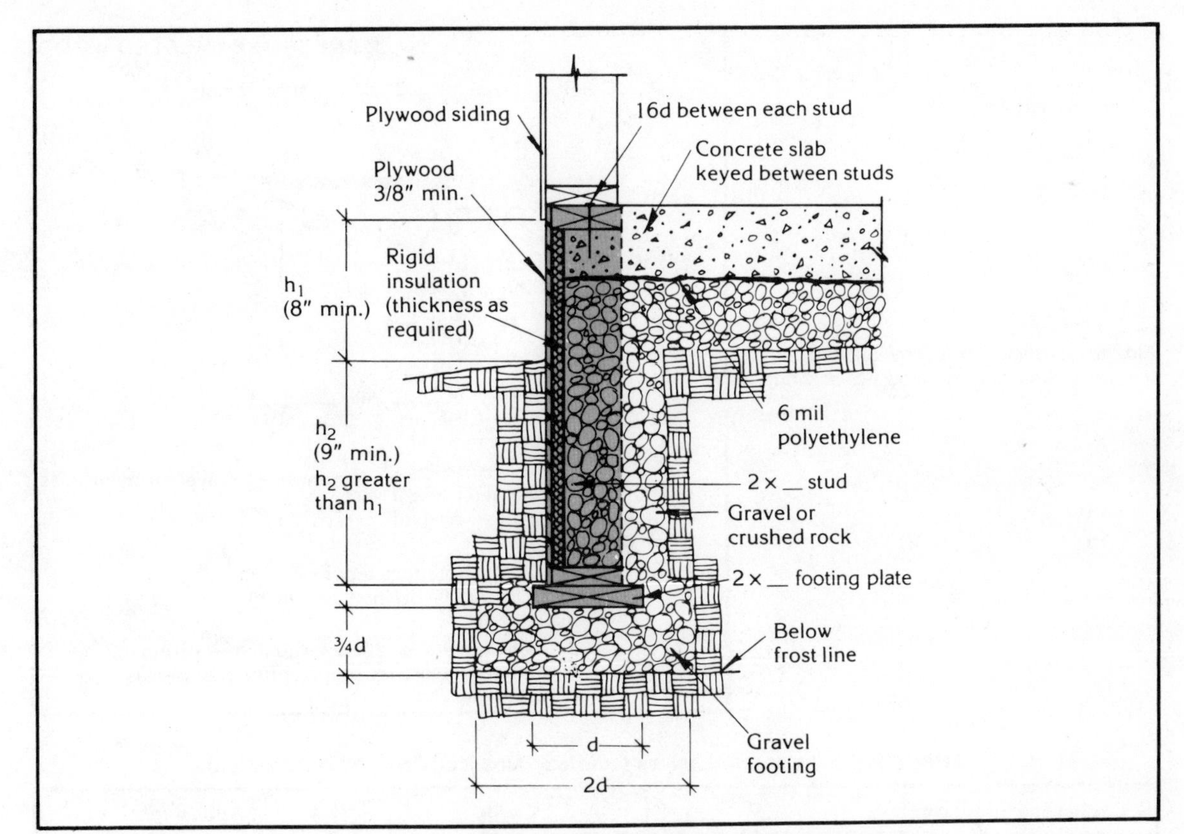

Fig. 6-14. PWF stem wall, concrete slab on grade (courtesy American Plywood Association).

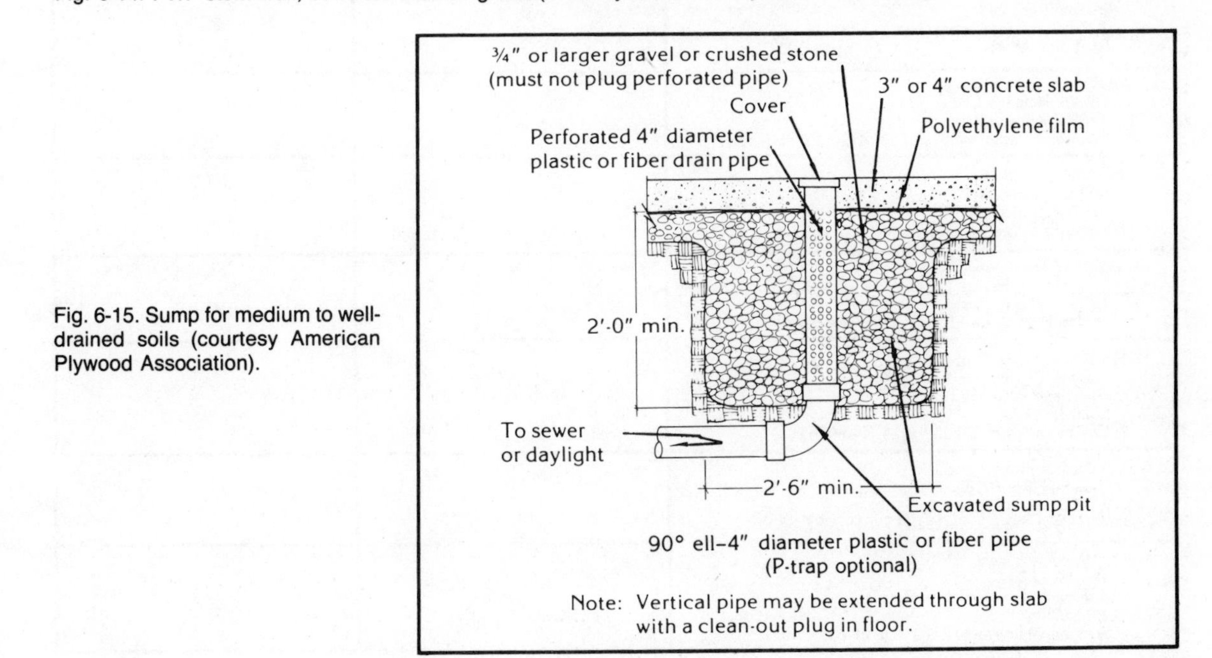

Fig. 6-15. Sump for medium to well-drained soils (courtesy American Plywood Association).

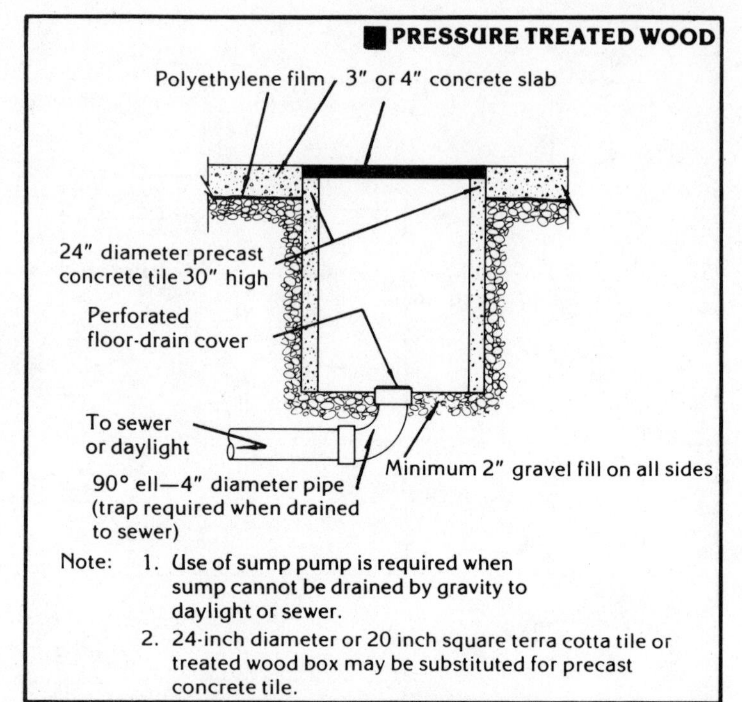

Fig. 6-16. Sump for poorly drained soils (courtesy American Plywood Association).

Table 6-9. Caulks and Adhesives (courtesy American Plywood Association).

Caulks and Adhesives	Caulk	Adhesive
Presstite #579.6 Plastic Sealer (Tape) Inmont Corporation 1218 Central Industrial Drive St. Louis, Missouri 63110 (314) 664-6000	X	X
Eternaflex Hypalon Sealant Gibson-Homans Co. 1755 Enterprise Parkway Twinsburg, Ohio 44087 (216) 425-3255	X	
Clear Butyl Sealant Wilhold Glues, Inc. 8707 Millergrove Drive Santa Fe Springs, California 90670 (213) 692-0911	X	X
Presstite #579.64 Plastic Sealer (Caulk) Inmont Corporation 1218 Central Industrial Drive St. Louis, Missouri 63110 (314) 664-6000		X
#5411 Acrylic Latex Caulk Franklin Chemical Industries, Inc. Consumer Products Div. 2020 Bruck St. Columbus, Ohio 43207 (614) 443-0241	X	
Silpruf Weatherproofing Sealant General Electric Company RTV Products Department Waterford, New York 12188 (518) 237-3330	X	
#5230 Wood Adhesive 3M Company Adhesives, Coatings & Sealers Div. 3M Center St. Paul, Minnesota 55101 (612) 733-1110	X	

Table 6-11. Minimum APA-Rated Sheathing Plywood Requirements for Basement Construction (courtesy American Plywood Association).

Height of Fill (Inches)	Stud Spacing (Inches)	Face Grain Across Studs[b]			Face Grain Parallel to Studs		
		Grade[c]	Minimum Thickness[a]	Span Rating	Grade[c]	Minimum Thickness[a][d]	Span Rating
24	12	B	15/32	32/16	A B	15/32 15/32[e]	32/16 32/16
	16	B	15/32	32/16	A B	15/32[e] 19/32[e] (4-ply)	32/16 40/20
36	12	B	15/32	32/16	A B B	15/32 15/32[e] (4-ply) 19/32 (4-ply)	32/16 32/16 40/20
	16	B	15/32[e]	32/16	A B	19/32 23/32	40/20 48/24
48	12	B	15/32	32/16	A B	15/32[e] 19/32[e] (4-ply)	32/16 40/20
	16	B	19/32	40/20	A A	19/32[e] 23/32	40/20 48/24
60	12	B	15/32	32/16	A B B	19/32 19/32[e] (5-ply) 23/32	40/20 40/20 48/24
	16	B	19/32[e]	40/20	A	23/32[e]	48/24
72	12	B	15/32[e]	32/16	A B	19/32 23/32[e]	40/20 48/24
	16	B	23/32[e]	48/24	—	—	—
86	12	B	19/32	40/20	A A	19/32[e] 23/32	40/20 48/24
	16	B	23/32[e]	48/24	—	—	—

(a) Minimum thickness 15/32-inch, except crawl space sheathing may be 3/8-inch for face grain across studs 16 inches on center and maximum 2-foot depth of unequal fill.

(b) Minimum 2-inch blocking between studs required at all horizontal panel joints more than 4 feet below adjacent ground level (also where noted in construction details).

(c) Minimum all-veneer plywood grades are:
A. APA Structural I RATED SHEATHING Exposure 1
B. APA RATED SHEATHING Exposure 1 marked PS 1-83.

If a major portion of the wall is exposed above ground, a better appearance may be desired. The following Exterior grades would be suitable:
A. APA Structural I A-C, APA Structural I B-C or APA Structural I C-C (Plugged)
B. APA A-C Exterior Group 1, APA B-C Exterior Group 1, APA C-C (Plugged) Exterior Group 1, APA MDO Exterior Group 1, or ungrooved APA 303 Siding Group 1.

(d) When face grain is parallel to studs, all-veneer plywood panels of the required thickness, grade and Span Rating may be of any construction permitted except as noted in the table for minimum number of plies required.

(e) For this fill height, thickness and grade combination, panels which are continuous over less than three spans require blocking 16 inches above the bottom plate. Offset adjacent blocks and fasten through studs with two 16d corrosion resistant nails at each end.

Table 6-10 lists the minimum structural requirements for houses of various widths. Minimum plywood thicknesses are shown in Table 6-11 for backfill heights up to 86 inches.

BASEMENT DESIGN DETAILS

The permanent wood foundation adapts easily to daylight configurations, even on sloping sites, and it allows for windows or doors much more easily than conventional foundation construction. It also easily accommodates regular lumber joists or the variety of trusses now available throughout the country.

You can construct beam pockets (Fig. 6-17)

Table 6-10. Minimum Structural Requirements for Basement Walls (courtesy American Plywood Association).

			Uniform load conditions					
			Roof—40 psf live; 10 psf dead Ceiling—10 psf 1st floor—50 psf live and dead 2nd floor—50 psf live and dead			Roof—30 psf live; 10 psf dead Ceiling—10 psf 1st floor—50 psf live and dead 2nd floor—50 psf live and dead		
Construction	House width (feet)	Height of fill (inches)	Lumber species and grade	Stud and plate size (nominal)	Stud spacing (inches)	Lumber species and grade	Stud and plate size (nominal)	Stud spacing (inches)
2 Stories	32 or less	24	D	2×6	16	D	2×6	16
		48	D	2×6	16	D	2×6	16
		72	A B C D	2×6 2×6 2×8 2×8	16 12 16 12	A B C	2×6 2×6 2×8	16 12 16
		86	A* B C	2×6 2×8 2×8	12 16 12	A* B C	2×6 2×8 2×8	12 16 12
	24 or less	24	D	2×6	16	D	2×6	16
		48	D	2×6	16	D	2×6	16
		72	C D	2×6 2×8	12 16	C	2×6	12
		86	A* B C	2×6 2×8 2×8	12 16 12	D	2×8	12
1 Story	32 or less	24	B D D	2×4 2×4 2×6	16 12 16	B D D	2×4 2×4 2×6	16 12 16
		48	D	2×6	16	D	2×6	16
		72	A B D	2×6 2×6 2×8	16 12 16	A C D	2×6 2×6 2×8	16 12 16
		86	A* B C	2×6 2×8 2×8	12 16 12	A* B D	2×6 2×8 2×8	12 16 12
	28 or less	24	B D D	2×4 2×4 2×6	16 12 16	D	2×4	16
		48	D	2×6	16	B	2×4	12
		72	C	2×6	12	C	2×6	12
		86	D	2×8	12	A* B D	2×6 2×8 2×8	12 16 12
	24 or less	24	D	2×4	16	D	2×4	16
		48	B	2×4	12	B	2×4	12
		72	C	2×6	12	C	2×6	12
		86	D	2×8	12	A* B D	2×6 2×8 2×8	12 16 12

Wall height—8 feet. Roof supported on exterior walls. Floors supported on interior and exterior bearing walls.[1,2] 30 lbs. per cu. ft. equivalent-fluid density soil pressure—2000 lbs. per sq. ft. allowable soil bearing pressure.

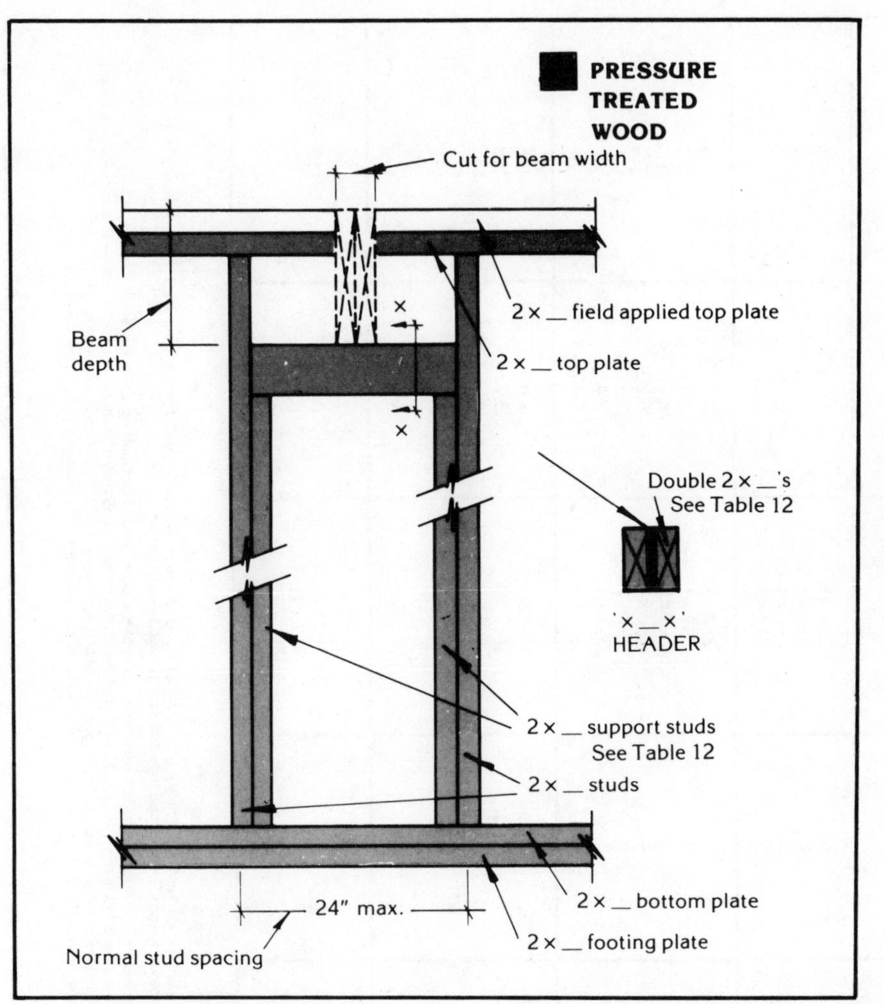

Fig. 6-17. Beam pockets in basement or crawl space end wall (courtesy American Plywood Association).

quickly with support studs and a header (Table 6-12). Figures 6-18 and 6-19 illustrate typical pressure-treated wood basement wall construction.

For brick up to 18 feet high, minimum knee wall framing is 2- × -4 studs 16 inches on center, with a 1- × -4 bottom plate and a 2- × -6 top plate (Fig. 6-20). Lumber can be species grade combination "D" if brick is no more than 16 1/2 feet high. Provide double studs under all butt joints in the top plate. Footing plates for knee walls must be 2 × 10s where 2- × -4 studs are called out in Table 6-10 and 2 × 12s where 2- × -6 studs are required.

Alternatively, you can fabricate a footing plate from two widths of lumber with a joint offset from permanent wood foundation and knee wall bottom plates. Fasten the treated plywood plate to the bottom of a two-piece footing plate. (See footnote 2, Table 6-7 for details.) If you are using a polyethylene film, place it between the foundation sheathing and knee wall framing before you attach the knee wall to the foundation. Figures 6-21 through 6-25 offer additional cross-sectional views of permanent wood foundation construction.

In daylight basements where backfill is against only one to three sides of the building, pour the slab to "key" between the studs of the end walls, as shown in Fig. 6-26. Where there is more than 4 feet of difference in height of the backfill on opposite sides of the house, you might need to individually design the wall nailing using the *Design, Fabrica-*

Table 6-12. Beam Pockets in Basement or Crawl-Space Walls (courtesy American Plywood Association).

Minimum soil bearing 2000 lb./sq. ft. Stud spacing 12 to 24 inches.

Beam pocket load (pounds)	Species and grade of lumber required[1]	Nominal size and no. of support studs	Nominal footing plate width (inches)	Nominal size and no. of header laminations[2]	Minimum width of beam bearing (inches)	Capacity of footing plate and gravel to support additional load[3] (lbs./lin. ft.)
2400	D	2-2×4	6	2-2×6	3.03	365
3100	D	2-2×6	6	3-2×6	2.56	135
3400	D	4-2×4	6	2-2×6	4.45	35
3500	B	2-2×4	6	2-2×6	2.66	—
3700	B	2-2×4	8	2-2×6	2.83	365
3700	D	2-2×6	8	3-2×6	3.12	365
3900	C	4-2×4	8	2-2×6	4.93	300
4700	D	4-2×4	8	2-2×8	6.29	35
4800	B	4-2×4	8	2-2×6	3.78	—
4800	B	2-2×6	8	3-2×6	2.40	—
4800	D	4-2×6	8	3-2×6	4.16	—
4900	B	4-2×4	10	2-2×6	3.87	465
5700	B	2-2×6	10	3-2×6	2.92	200
5800	A	4-2×4	10	2-2×6	4.65	165
5900	B	4-2×4	10	2-2×8	4.73	135
5900	D	4-2×6	10	3-2×6	5.20	135
6200	C	4-2×6	10	3-2×6	5.25	35
6300	D	4-2×6	10	3-2×8	5.58	—
7100	B	4-2×6	12	3-2×6	3.72	235
7300	D	4-2×6	12	3-2×8	6.53	165
7600	C	4-2×6	12	3-2×8	6.52	65
7800	B	4-2×6	12	3-2×8	4.13	—

[1] See Table 10 for minimum properties of lumber species and grade combinations.

[2] Headers having two laminations of 2 inch (nominal) thickness lumber shall have a 15/32 inch plywood spacer with grain parallel to lumber grain. Headers having three lumber laminations shall have two 15/32 inch plywood spacers. Lumber and plywood shall be well spiked together.

[3] Some foundations may carry contributory brick veneer and/or floor and ceiling loads in addition to the beam load. The tabulated additional loads (column 7) may be supported by the footing plate and gravel without increased foundation size. For heavier contributory loads, reduce allowable beam pocket load or increase size of footing For each pound per linear foot that contributory loads exceed the values shown, reduce allowable beam pocket load by three pounds (Brick veneer is estimated 300 lb. per linear foot of wall per 8 foot height of brick.) An increase of 2 inches in footing plate width is also equivalent to a 500 lb. per linear foot increase in this capacity to support additional loads.

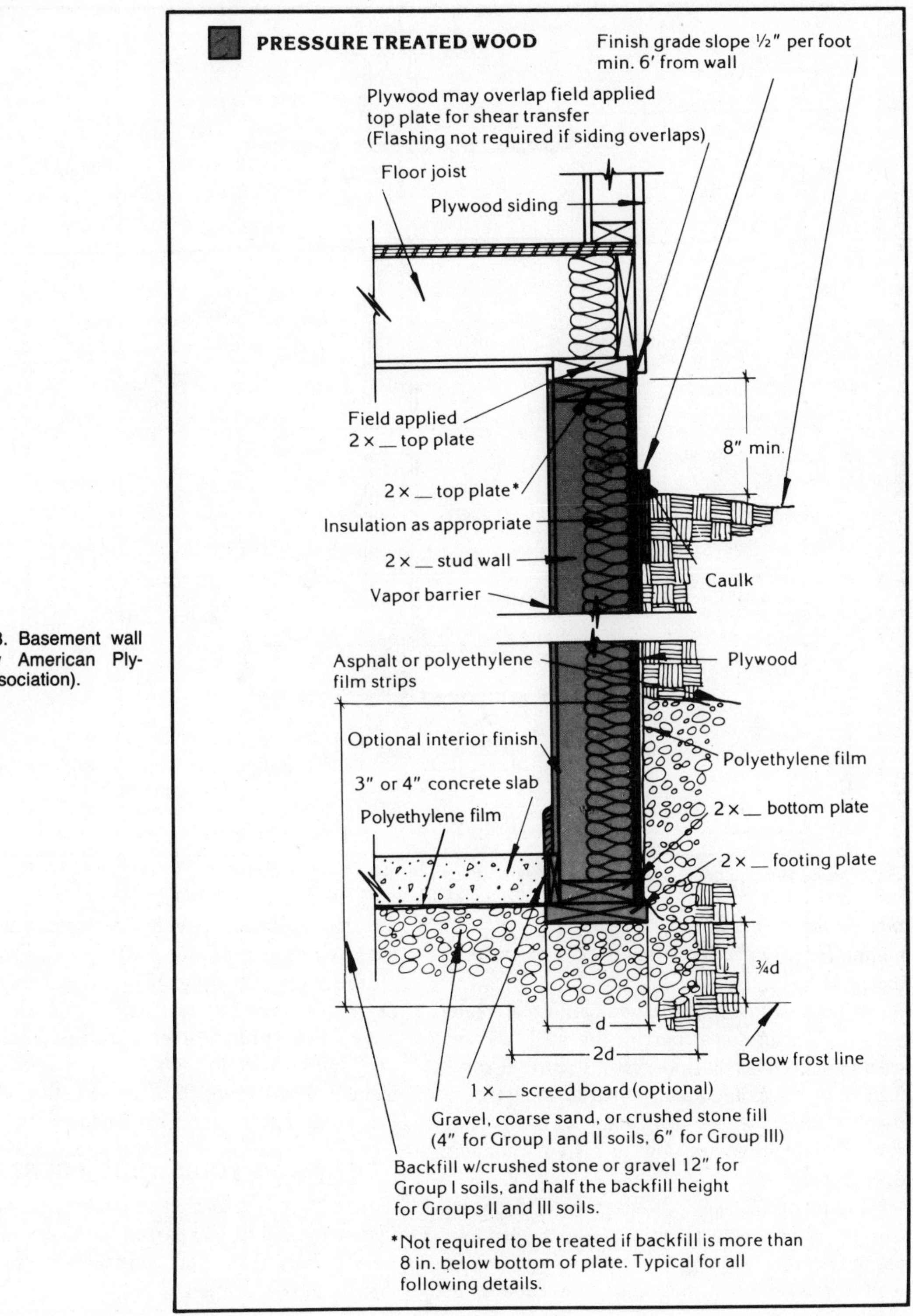

Fig. 6-18. Basement wall (courtesy American Plywood Association).

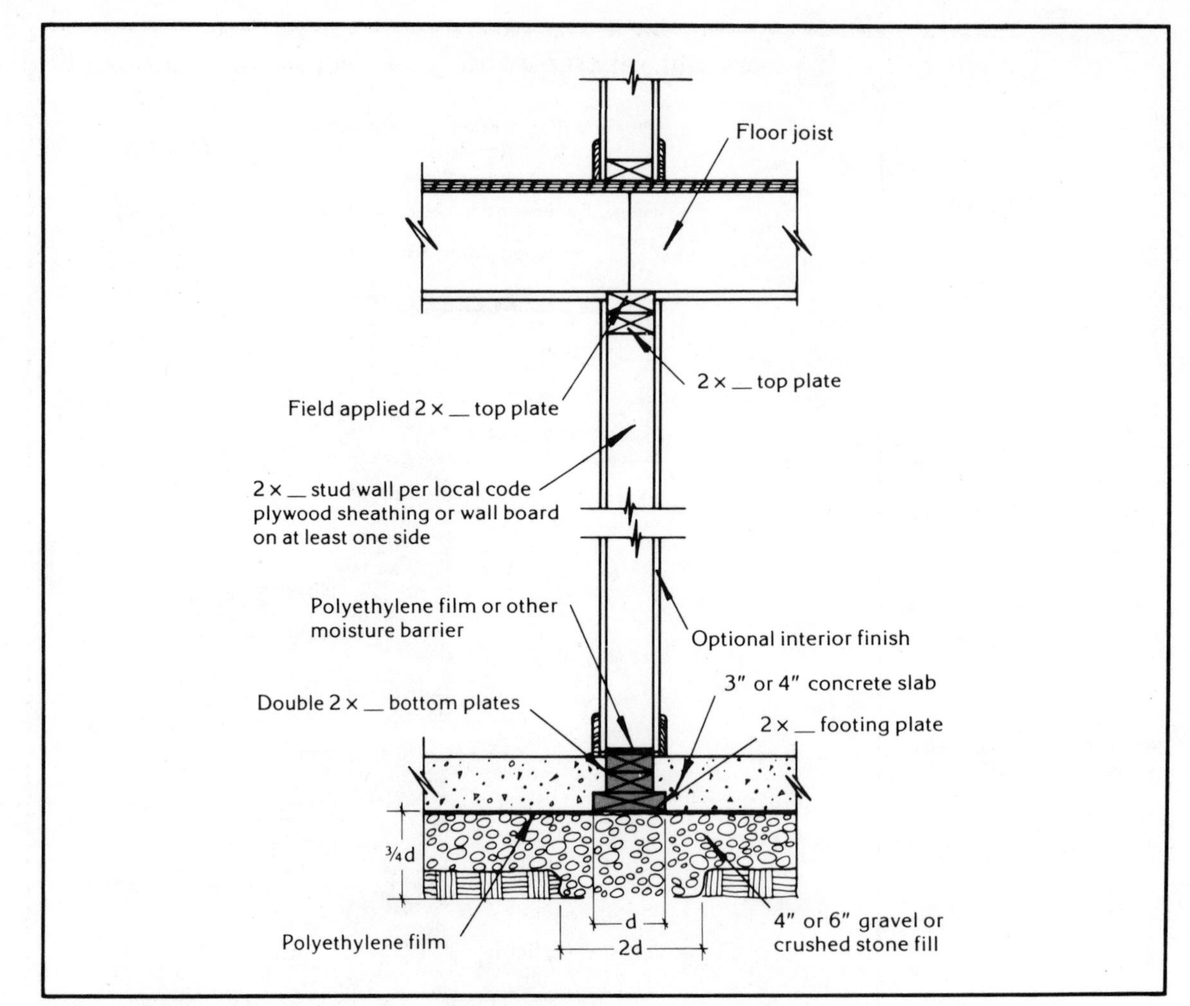

Fig. 6-19. Basement bearing partition (courtesy American Plywood Association).

tion, Installation Manual. Table 6-1 gives required nailing for plywood for 4-foot maximum differential in fill.

If there is low backfill, you can use a treated stub wall in the lower part of the wall, with untreated material in the upper part. Figures 6-26 and 6-28 show ways of reducing costs by using treated material only where necessary. Figures 6-29 through 6-31 illustrate how to fasten walls with deep backfill.

Figures 6-32 and 6-33 show how clear-span trusses can be used with the permanent wood foundation. Because Table 6-10 assumes a load-bearing wall in the center of the house, the Table is not directly applicable when clear-span trusses are used. You can use Table 6-10 to select the framing for the nonload-bearing walls, but you must use the *Design, Fabrication, Installation Manual* to determine framing requirements for the other walls.

Figures 6-34 through 6-38 offer additional permanent wood foundation construction views for basement and garage foundations.

DESIGNING YOUR WOOD FOUNDATION

To review the selection of lumber, plywood, and fasteners for a typical permanent wood foundation, refer to Figs. 6-39, 6-40, and other illustrations discussed in this section.

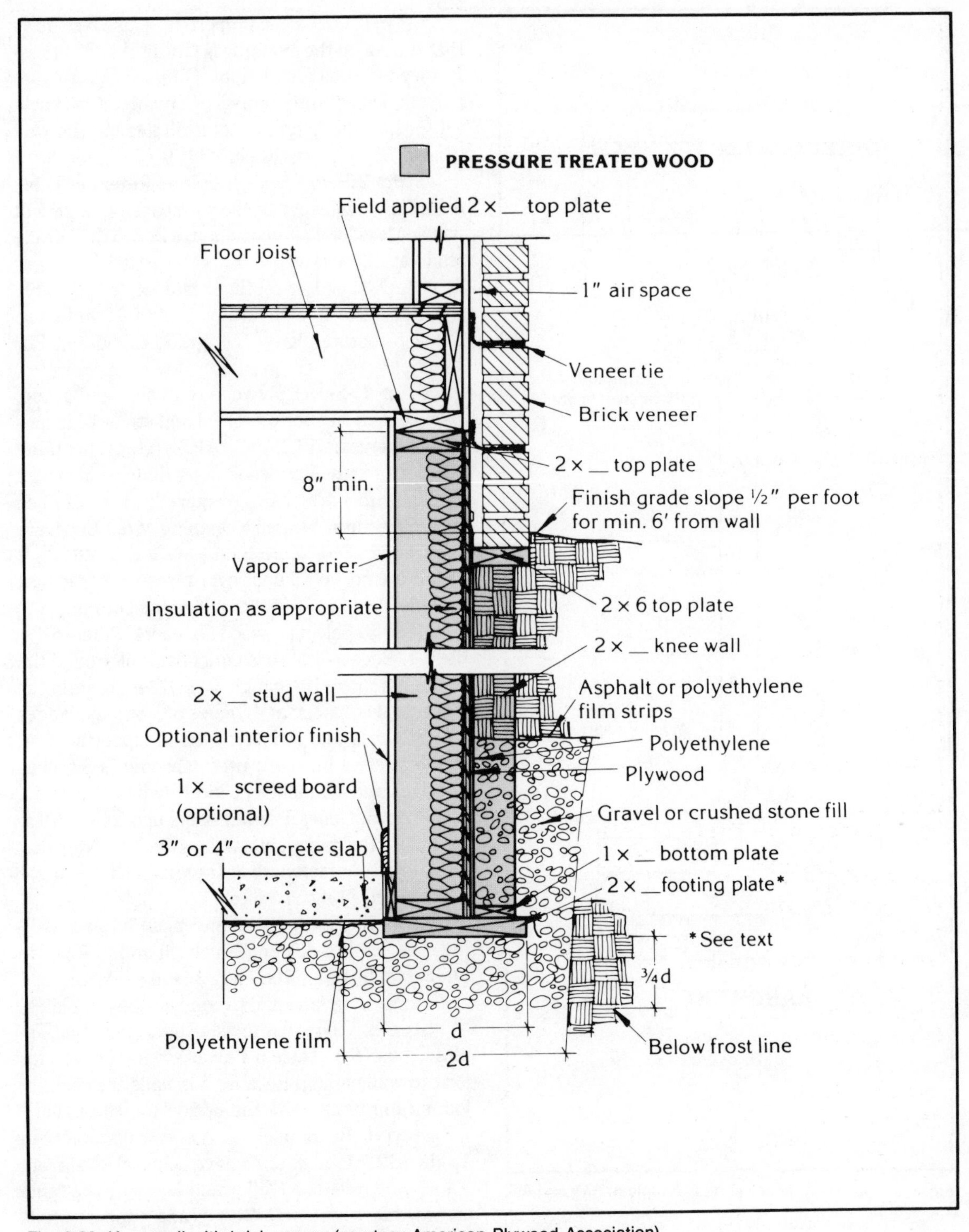

Fig. 6-20. Knee wall with brick veneer (courtesy American Plywood Association).

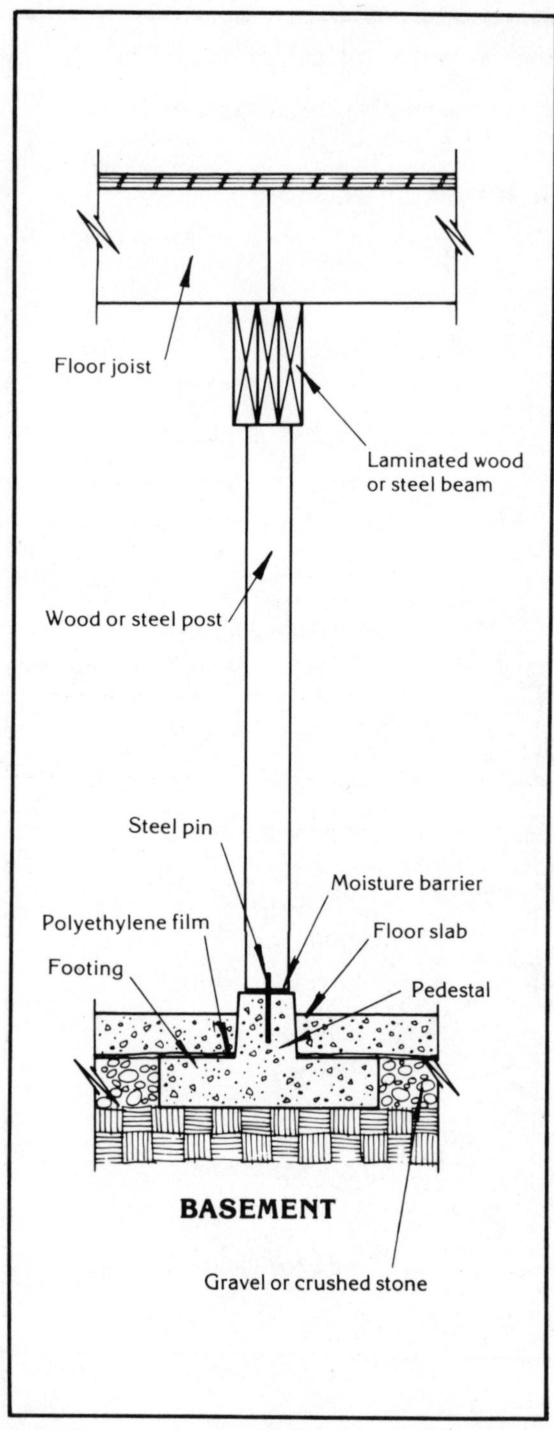

Fig. 6-21. Basement post (courtesy American Plywood Association).

Step 1. Check soil properties. Table 6-6 shows that the soil in the example is Group II and is satisfactory for wood foundations. Group II soils require polyethylene film and gravel or crushed stone backfill for half the height of total fill against the wall (Fig. 6-18). Use sump per Fig. 6-15 and 6-16.

Step 2. Select wall framing. Refer to Table 6-10. For a one-story house with basement, 28 feet wide or less, and a 6-foot fill, use 2- × -6 "C" grade studs at 12 inches.

Step 3. Select footing plate. Refer to Table 6-7. For a one-story house up to 28 feet wide, use a 2- × -8 footing plate, "D" grade minimum. For simplicity, use "C" grade.

Step 4. Select plywood thickness and grade. Use Table 6-11. For 6-foot fill and studs 12 inches on center use: 15/32 inch APA-Rated Sheathing 32/16 Exposure 1 plywood (long dimension across studs), with no blocking required; or 19/32 inch APA Structural I-Rated Sheathing 40/20 Exposure 1 plywood (long dimension parallel to studs). If panels are not continuous over three-stud spacings, provide blocking 16 inches above the bottom plate.

Step 5. Select plywood fasteners. Refer to Table 6-1. Because there is equal backfill around the perimeter, use 16- gauge- × -1 1/2-inch stainless steel staples spaced at 4 inches o.c. at panel edges and 8 inches o.c. at intermediate supports. If 8d stainless steel nails are used, spacing is 6 inches at edges and 12 inches at intermediate supports.

Step 6. General nailing schedule. Refer to Table 6-2 for the minimum nailing schedule. Note that more or larger fasteners or framing anchors might be required in some cases.

Step 7. Plate-to-stud and plate-to-plate nailing. See Table 6-3. For 72-inch fill and "C" grade lumber, select four 16d end nails in studs and 10d at 4 inches o.c.; face-nail to connect the top plates.

Step 8. Floor-joist-to-wall connection. See Tables 6-4 and 6-5. Table 6-4 gives several options for joist-to-wall nailing. Because 10d nails are used for joining top plates, use the option for 10d nails.

Step 9. Beam pockets in end walls. Refer to Table 6-12. Member sizes have been selected from Table 6-10 based on "C" grade lumber. The table shows that two 2- × -4 "B" grade support studs

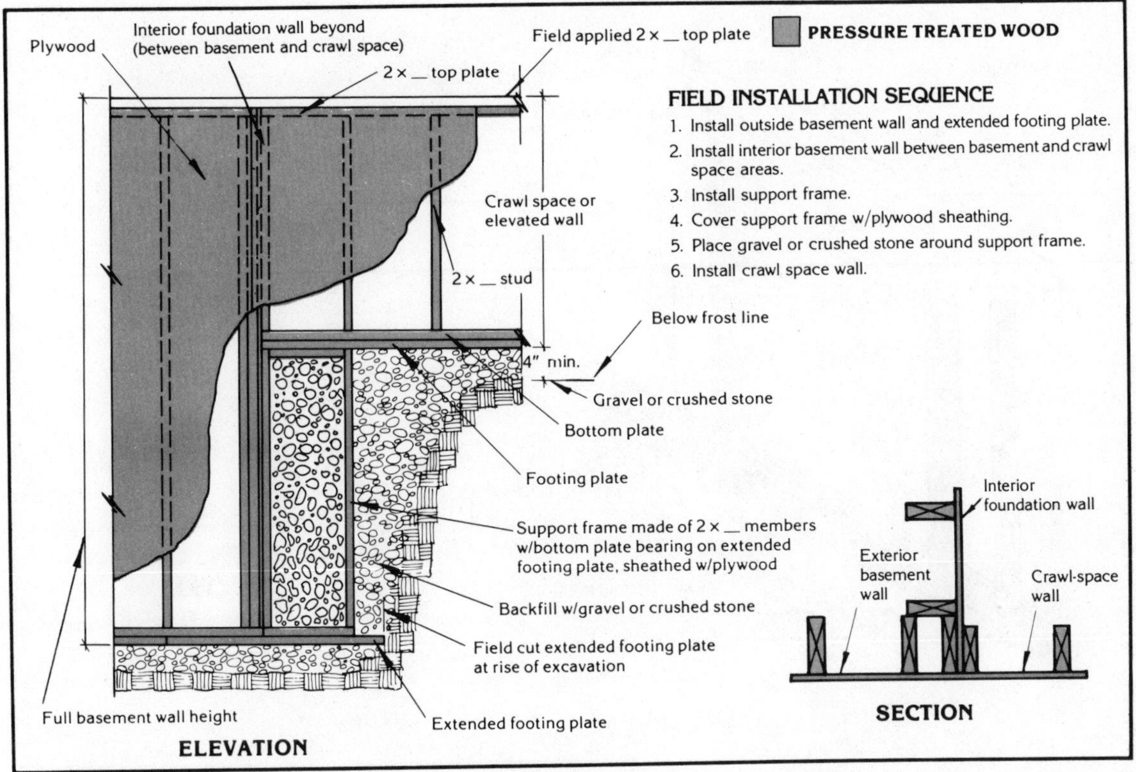

Fig. 6-22. Stepped footing (courtesy American Plywood Association).

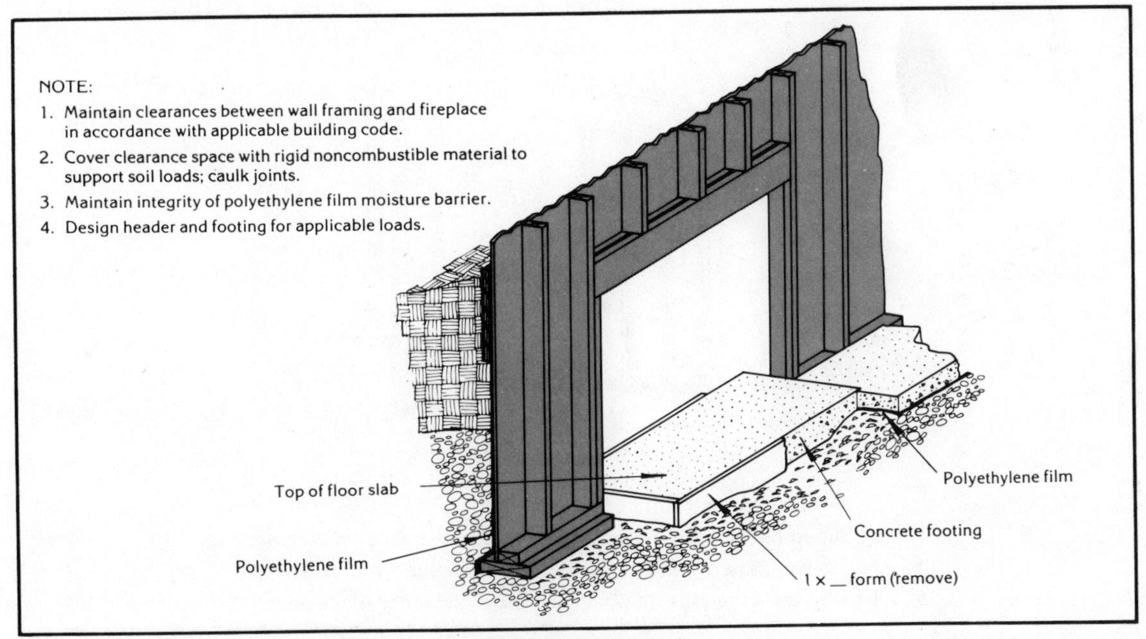

Fig. 6-23. Fireplace opening (courtesy American Plywood Association).

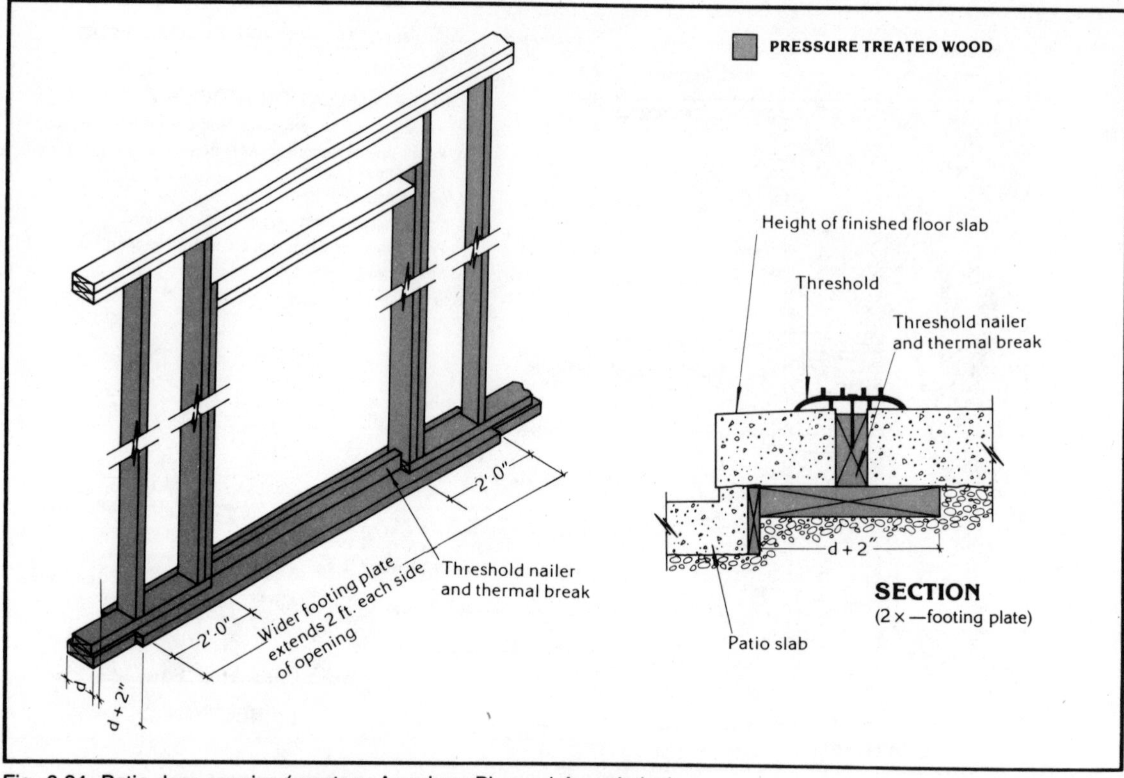

Fig. 6-24. Patio door opening (courtesy American Plywood Association).

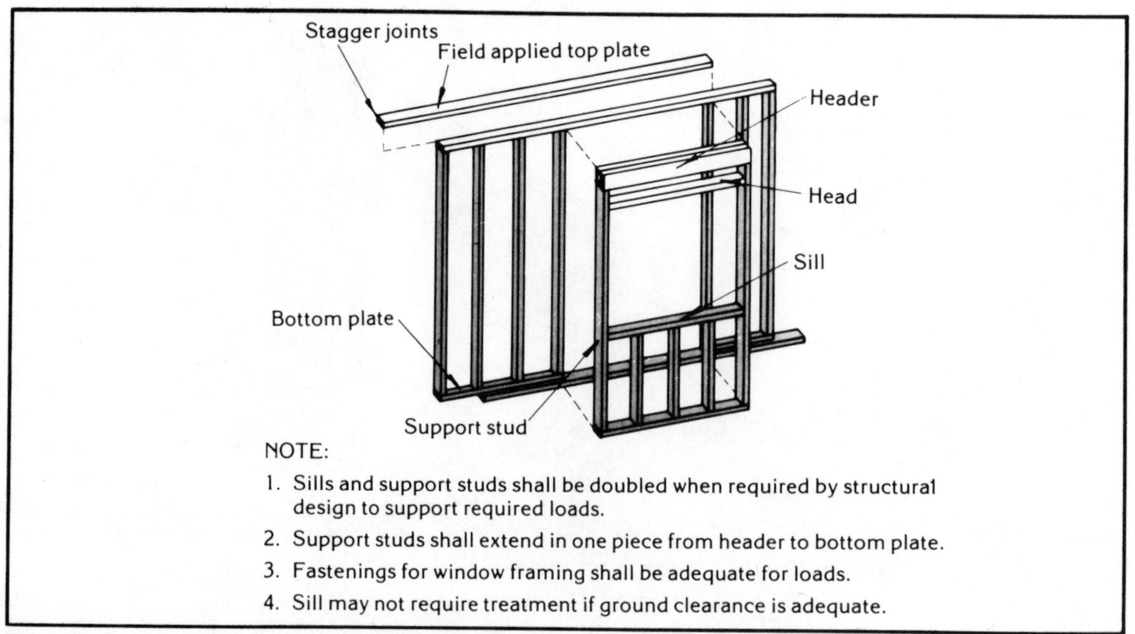

NOTE:

1. Sills and support studs shall be doubled when required by structural design to support required loads.
2. Support studs shall extend in one piece from header to bottom plate.
3. Fastenings for window framing shall be adequate for loads.
4. Sill may not require treatment if ground clearance is adequate.

Fig. 6-25. Typical window opening framing (courtesy American Plywood Association).

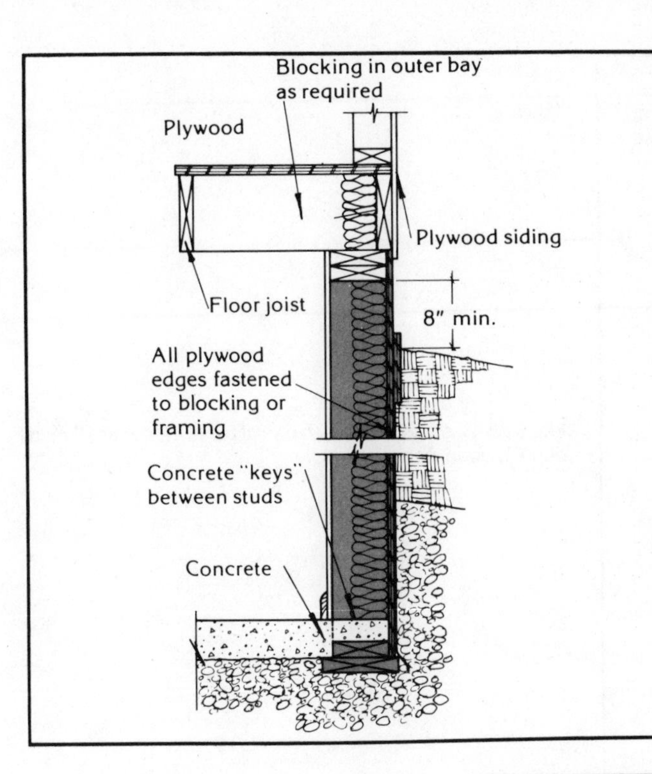

Fig. 6-26. Daylight basement end walls (courtesy American Plywood Association).

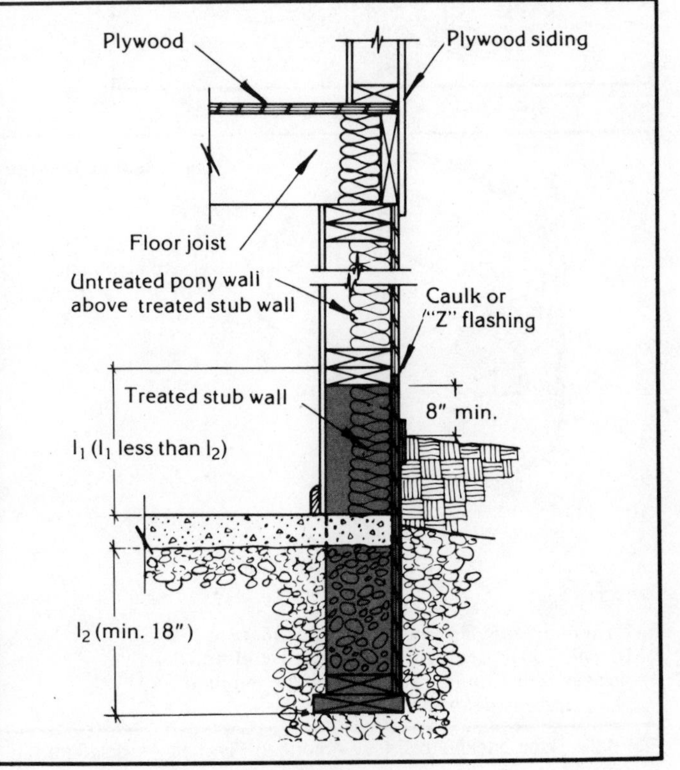

Fig. 6-27. Basement stub wall (courtesy American Plywood Association).

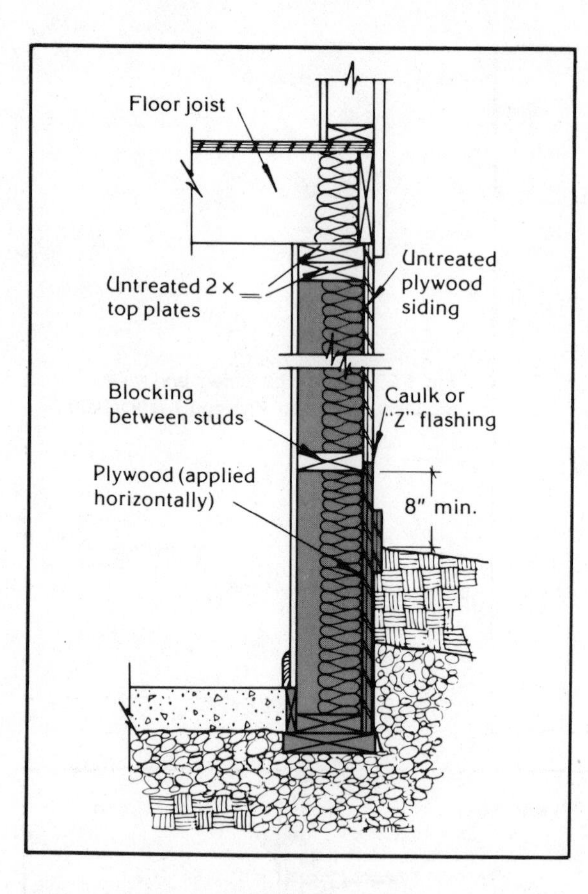

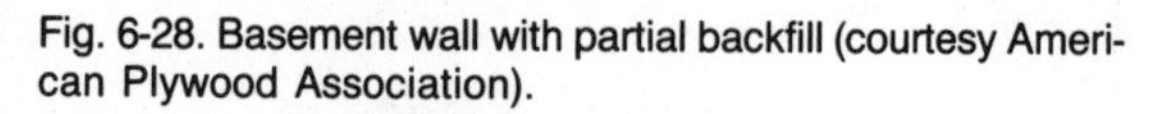
Fig. 6-28. Basement wall with partial backfill (courtesy American Plywood Association).

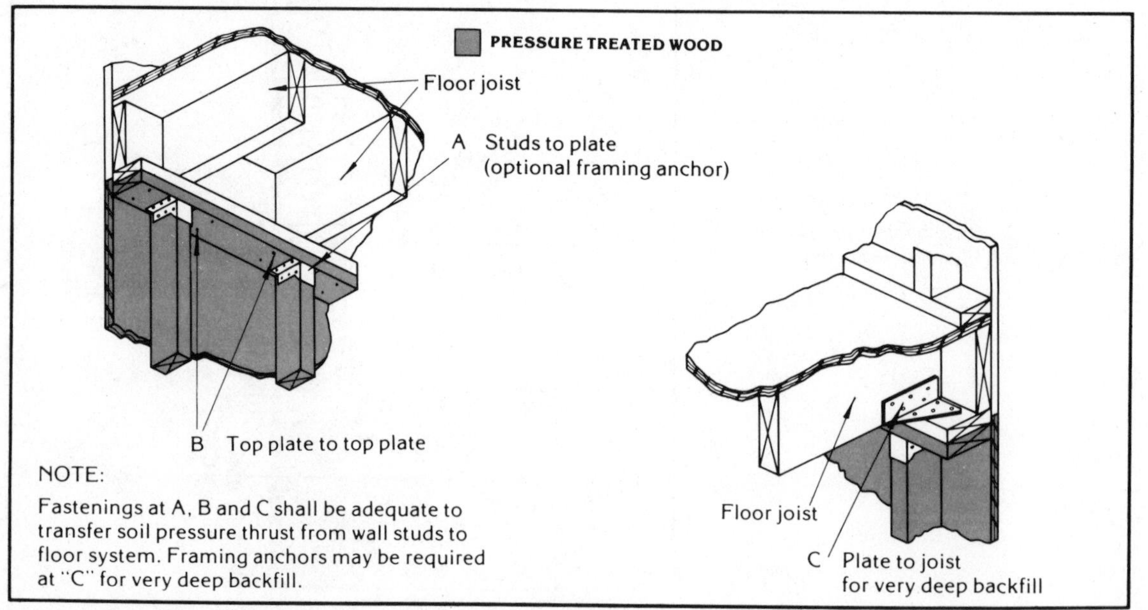

Fig. 6-29. Deep backfill (courtesy American Plywood Association).

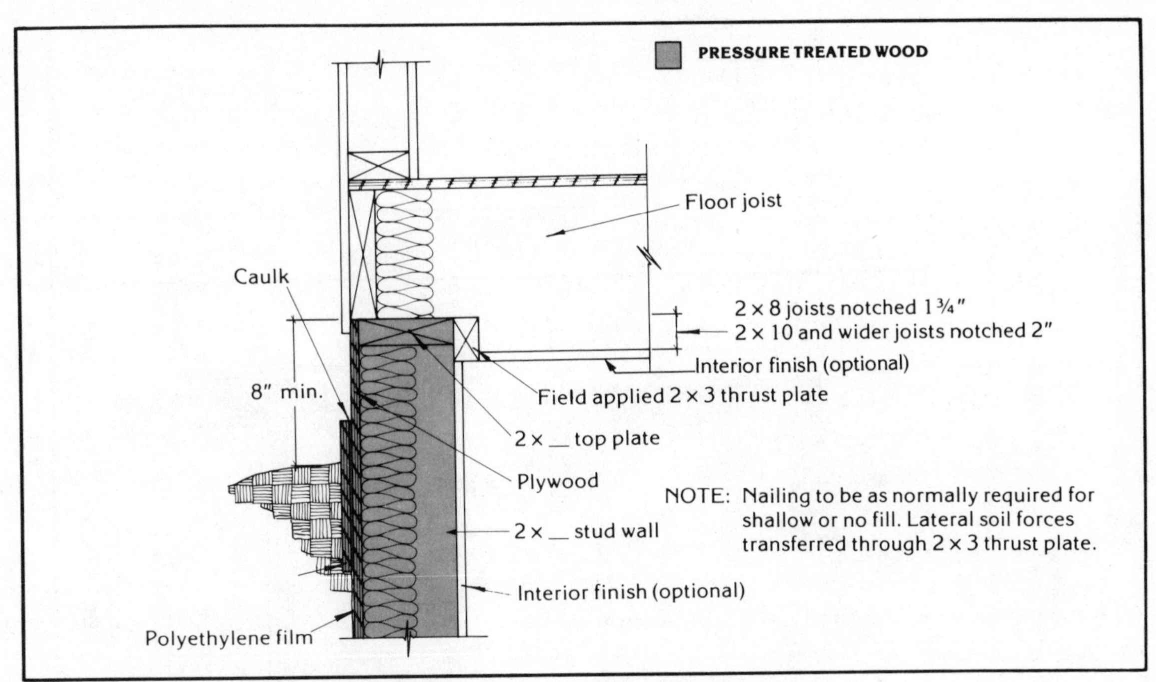

Fig. 6-30. Another view of deep backfill (courtesy American Plywood Association).

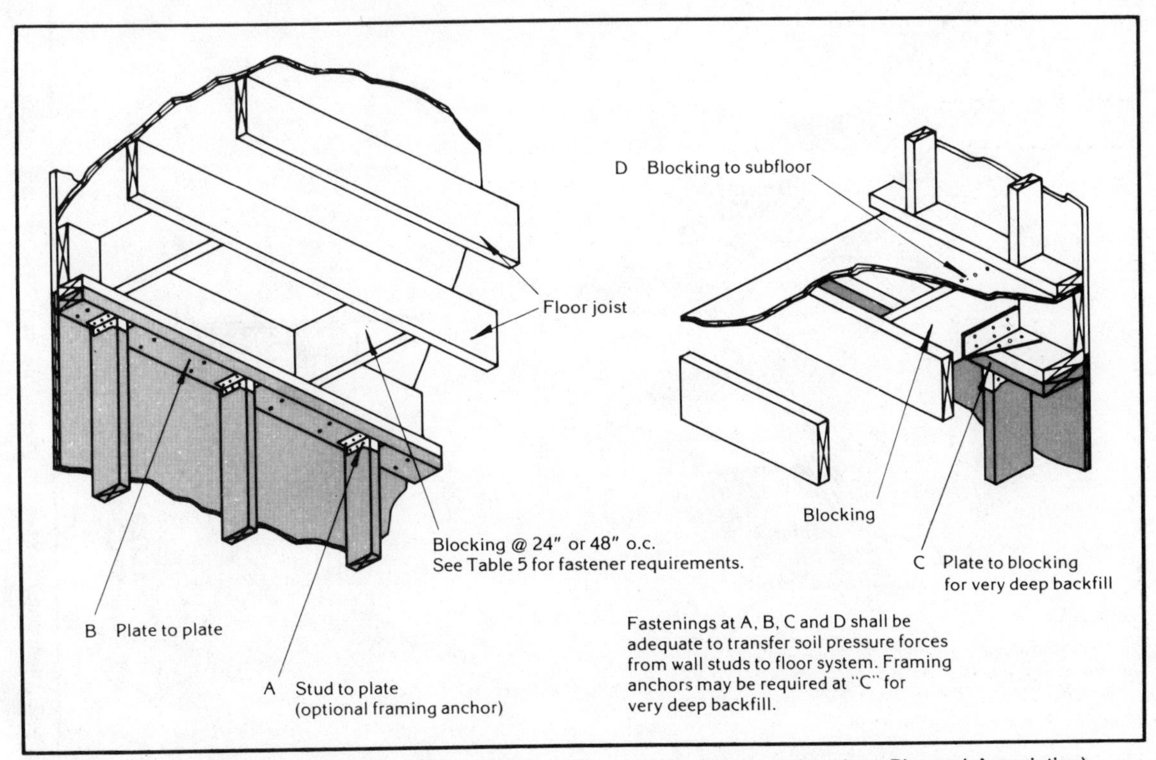

Fig. 6-31. Deep backfill fastening foundation end walls to floor system (courtesy American Plywood Association).

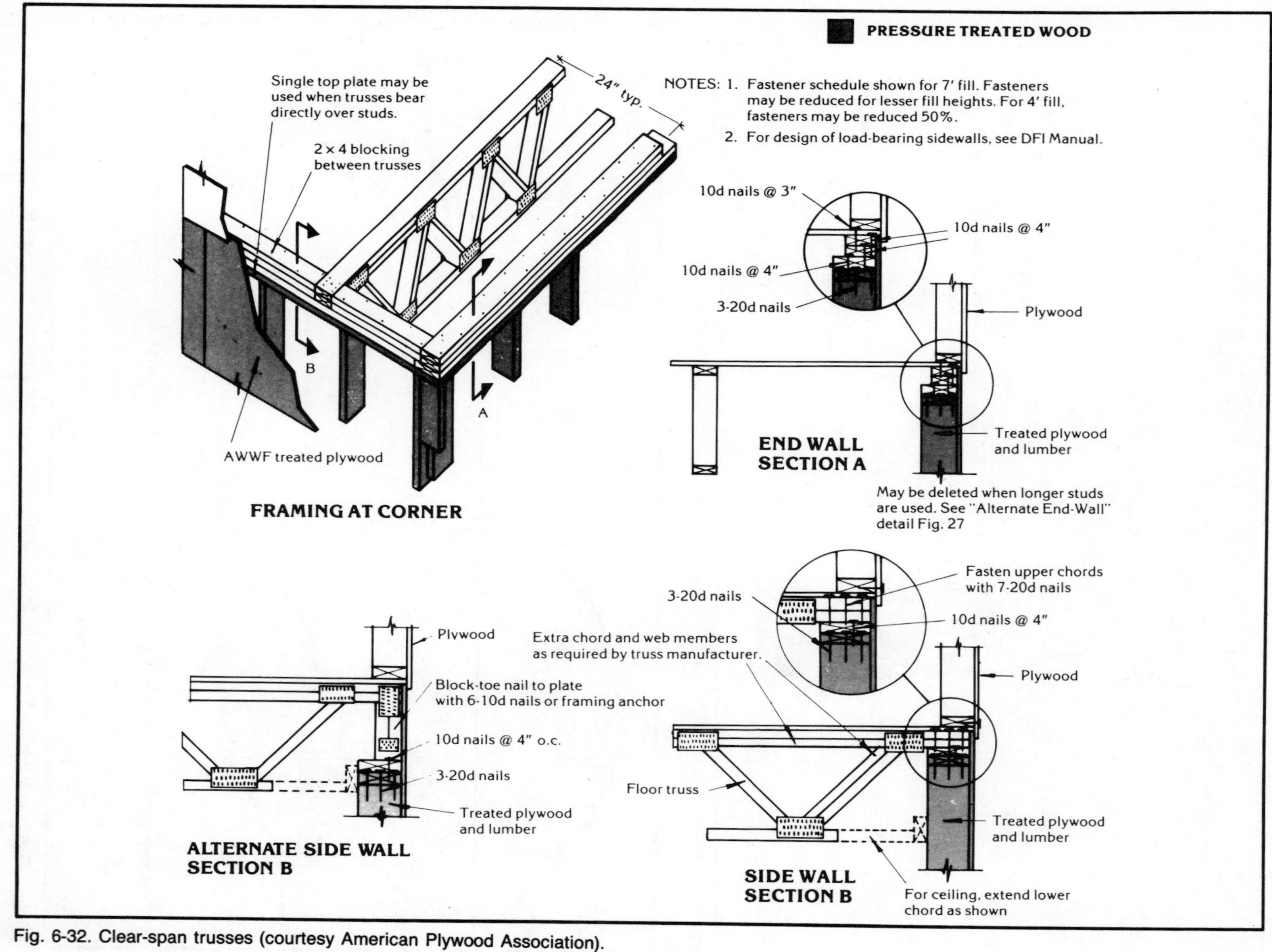

Fig. 6-32. Clear-span trusses (courtesy American Plywood Association).

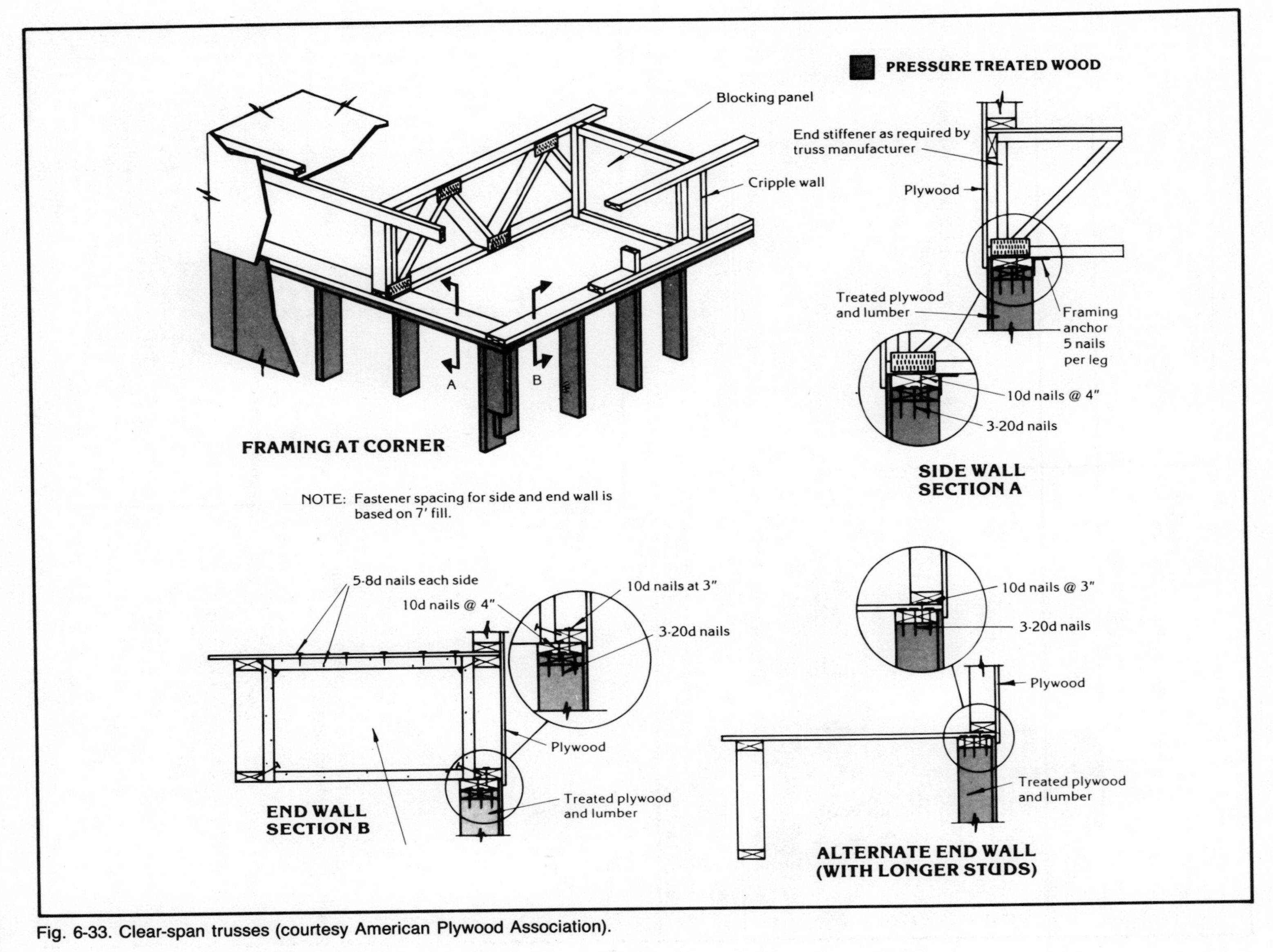

Fig. 6-33. Clear-span trusses (courtesy American Plywood Association).

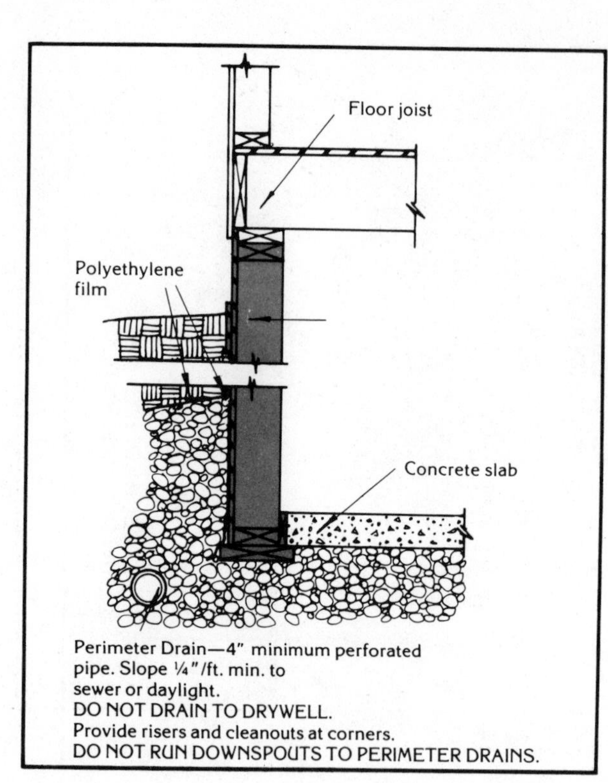

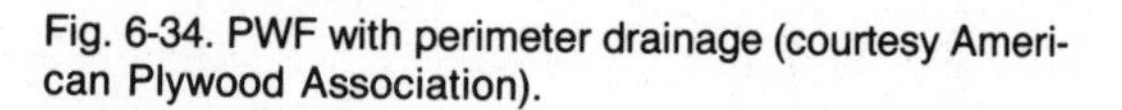

Fig. 6-34. PWF with perimeter drainage (courtesy American Plywood Association).

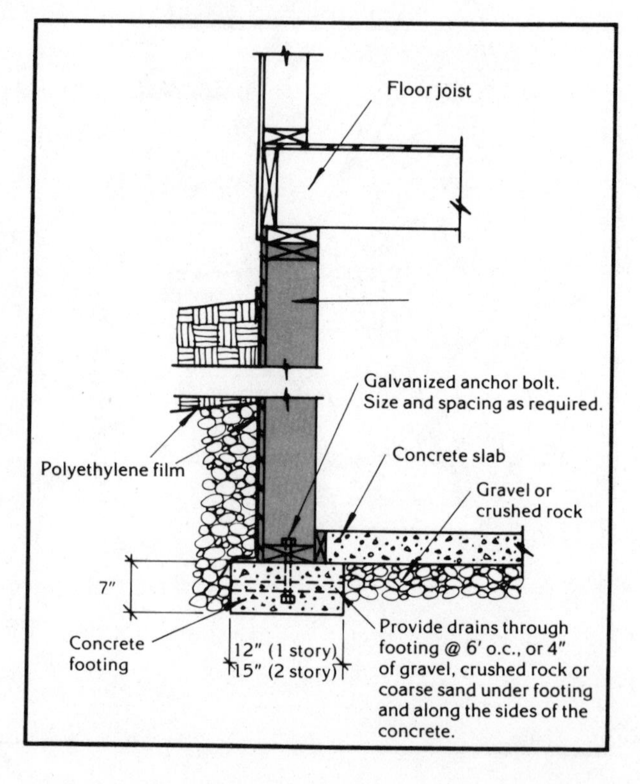

Fig. 6-35. Basement PWF on concrete footing (courtesy American Plywood Association).

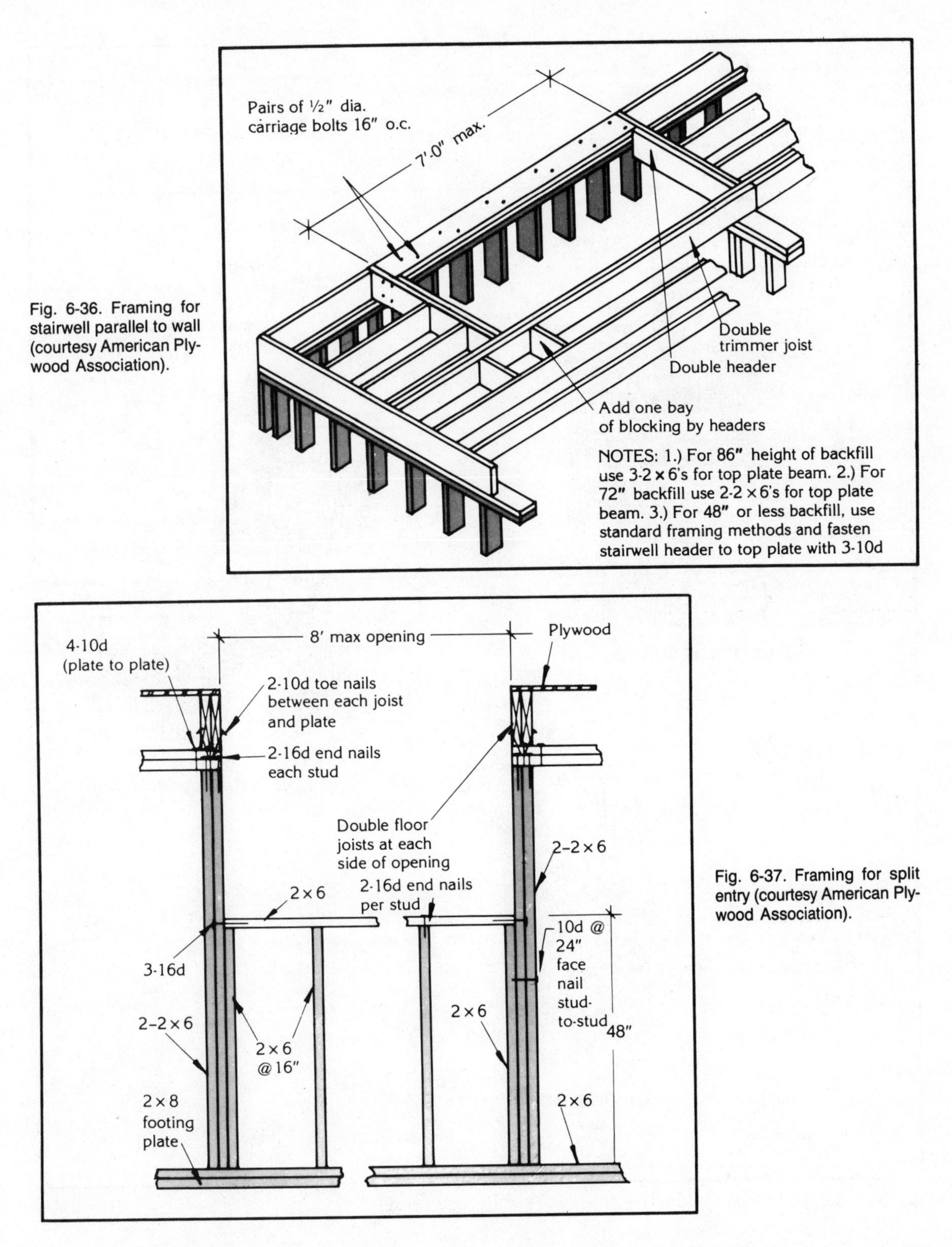

Fig. 6-36. Framing for stairwell parallel to wall (courtesy American Plywood Association).

Fig. 6-37. Framing for split entry (courtesy American Plywood Association).

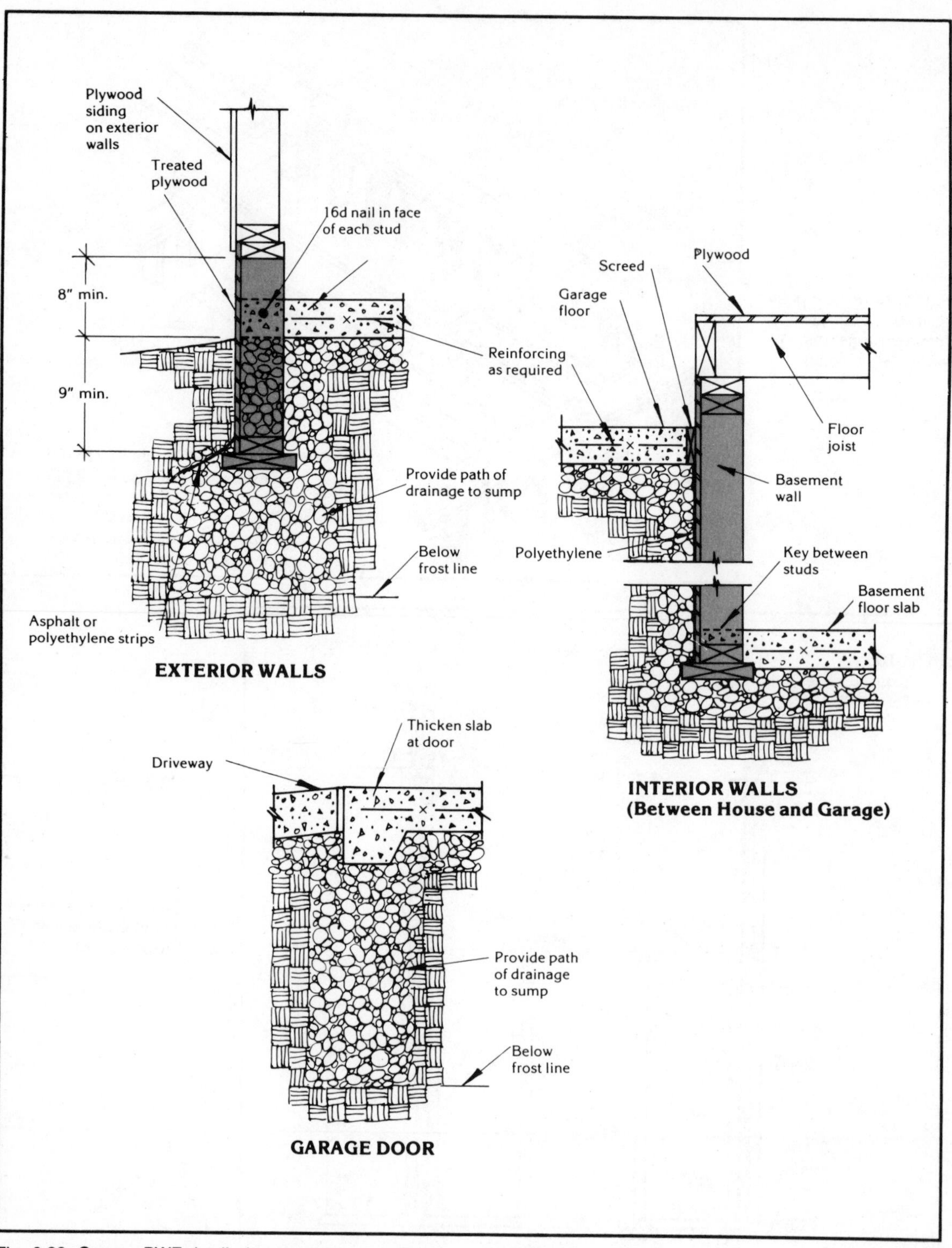

Fig. 6-38. Garage PWF details (courtesy American Plywood Association).

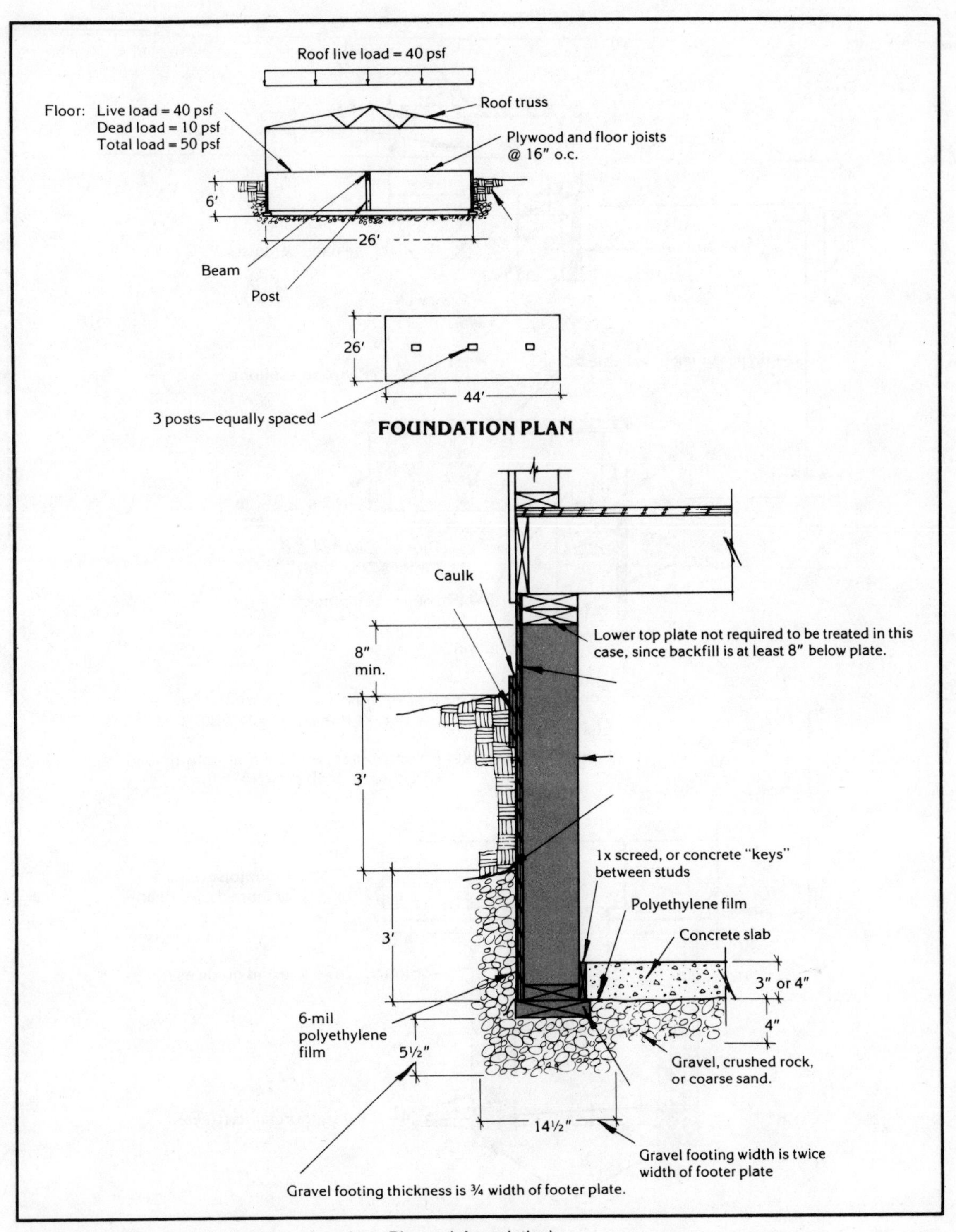

Fig. 6-39. Design example (courtesy American Plywood Association).

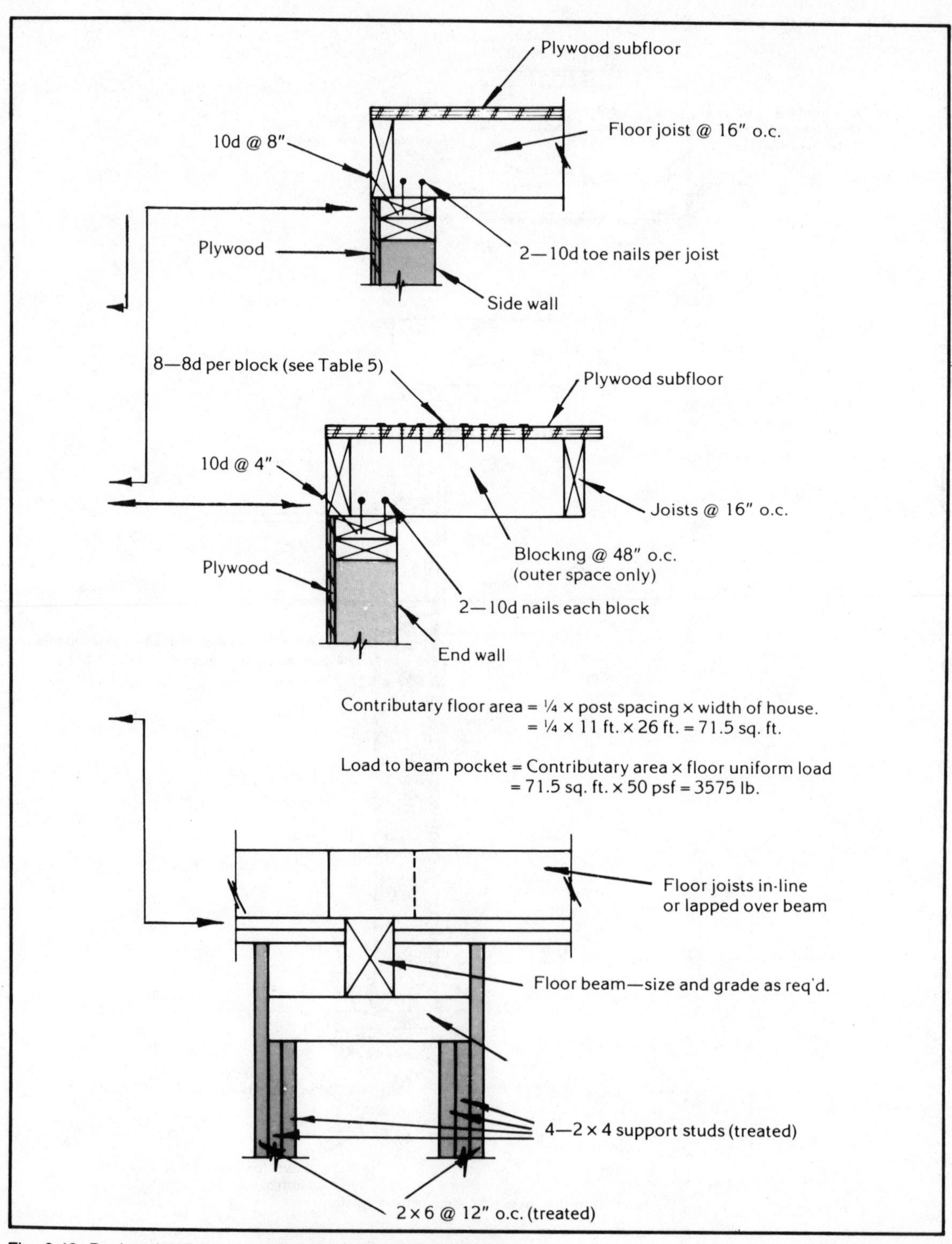

Fig. 6-40. Design detail (courtesy American Plywood Association).

could also be used. To avoid using two lumber grades, it is more practical to overdesign the support studs.

Step 10. Post and piers at center of house. Refer to Fig. 6-21. These can be the same as the posts and piers used for conventional foundations.

Using these steps, you can select the materials you will need for your own permanent wood foundation.

Chapter 7

Floor Framing

The floor framing in a wood-frame house consists specifically of the posts, beams, sill plates, joists, and subfloor. When these components are assembled properly on foundation, they form a level, anchored platform for the rest of the house (Fig. 7-1).

The posts and center beams of wood or steel, which support the inside ends of the joists, are sometimes replaced with a wood-frame or masonry wall when the basement area is divided into rooms. Wood-frame houses also can be constructed on a concrete floor slab or over a crawl space with floor framing similar to that used for a full basement.

One of the important factors in the design of a wood floor system is to equalize shrinkage and expansion of the wood framing at the outside walls and at the center beam. This equalization is usually accomplished through the use of approximately the same total depth of wood at the center beam as the outside framing. Thus, as beams and joists approach *moisture equilibrium,* or the moisture content they reach in service, there are only small differences in the amount of shrinkage. Equalization will minimize plaster cracks and prevent sticking doors and other inconveniences caused by uneven shrinkage. If there are 12 inches of wood at the foundation wall (including joists and sill plate), you should balance it with about 12 inches of wood at the center beam.

Moisture contents of beams and joists used in floor framing should not exceed 19 percent. A moisture content of about 15 percent is much more desirable, however. Dimension material can be obtained at these moisture contents when so specified. When moisture contents are in the higher ranges, it is good practice to allow joists and beams to approach their moisture equilibrium before you apply inside finish and trim, such as baseboard, base shoe, door jambs, and casings.

Grades of dimension lumber vary considerably by species. In general, the first grade is for a high or special use, the second for better than average, the third for average, and the fourth and fifth for more economical construction. Joists and girders are usually second-grade materials of a species, while sills and posts are usually of third or fourth grade. Before continuing with floor framing con-

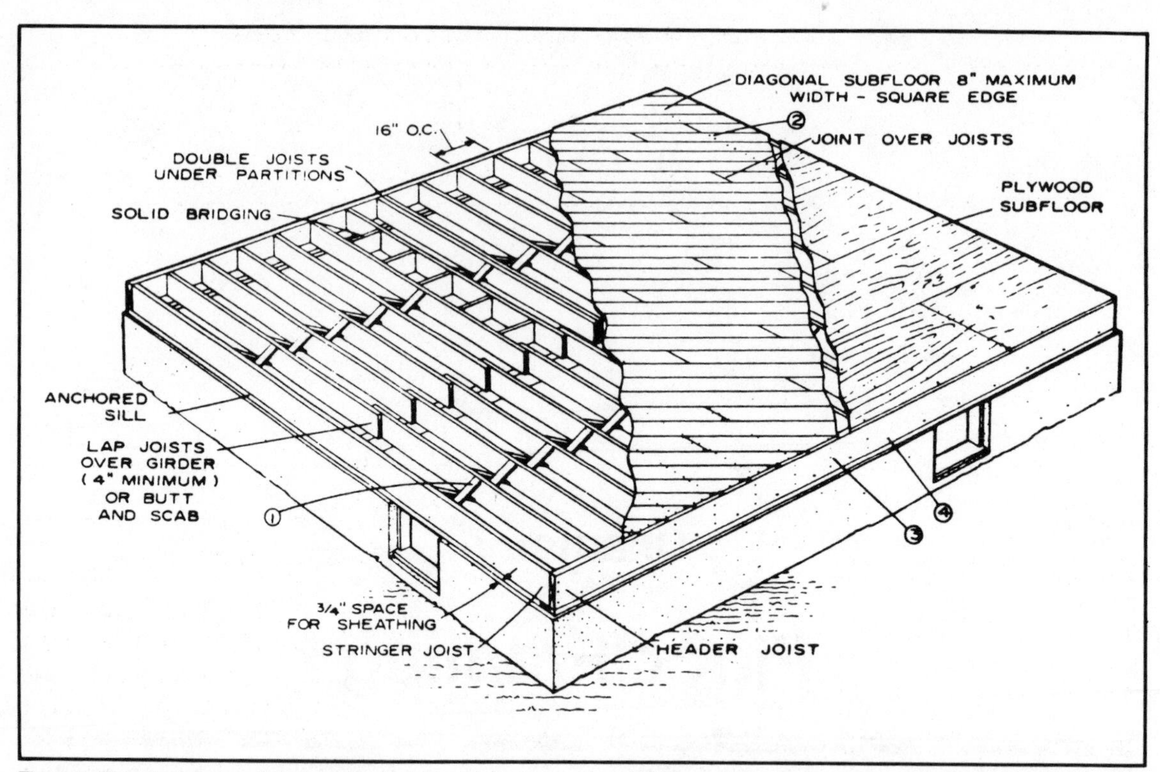

Fig. 7-1. Floor framing: nailing bridging to joists (1); nailing board subfloor to joists (2); nailing header to joists (3); toenailing header to sill (4).

struction techniques, let's consider how to select and buy construction lumber, emphasizing lumber used in floor framing.

BUYING CONSTRUCTION LUMBER

Two common terms in the lumber trade are *softwoods* and *hardwoods.* These terms do not refer to the hardness or softness of the wood itself, but are expressions denoting the botanical origins of the two major groups. Hardwoods are broad-leafed tree species that typically lose their leaves each fall (deciduous). Softwoods are cone-bearing species having needlelike leaves that usually remain on the tree year-round (evergreen). Most common construction lumber is manufactured from softwoods.

Lumber Grades

Lumber grades sort lumber products into their proper classes: size, shape, and quality. The basis of comparison is the usefulness of each piece. Naturally, the standards that determine usefulness must be universal for both buyer and seller so that the buyer knows what to order and the seller knows what is being ordered.

Construction lumber grades are primarily based on strength and appearance. In addition, some species are graded according to durability (heartwood vs. sapwood). Individual grades are a reflection of inherent characteristics of the species; the size, type, and distribution of natural growth characteristics (such as knots or pitch pockets); orientation of the grain, defects generated during drying; and moisture content.

The comparative characteristics of common softwood construction woods are shown in Table 7-1. Allowable size and types of knots vary with the different grades. Wood is strongest when the primary cell types are oriented parallel with the long axis of the piece. The slope of grain is restricted in any strength grades of lumber.

Table 7-1. Comparative Characteristics of Some Construction Woods.

Woods	Dry weight	Hardness	Workability	Nail holding	Paint holding	Freedom from shrinking	Freedom from warping	Freedom from pitch (resin)	Decay resistance (heartwood)	Bending strength	Stiffness
Douglas-fir-larch	H	H	M	H	L	M	M	M	M	H	H
Hem-fir	M	M	M	M	L	M	M	H	L	M	M
Englemann spruce-lodgepole pine	L	L	M	L	L	M	M	H	L	L	L
Eastern hemlock	M	M	M	M	M	H	M	H	L	M	M
Southern Pine	H	H	M	H	L	M	M	L	M	H	H
Idaho white pine	L	L	H	L	H	M	H	H	L	L	L
Spruce-pine-fir (Canada)	L	L	M	L	L	M	M	H	L	M	M
Western cedar	L	L	H	L	H	H	H	H	H	L	L
California redwood	L	M	H	M	H	H	H	H	H	M	M

H = high; M = medium; L = low

Grain also might refer to the direction of growth rings in lumber with respect to the wide face. Figure 7-2 shows flat-grain and edge-grain lumber. Drying defects, such as warp and checks, can reduce the utility of a board. Limitations are often specified in the grade. You should use only dry lumber in construction. As noted earlier, lumber used outdoors should be purchased with a moisture content of 15 to 19 percent. Interior wood should have a moisture content of 6 to 8 percent.

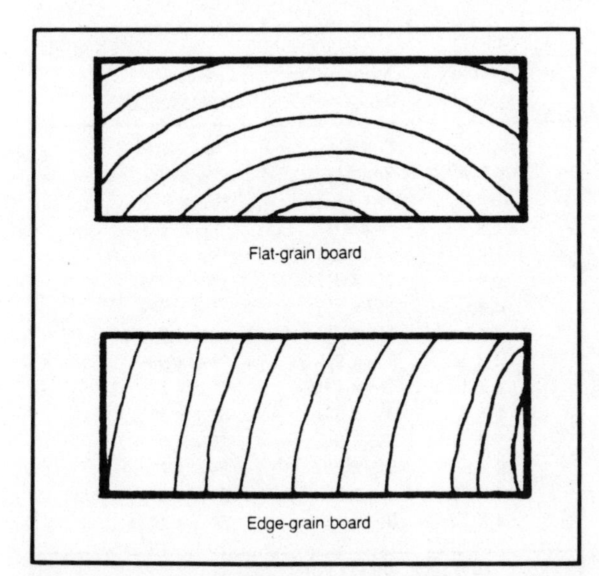

Fig. 7-2. End view of flat-grain and edge-grain boards showing orientation of annual rings.

Grade Stamps

How can the floor framer know that the right lumber is purchased for the project? The lumber grade mark, stamped on the surface of lumber by the manufacturer, is an aid. This mark signifies to the consumer that the lumber has been inspected under grading criteria established by a regional grading association or agency and conforms to American Lumber Standard Committee guidelines.

Specifically, each grade stamp shows the registered symbol of the certifying agency or association, sawmill identification, grade classification or name, lumber species or species group, and perhaps moisture content and size at time of shipment. An example of a grade mark is shown in Fig. 7-3.

Lumber Sizes

In addition to the classification system for softwood lumber, you should understand other characteristics of lumber before you examine specific grades. For many years, sawmills cut lumbers to various dimensions that, over the years, have developed into the nominal standard thickness of boards (1 inch), dimension (2 to 4 inches), and timbers (4 inches and greater).

The popular sizes of 1- × -4, 1- × -6, 2- × -4, and 2- × -8, to mention a few, have come to be known as product names rather than size descriptions. As lumber technology and machinery developed, saw-

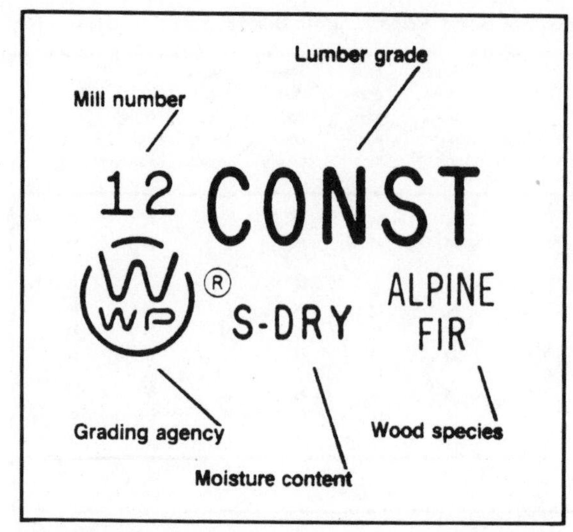

Fig. 7-3. Example of grade mark on construction lumber.

mills began to surface the rough-sawn lumber in a planing mill to produce a smooth, uniform piece of lumber. However, this operation reduced the size of a rough piece so that the popular "2 × 4" no longer measured 2 inches × 4 inches. Dressed lumber came into popular demand. Today it is virtually the only type of lumber used.

The wood of all living trees contains substantial percentages of water by weight. As soon as trees are cut into lumber, moisture is lost from the surface of the lumber, and the pieces of lumber shrink in size. To speed up the process of moisture removal, the lumber industry developed dry kilns so that the moisture content of the lumber could quickly be brought into equilibrium with the environment where it would be used. The final size of the piece might vary depending on the moisture content, as shown in Table 7-2.

Lumber is commonly measured in *board feet.* A board foot is a unit of measure represented by a piece of lumber 1 foot long, 1 foot wide, and 1 inch thick. In practice, the board-foot calculation usually is based on nominal thickness and width. To determine the amount of board feet in a piece of lumber, use this formula:

$$\text{board feet} = \frac{\underset{\text{(inches)}}{\text{thickness}} \times \underset{\text{(inches)}}{\text{width}} \times \textit{length}\ \text{(feet)}}{12}$$

For example, a piece of lumber 2 inches thick, 6 inches wide, and 12 feet long has 12 board feet.

When purchasing lumber for specific jobs, there is an organized way to proceed based on how the lumber is to be used. Most lumber can be classified into three end-use categories: finishing, framing, and general use. Floor framing uses framing lumber.

Lumber for Finishing

Finish lumber includes the clear (select) grades used for interior and exterior trim, porch flooring, cabinet work, or shelving. Lumber in this category is generally kiln dried and surfaced to a uniform thickness, and the surface and edges might be machined to a prescribed pattern. *Uppers* is the term used to describe kiln-dried, surfaced or patterned, clear lumber. *First lumber* is used as a synonym for clear but refers specifically to kiln-dried and surfaced clear grades that are not patterned. Clear grades include:

- ☐ *Grade B and Better (B & Btr.)* is suitable for natural finish; reverse side permits some imperfections. The highest grade you can purchase, it

Table 7-2. Guides to Lumber Sizes.

Product name	Dry or seasoned*	Green or unseasoned**
1 x 4	3/4 x 3 1/2	25/32 x 3 9/16
1 x 6	3/4 x 5 1/2	25/32 x 5 5/8
1 x 8	3/4 x 7 1/4	25/32 x 7 1/2
1 x 10	3/4 x 9 1/4	25/32 x 9 1/2
1 x 12	3/4 x 11 1/4	25/32 x 11 1/2
2 x 4	1 1/2 x 3 1/2	1 9/16 x 3 9/16
2 x 6	1 1/2 x 5 1/2	1 9/16 x 5 5/8
2 x 8	1 1/2 x 7 1/4	1 9/16 x 7 1/2
2 x 10	1 1/2 x 9 1/4	1 9/16 x 9 1/2
2 x 12	1 1/2 x 11 1/4	1 9/16 x 11 1/2
4 x 4	3 1/2 x 3 1/2	3 9/16 x 3 9/16
4 x 6	3 1/2 x 5 1/2	3 9/16 x 5 5/8
4 x 8	3 1/2 x 7 1/4	3 9/16 x 7 1/2
4 x 10	3 1/2 x 9 1/4	3 9/16 x 9 1/2
4 x 12	3 1/2 x 11 1/4	3 9/16 x 11 1/2

*19% moisture content or less
**more than 19% moisture content

is practically clear and almost knot-free, which makes it an excellent choice for fine furniture, wall paneling, interior and exterior trim and flooring, and applications where grain and color are to be featured.

- ☐ *Grade C* has limited imperfections that can be covered by paint. Small knots are permitted. This grade is useful for cabinets, cornices, porch flooring, and columns.
- ☐ *Grade D* is suitable for paint finish. Make allowances for imperfections that can be cut out. It can be finished clear or stained to emphasize the knots, and is useful for shelving and molding.
- ☐ *Grade E* is a utility grade in certain patterned items, such as flooring and ceiling boards. It permits wasting of one-fourth of the board length.

Lumber for Framing

Framing lumber includes dimension lumber grades that are determined by size and location of knots and in which strength is an important criterion. The number of knots per piece is not limited as in the clear grades.

Framing dimension, or simply *dimension,* is used to describe construction or structural framing lumber 2 to 4 inches in thickness, principally 2-inch material. Dimension lumber is used in the framing of floors, walls, and roofs, and is classified into two widths and five categories.

Dimensions up to 4 inches wide is classified as *structural light framing, light framing,* and *studs.* Dimension 6 inches and wider is classified as *structural joists and planks.* There is also an *appearance framing* grade for applications where high strength and appearance are required. For general construction framing, light framing, stud, and structural joists and planks dimension lumber is most commonly used.

Light Framing Dimension Grades. The light framing grades are designed to provide dimension of good appearance and lower design load levels for uses where high design load levels coupled with high appearance characteristics are not needed.

- ☐ *Construction* grade is recommended for general framing purposes. It has good appearance and is graded primarily for strength and serviceability.
- ☐ *Standard* is used for most normal framing where appearance is not a factor. It provides good strength and serviceability.
- ☐ *Utility* is used where a combination of good strength and economical construction are needed. Applications are studding, bracing, blocking, and rafters.
- ☐ *Economy* is used for temporary or lower cost construction in applications such as blocking, bracing, cribbing, and for purposes such as railroad dunnage and machinery crating.

Studs. Studs are the vertical support members in most light-frame wall construction. There is only one grade. Dimension lumber in this grade is limited by characteristics such as knots that affect strength and stiffness. This grade is satisfactory for load-bearing walls. Studs are generally precision-trimmed to lengths of 10 feet or less.

Structural Joist and Plank Dimension Grades. Structural joist and plank grades are designed especially to fit engineering applications for lumber 6 inches and wider.

- ☐ *Select Structural* grade is recommended for use where both high strength and good appearance are required.
- ☐ *No. 1* is for use where high strength, stiffness, and good appearance are desired. This grade is slightly lower in bending strength than select structural.
- ☐ *No. 2* grade is recommended for most general construction.
- ☐ *No. 3* grade is the least restrictive for those characteristics that affect appearance and strength. Much of the lumber is classified in this grade because of one characteristic, however. It is usually recommended for use in general construction where appearance is not important.

Lumber for General Use

This is the *common board* category and covers lumber less than 2 inches in nominal or rough thickness. This category of lumber plays an important role in modern construction. It is suitable for general utility and construction purposes where appearance is not a crucial factor. It is used to brace the framing members of a building when applied as subflooring and roof decking or sheathing. At the same time, it furnishes insulation and acts as a base for the finished surface.

Boards are nominally 1 inch thick and can be purchased rough or surfaced to 3/4 inch. Boards are graded based on the characteristics of the best surface, but features affecting surface tightness are also considered.

There are basically five grades of general-use boards, but each manufacturers' grading association retains its own grading nomenclature, even though the grade specifications are fairly well standardized. To facilitate board-grade identification, three commonly used names are listed for each grade.

- ☐ *Select Merchantable, 1 Common, Colonial* is recommended for interior and exterior use when serviceability and fine appearance of knotty material are required.
- ☐ *Construction, 2 Common, Sterling* is used for paneling, shelving, subfloors, and roof and wall sheathing where knotty lumber with a fine appearance is required.
- ☐ *Standard, 3 Common, Standard* is used for a wide range of construction purposes where appearance and strength are important. Applications include shelving, sheathing, subflooring, paneling, and fencing.
- ☐ *Utility, 4 Common, Utility* is the most widely used lumber for general construction purposes when serviceability, not appearance, is the most important feature. Used for subflooring, roof and wall sheathing, low-cost fencing, and concrete forms.
- ☐ *Economy, 5 Common, Industrial* is lumber suitable for low-grade sheathing and other applications where appearance and strength are not basic requirements.

Softwood floor framing lumber will perform properly only if the right species, size, and grade are used for appropriate applications. Use only seasoned lumber and use durable or preservative-treated woods where decay hazards exist. Remember that the lowest quality or grade of lumber that will satisfy your use requirement is the economical choice. It is also the best choice in conserving the supply of quality lumber.

NAILING

With that view of building materials used in floor framing, let's consider the methods used to fasten the various wood members together. These connections are most commonly made with nails, but on occasions metal strips, lag screws, bolts, and adhesives are used.

Proper fastening of frame members and covering materials provides the rigidity and strength to resist severe windstorms and other hazards. Good nailing is also important from the standpoint of normal performance of wood parts. For example, proper fastening of intersecting walls usually reduces plaster cracking at the inside corners.

The schedule in Table 7-3 outlines good nailing practices for the framing and sheathing of a well-constructed, wood-frame house. Sizes of common wire nails are shown in Fig. 7-4.

When houses are located in hurricane areas, they should have supplemental fasteners. Refer to building material retailers in these areas for more specifics.

POSTS AND GIRDERS

Wood or steel posts are generally used in the basement or to support wood girders or steel beams. Masonry piers also might be used for this purpose and are commonly employed in crawl-space houses.

You can use round steel posts to support both wood girders and steel beams. These posts are normally supplied with a steel bearing plate at each end. Secure anchoring to the girder or beam is important (Fig. 7-5).

Wood posts should be solid and not less than 6 × 6 inches in size for freestanding use in a basement. When combined with a framed wall, they can be 4 × 6 inches to conform to the depth of the

Table 7-3. Recommended Schedule for Nailing the Framing and Sheathing.

Joining	Nailing Method	Nails		
		Number	Size	Placement
Header to joist	End-nail	3	16d	
Joist to sill or girder	Toenail	2 3	10d or 8d	
Header and stringer joist to sill	Toenail		10d	16 in. on center
Bridging to joist	Toenail each end	2	8d	
Ledger strip to beam, 2 in. thick		3	16d	At each joist
Subfloor, boards:				
1 by 6 in. and smaller		2	8d	To each joist
1 by 8 in.		3	8d	To each joist
Subfloor, plywood:				
At edges			8d	6 in. on center
At intermediate joists			8d	8 in. on center
Subfloor (2 by 6 in., T&G) to joist or girder	Blind-nail (casing) and face-nail	2	16d	
Soleplate to stud, horizontal assembly	End-nail	2	16d	At each stud
Top plate to stud	End-nail	2	16d	
Stud to soleplate	Toenail	4	8d	
Soleplate to joist or blocking	Face-nail		16d	16 in. on center
Doubled studs	Face-nail, stagger		10d	16 in. on center
End stud of intersecting wall to exterior wall stud	Face-nail		16d	16 in. on center
Upper top plate to lower top plate	Face-nail		16d	16 in. on center
Upper top plate, laps and intersections	Face-nail	2	16d	
Continuous header, two pieces, each edge			12d	12 in. on center
Ceiling joist to top wall plates	Toenail	3	8d	
Ceiling joist laps at partition	Face-nail	4	16d	
Rafter to top plate	Toenail	2	8d	
Rafter to ceiling joist	Face-nail	5	10d	
Rafter to valley or hip rafter	Toenail	3	10d	
Ridge board to rafter	End-nail	3	10d	
Rafter to rafter through ridge board	Toenail Edge-nail	4 1	8d 10d	
Collar beam to rafter:				
2 in. member	Face-nail	2	12d	
1 in. member	Face-nail	3	8d	
1-in. diagonal let-in brace to each stud and plate (4 nails at top)		2	8d	
Built-up corner studs:				
Studs to blocking	Face-nail	2	10d	Each side
Intersecting stud to corner studs	Face-nail		16d	12 in. on center
Built-up girders and beams, three or more members	Face-nail		20d	32 in. on center, each side
Wall sheathing:				
1 by 8 in. or less, horizontal	Face-nail	2	8d	At each stud
1 by 6 in. or greater, diagonal	Face-nail	3	8d	At each stud
Wall sheathing, vertically applied plywood:				
3/8 in. and less thick	Face-nail		6d	6 in. edge
1/2 in. and over thick	Face-nail		8d	12-in. intermediate
Wall sheathing, vertically applied fiberboard:				
1/2 in. thick	Face-nail		1 1/2 in. roofing nail	3 in. edge and
25/32 in. thick	Face-nail		1 3/4 in. roofing nail	6 in. intermediate
Roof sheathing, boards, 4-, 6-, 8-in. width	Face-nail	2	8d	At each rafter
Roof sheathing, plywood:				
3/8 in. and less thick	Face-nail		6d	6 in. edge and 12 in. intermediate
1/2 in. and over thick	Face-nail		8d	

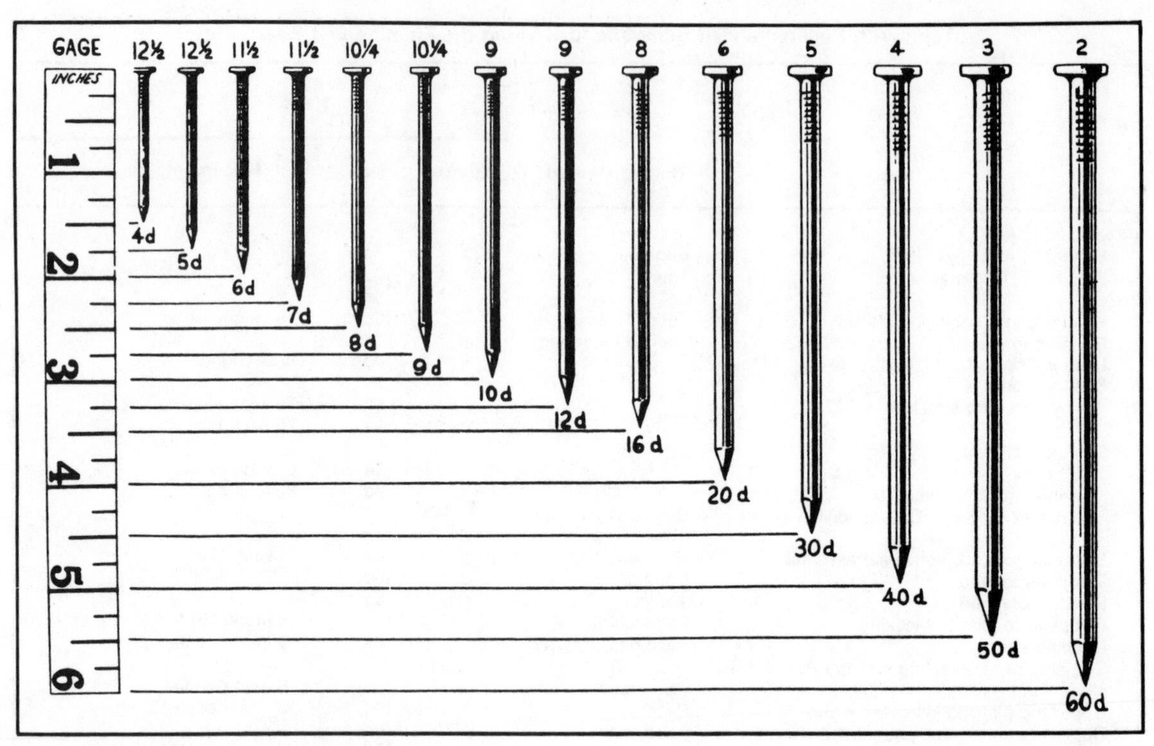

Fig. 7-4. Sizes of common wire nails.

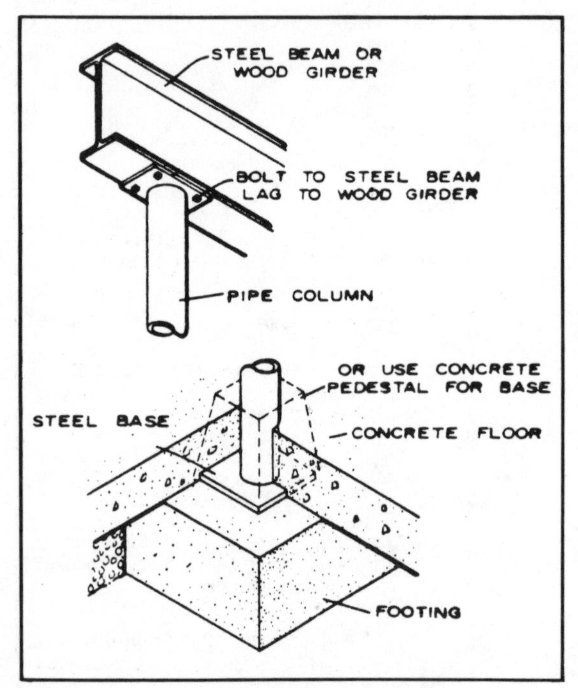

Fig. 7-5. Steel post for wood or steel girder.

studs. Square wood posts at both ends, and securely fasten them to the girder (Figs. 7-6 and 7-7). The bottom of the post should rest on and be pinned to a masonry pedestal 2 to 3 inches above the finish floor. In moist or wet conditions, it is good practice to treat the bottom end or use a moistureproof covering over the pedestal.

Both wood girders and steel beams are used in present-day house construction. The standard I beam and the wide flange beam are the most commonly used steel-beam shapes. Wood girders are of two types: solid and built-up. The built-up beam is preferred because it can be made up from drier dimension material and is more stable. Commercially available glue-laminated beams might be desirable where exposed in finished basement rooms.

The built-up girder (Fig. 7-8) is usually made up of two or more pieces of 2-inch dimension lumber spiked together, the ends of the pieces joining over a supporting post. Nail a two-piece girder from one side with 10d nails—two at the end of each

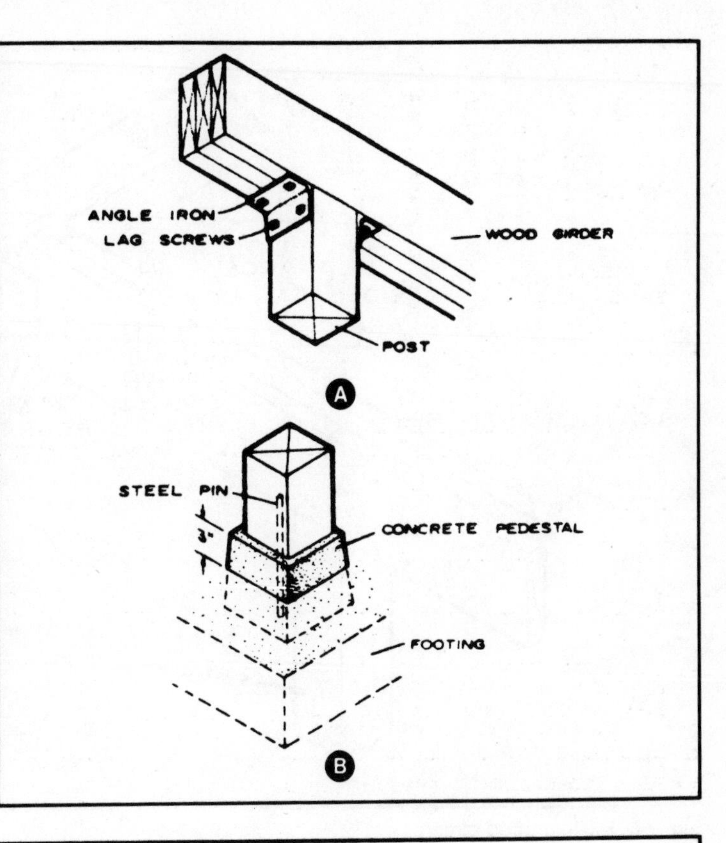

Fig. 7-6. Wood post for wood girder; connection to girder (A), and base (B).

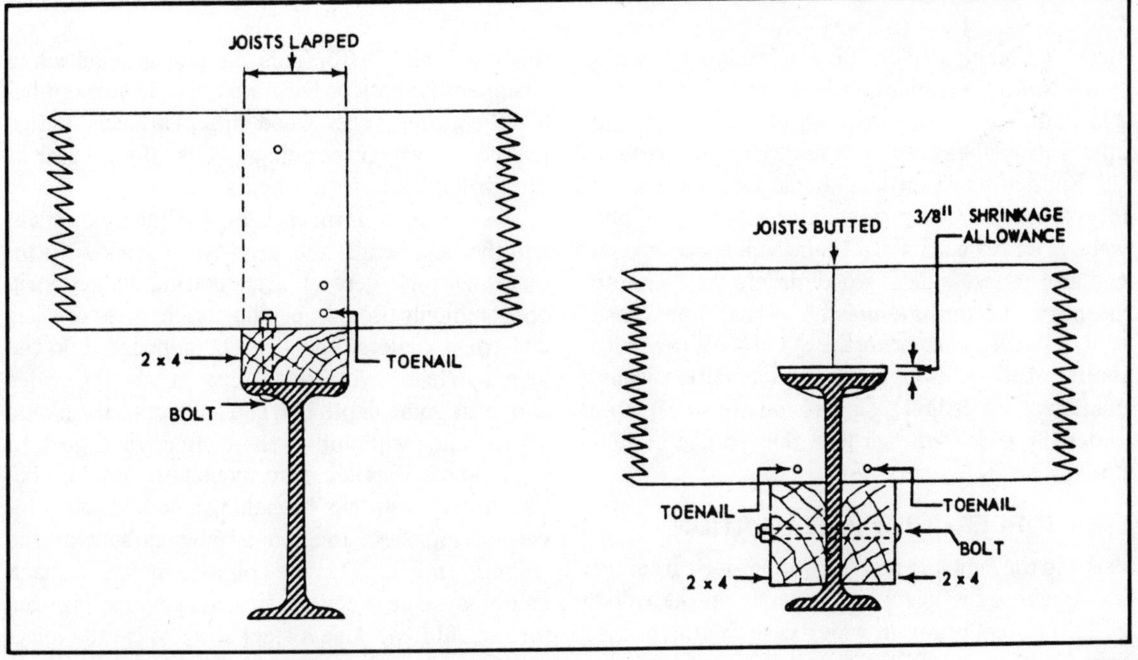

Fig. 7-7. Methods of framing joists to steel girders.

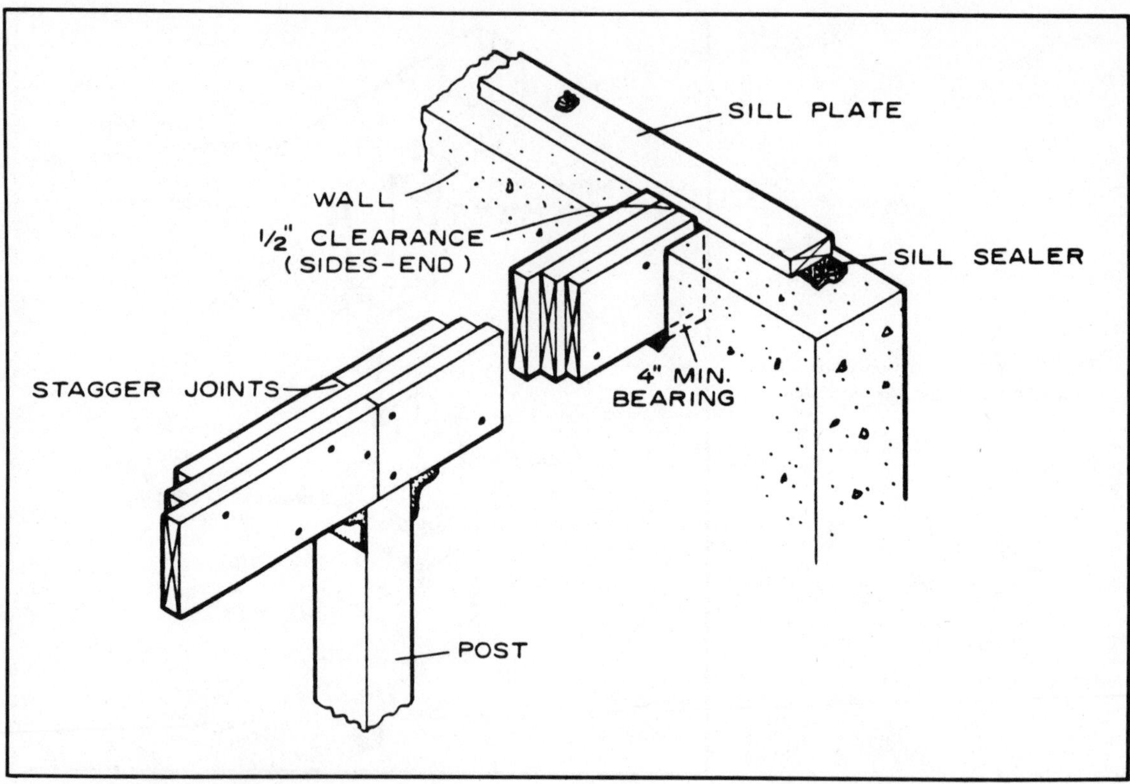

Fig. 7-8. Built-up wood girder.

piece and others driven stagger fashion 16 inches apart. Nail a three-piece girder from each side with 20d nails—two near each end of each piece and others driven stagger fashion 32 inches apart.

Ends of wood girders should beam at least 4 inches on the masonry walls or pilasters. When wood is untreated, leave a 1/2-inch air space at each end and at each side of wood girders framing into masonry. In termite-infested areas, line these pockets with metal. Level the top of the girder with the top of the sill plates on the foundation walls, unless you use ledger strips. If you use steel plates under the ends of the girders, they should be full-bearing size.

GIRDER-JOIST INSTALLATION

Perhaps the simplest method of floor-joist framing is one where the joists bear directly on the wood girder or steel beam, in which case the top of the beam coincides with the top of the anchored sill (Figs. 7-9 and 7-10). This method is used when basement heights provide adequate headroom below the girder. When wood girders are used in this manner, however, shrinkage is usually greater at the girder than at the foundation.

For more uniform shrinkage at the inner beam and the outer wall, and to provide greater headroom, joist hangers or a supporting ledger strip are commonly used. Depending on the sizes of joists and wood girders, joists can be supported on the ledger strip in several ways. Each method provides about the same depth of wood subject to shrinkage at the outer wall and at the center wood girder.

You can obtain a continuous horizontal tie between exterior walls by nailing notched joists together (Fig. 7-9). Joists must always bear on the ledgers. In Fig. 7-11, the connecting scab at each pair of joists provides this tie and also a nailing area for the subfloor. Use a steel strap to tie the joists together when the tops of the beam and the joists

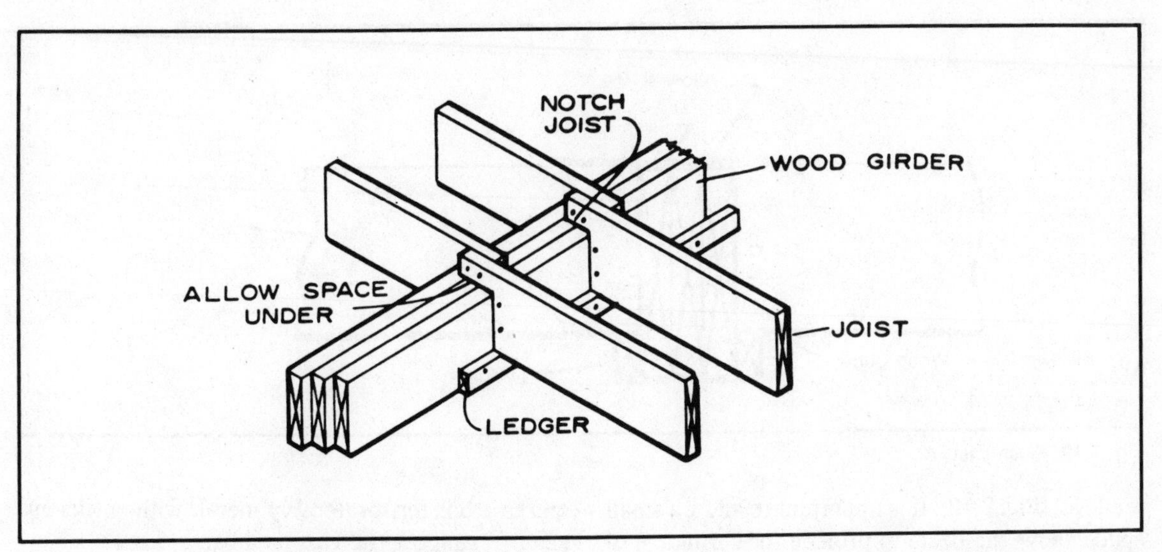

Fig. 7-9. Ledger on center wood girder; notched joist.

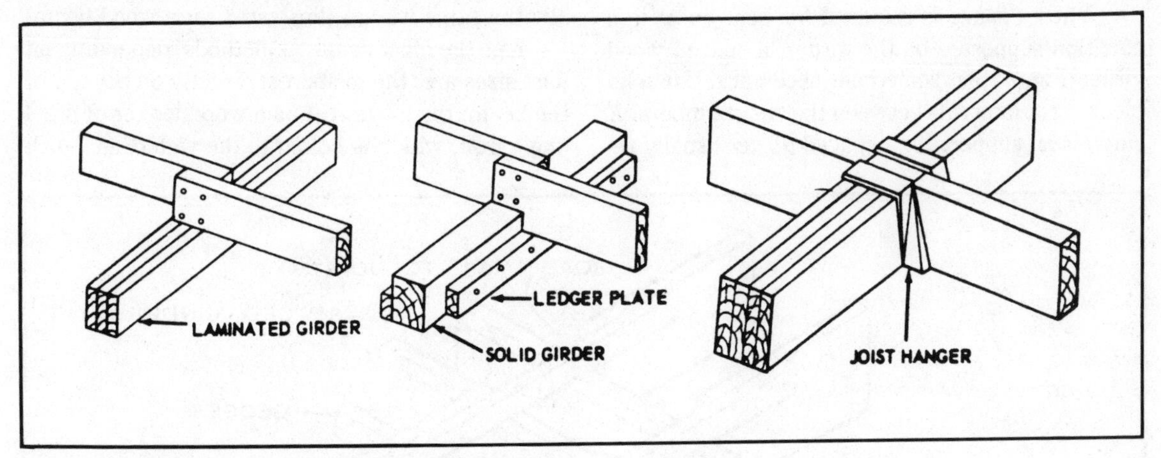

Fig. 7-10. Methods of framing joists to wooden girders.

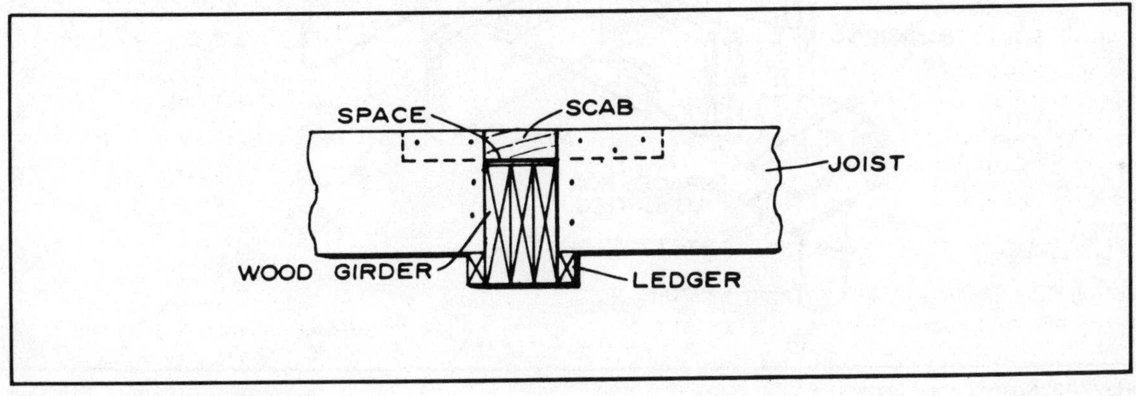

Fig. 7-11. Scab tie between joists.

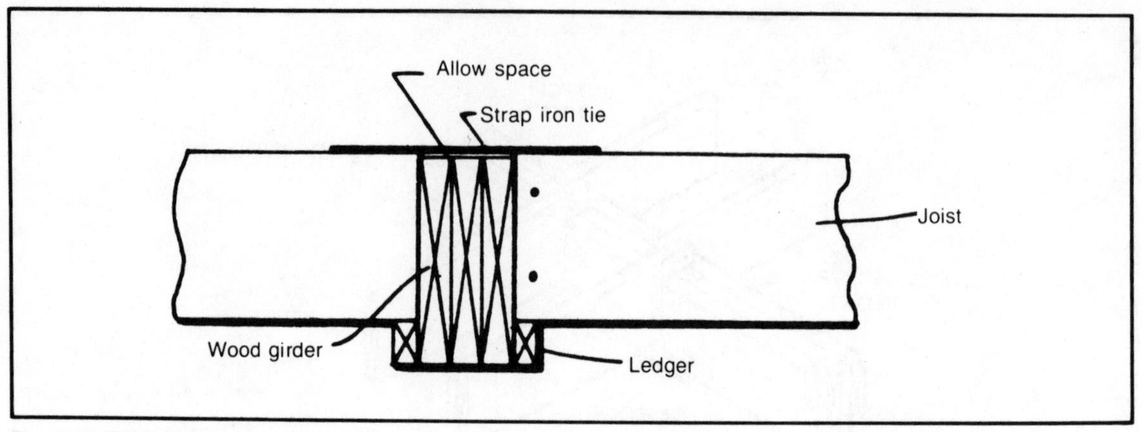

Fig. 7-12. Flush joist.

are level (Fig. 7-12). It is important to allow a small space above the beam to provide for shrinkage of the joists.

When a space is required for heat ducts in a partition supported on the girder, a spaced wood girder (Fig. 7-13) is sometimes necessary. Use solid blocking at intervals between the two members. A single post support for a spaced girder usually requires a bolster, preferably metal, with sufficient span to support the two members.

You can arrange joists with a steel beam generally the same way as illustrated for a wood beam. Perhaps the most common methods, depending on joist sizes are: the joists rest directly on the top of the beam; the joists rest on a wood ledger or steel angle iron, which is bolted to the web (Fig. 7-14);

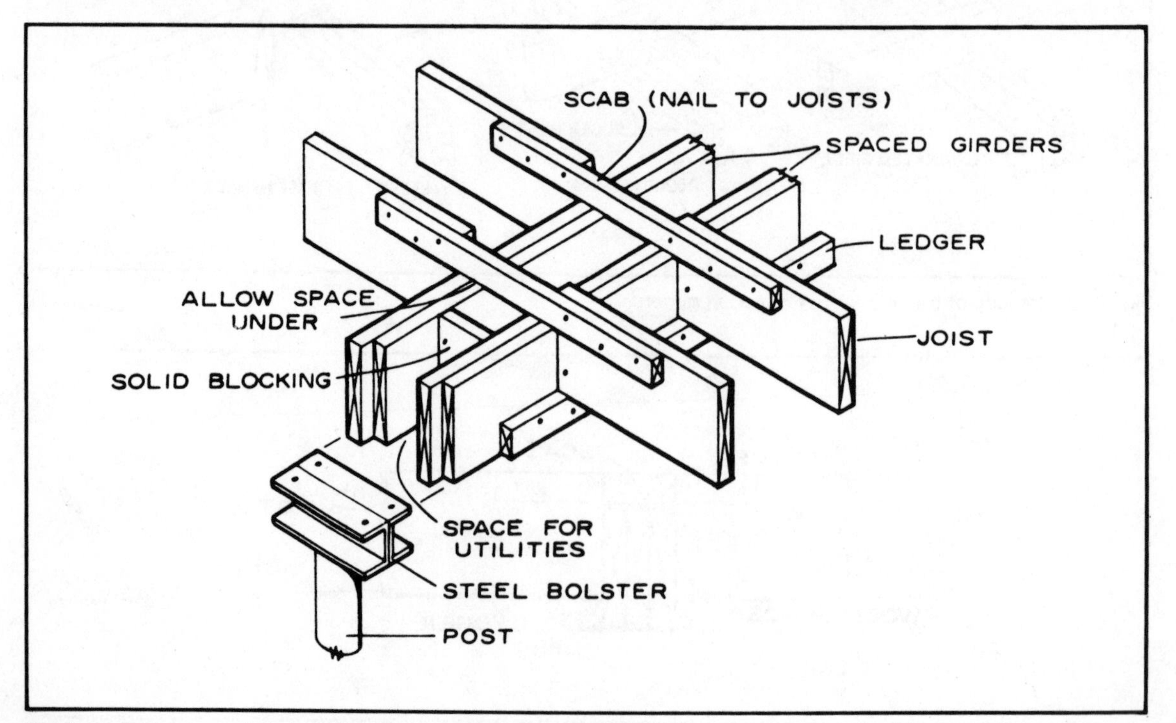

Fig. 7-13. Spaced wood girder.

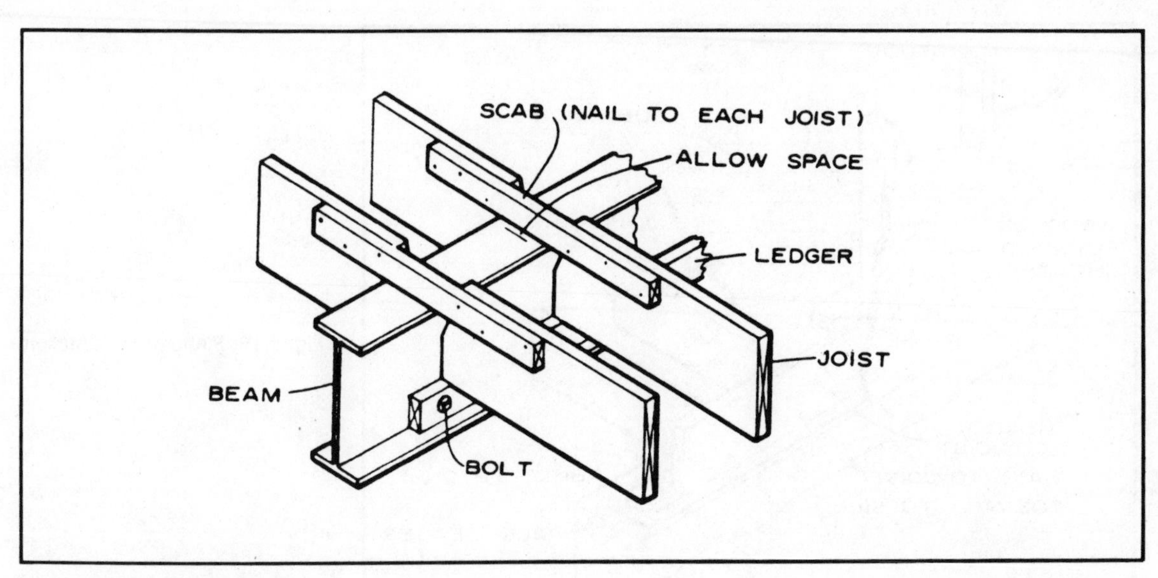

Fig. 7-14. Bearing on ledger.

or the joists bear directly on the flange of the beam (Fig. 7-15). In the third method, wood blocking is required between the joists near the beam flange to prevent overturning.

WOOD SILL CONSTRUCTION

The two types of wood sill construction used over the foundation wall conform either to platform or balloon framing. The box sill is commonly used in *platform construction.* It consists of a 2-inch or thicker plate anchored to the foundation wall over a sill sealer that provides support and fastening for the joists and header at the ends of the joists (Figs. 7-16 through 7-18). Some houses are constructed without benefit of an anchored sill plate, although this is not entirely desirable. In these cases, anchor the floor framing with metal strapping installed during pouring operations.

Balloon-frame construction uses a nominal 2-inch or thicker wood sill upon which the joists rest. The studs also bear on this member. Nail the studs both to the floor joists and the sill. Lay the subfloor diagonally or at right angles to the joists and add a fire-stop between the studs and the floor line (Fig. 7-19). When a diagonal subfloor is used, a nailing member is normally required between joists and studs at the wall lines.

There is less potential shrinkage in exterior walls with balloon framing than in the platform type. Therefore balloon framing is usually preferred over the platform in full two-story brick or stone veneer houses.

FLOOR JOISTS

Floor joists are selected primarily to meet strength and stiffness requirements. Strength requirements depend upon the loads to be carried. Stiffness requirements place an arbitrary control on deflection under load. Stiffness is also important in limiting vibrations from moving loads—often a cause of an-

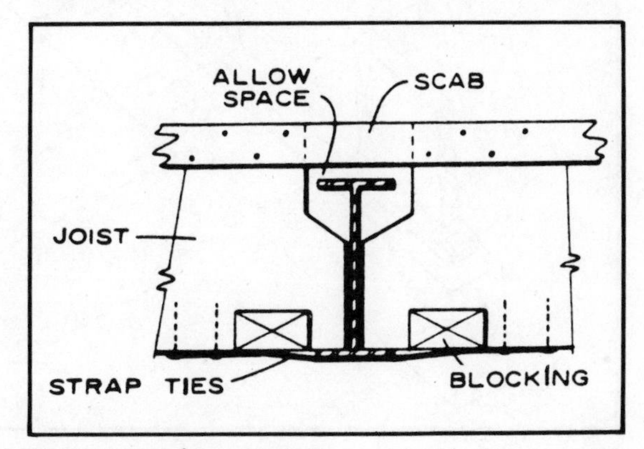

Fig. 7-15. Bearing on flange.

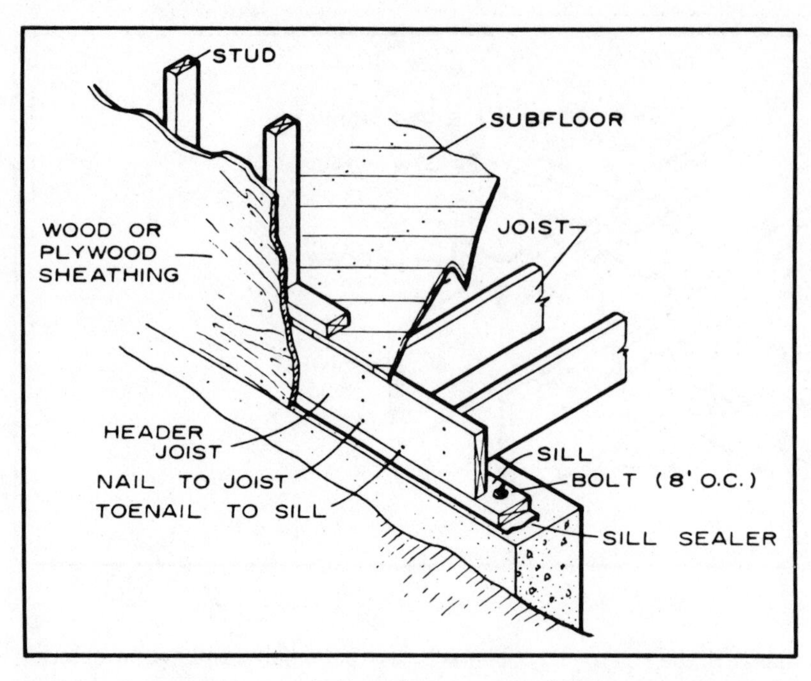

Fig. 7-16. Platform construction.

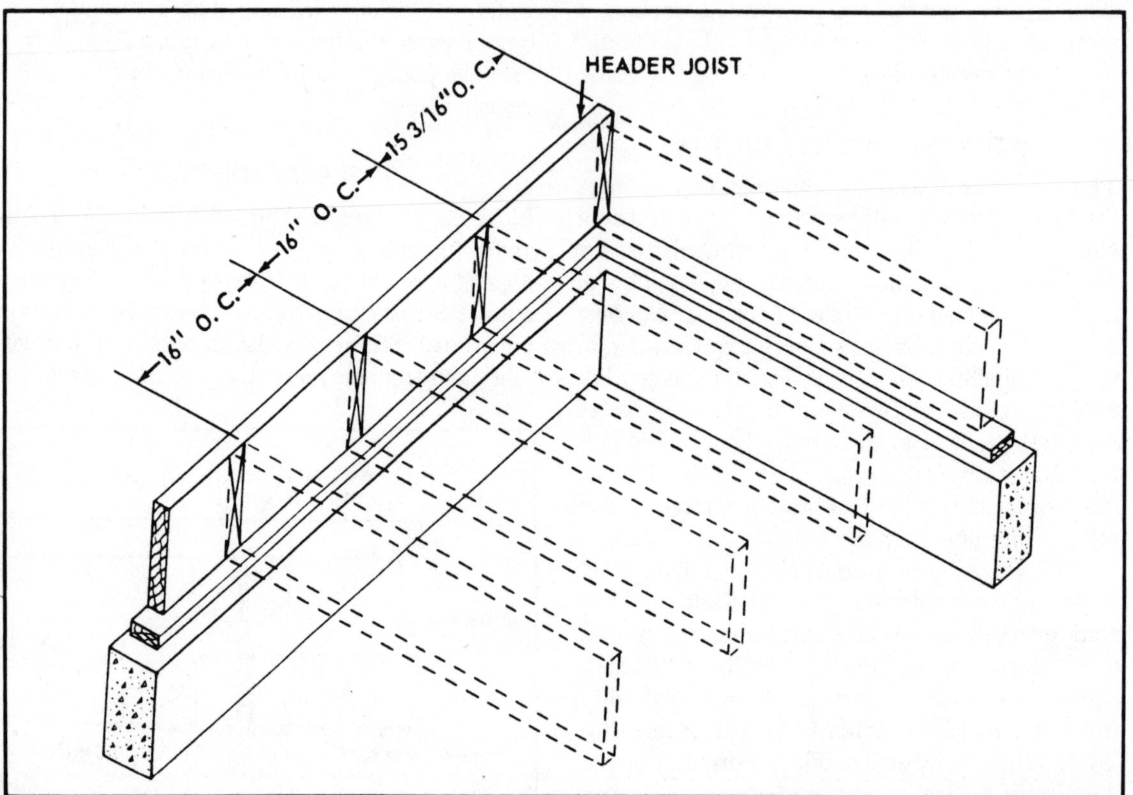

Fig. 7-17. Joist location layout.

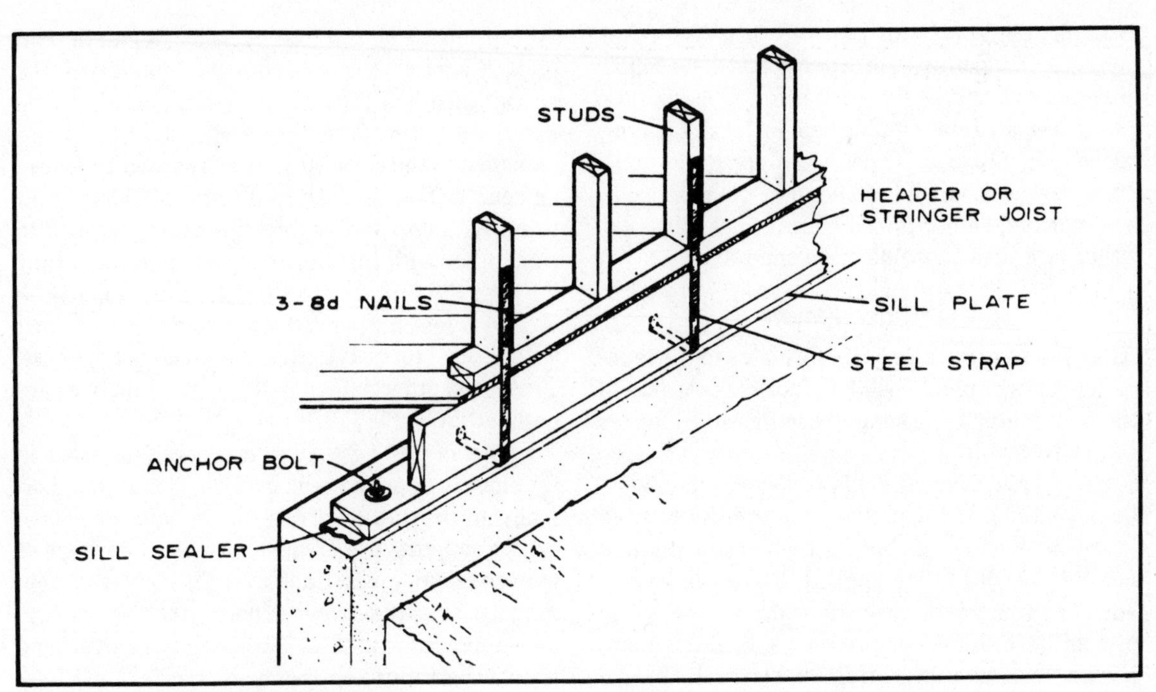

Fig. 7-18. Anchoring wall to floor framing.

noyance to occupants. Other desirable qualities for floor joists are good nail-holding ability and freedom from warp.

Wood joists are generally of 2-inch (nominal) thickness and of 8-, 10-, or 12-inch (nominal) depth. The sizes depend upon the loading, length of span, spacing between joists, and species and grade of lumber used. Grades in species vary greatly. For

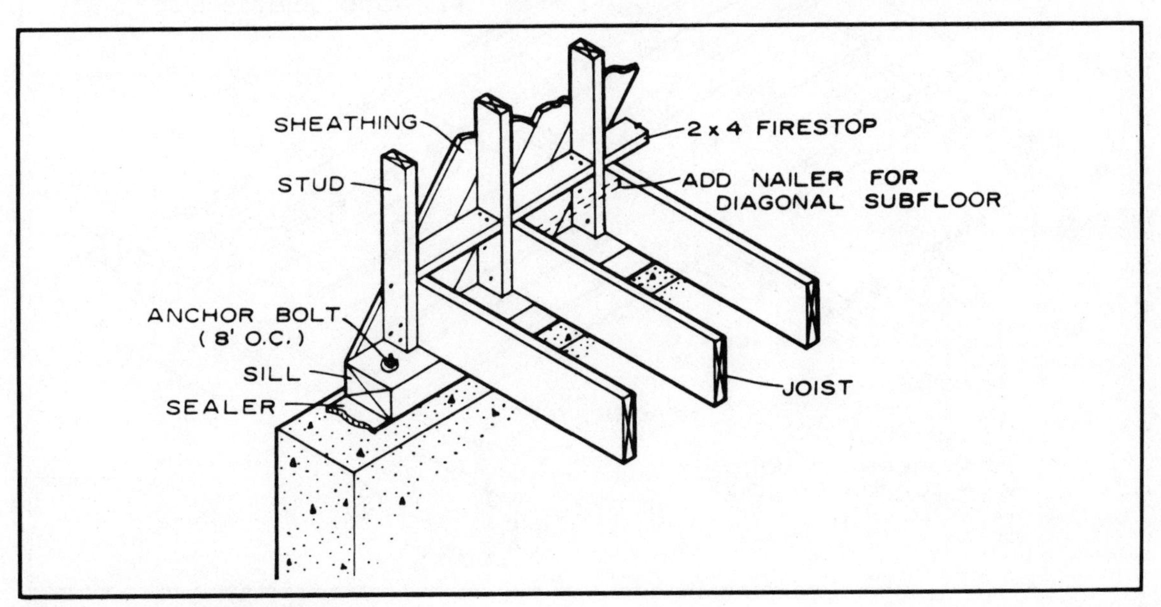

Fig. 7-19. Sill for balloon framing.

example, the grades for joists are usually "Standard" for Douglas fir, "No. 2 or No. 2KD" for southern pine, and comparable grades for other species.

Use span tables published in local building codes as guidelines. These sizes are often minimum, however, and it is sometimes the practice in medium- and higher-priced houses to use the next larger size than those listed in the tables.

Joint Installation

After you have anchored the sill plates to the foundation walls or piers, locate the joists according to the house design. Spacing is commonly 16 inches center to center.

Any joists having a slight bow edgewise should be so placed that the crown is on top. A crowned joist will tend to straighten out when subfloor and normal floor loads are applied. Place the largest edge knots on top because knots on the upper side of a joist are on the compression side of the member and will have less effect on strength.

Fasten the header joist by nailing it into the end of each joist with three 16d nails. In addition, the header joist and the stringer joists parallel to the exterior walls in platform construction (Fig. 7-1) are toenailed to the sill with 10d nails spaced 16 inches on center. Toenail each joist to the sill and center beam with two 10d or three 8d nails; then nail to each other with three or four 16d nails when they lap over the center beam. If a nominal 2-inch scab is used across butt-ended joists, nail it to each joist with at least three 16d nails at each end of the joist. These and other nailing patterns and practices are outlined in Table 7-2.

The *in-line joist splice* is sometimes used in framing for floor and ceiling joists. This system normally allows the use of one smaller joist size when center supports are present. Briefly, it consists of uneven length joists. Cantilever the long overhanging joist over the center support, then splice it to the supported joist (Fig. 7-20). Alternate overhang joists. Depending on the span, species, and joist

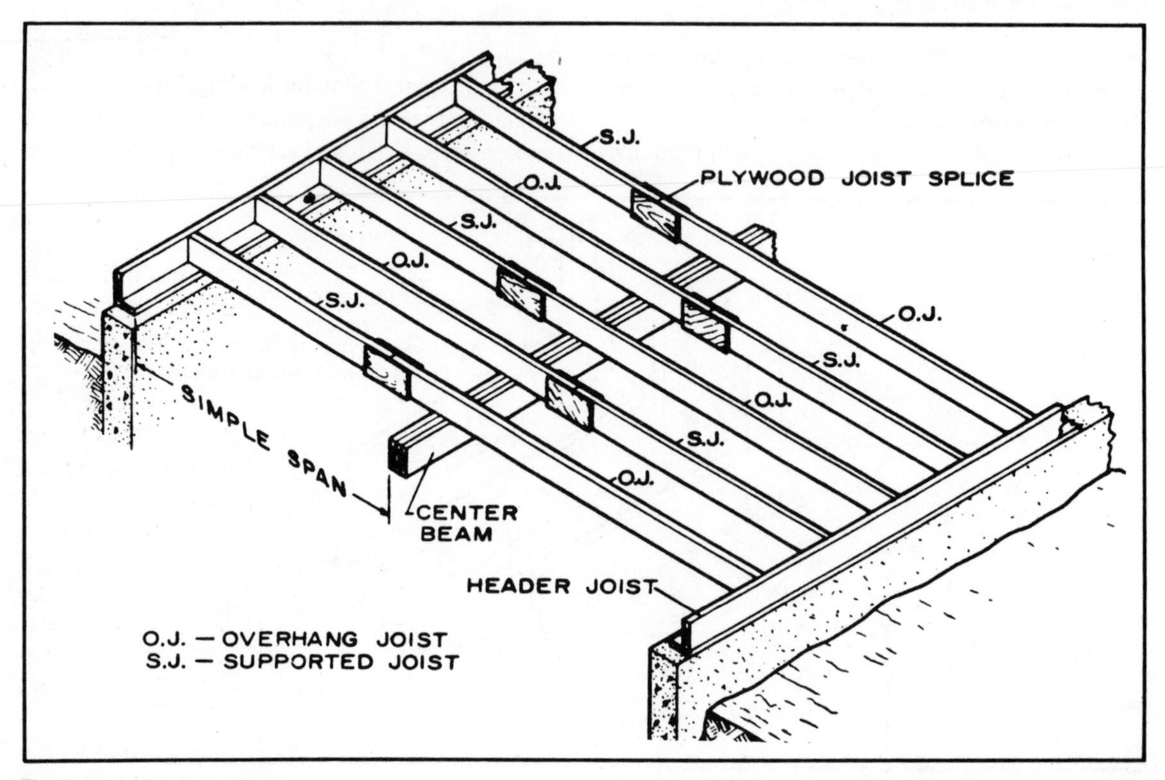

Fig. 7-20. In-line joist system.

size, the overhang varies between about 1 foot 10 inches and 2 feet 10 inches. Use plywood splice plates on each side of the end joists.

It is good practice to double joists under all parallel bearing partition walls. If space is required for heat ducts, use solid blocking between the joists.

BRIDGING

Cross-bridging between wood joists often has been used in house construction, but research by several laboratories has questioned the benefits of bridging in relation to its cost, especially in normal house construction. Even with tight-fitting, well-installed bridging, there is no significant ability to transfer loads after the subfloor and finish floor are installed. Some buildings require the use of cross-bridging or solid bridging, however.

Solid bridging is often used between joists to provide a more rigid base for partitions located above joist spaces. Well-fitted solid bridging securely nailed to the joists will aid in supporting partitions above them. Load-bearing partitions should be supported by doubled joists.

Figures 7-21 and 7-22 illustrate solid and cross bridging. Figures 7-23 through 7-26 show how to frame around floor openings.

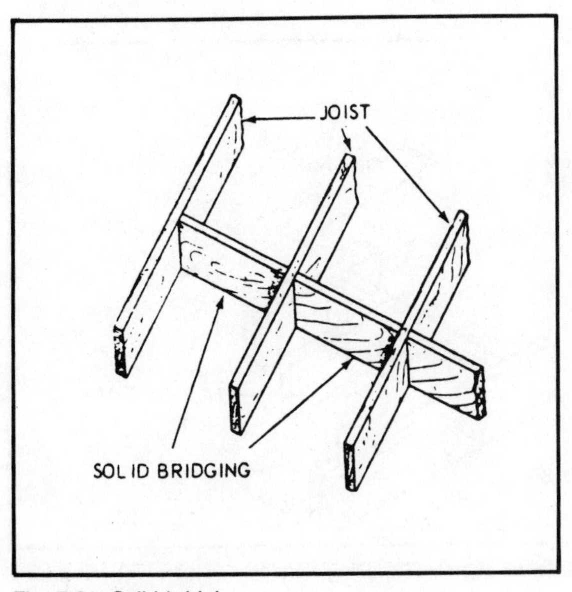

Fig. 7-21. Solid bridging.

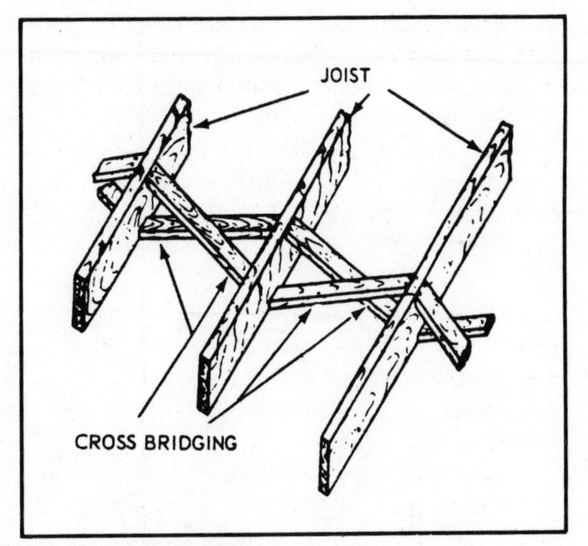

Fig. 7-22. Cross bridging.

SUBFLOOR

Subflooring is used over the floor joists to form a working platform and base for finish flooring. It usually consists of square-edged to tongue-and-grooved boards no wider than 8 inches and not less than 3/4 inch thick, or plywood 1/2 to 3/4 inch thick, depending on species, type of finish floor, and spacing of joists. (Refer to Fig. 7-1.)

Boards

You can apply subflooring either diagonally (most common) or at right angles to the joists. If you place subflooring at right angles to the joists, you should lay the finish floor at right angles to the subflooring. Diagonal subflooring permits finish flooring to be laid either parallel or at right angles (most common) to the joists. Always place end joints of the boards directly over the joists. Nail the subfloor to each joist with two 8d nails for 8-inch widths.

Joist spacing should not exceed 16 inches on center when finish flooring is laid parallel to the joists or when parquet finish flooring is used. Spacing should not exceed 24 inches on center when finish flooring at least 25/32 inch thick is at right angles to the joists.

Where balloon framing is used, install blocking between the ends of the joists at the wall for

Fig. 7-23. Framing around floor openings.

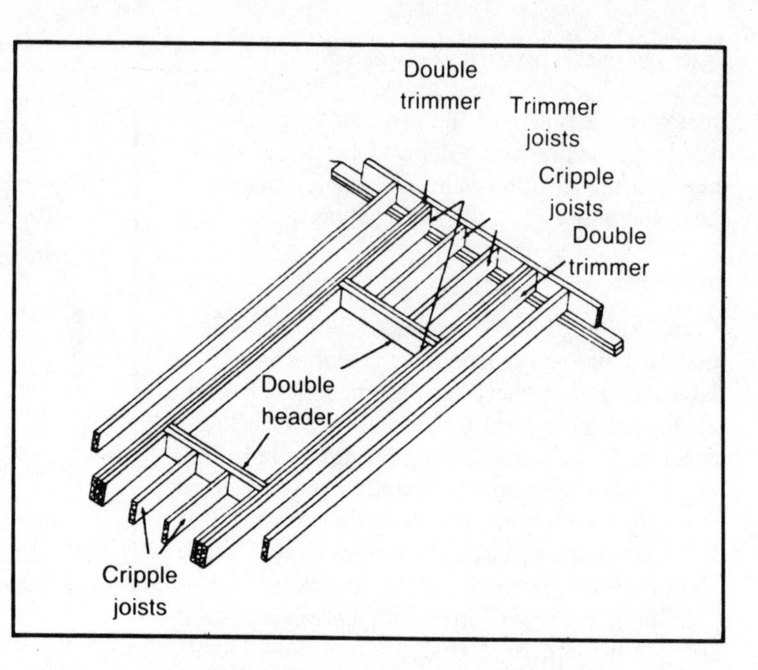

nailing the ends of diagonal subfloor boards. Refer to Fig. 7-19.

Plywood

Plywood comes in a number of grades designed to meet a broad range of end-use requirements. All interior-type grades are also available with fully waterproof adhesive identical with those used in exterior plywood. This type is useful where a hazard of prolonged moisture exists, such as in underlayments or subfloors adjacent to plumbing fixtures and for roof sheathing that might be exposed for long periods during construction. Under normal conditions and for sheathing used on walls, Standard sheathing grades are satisfactory.

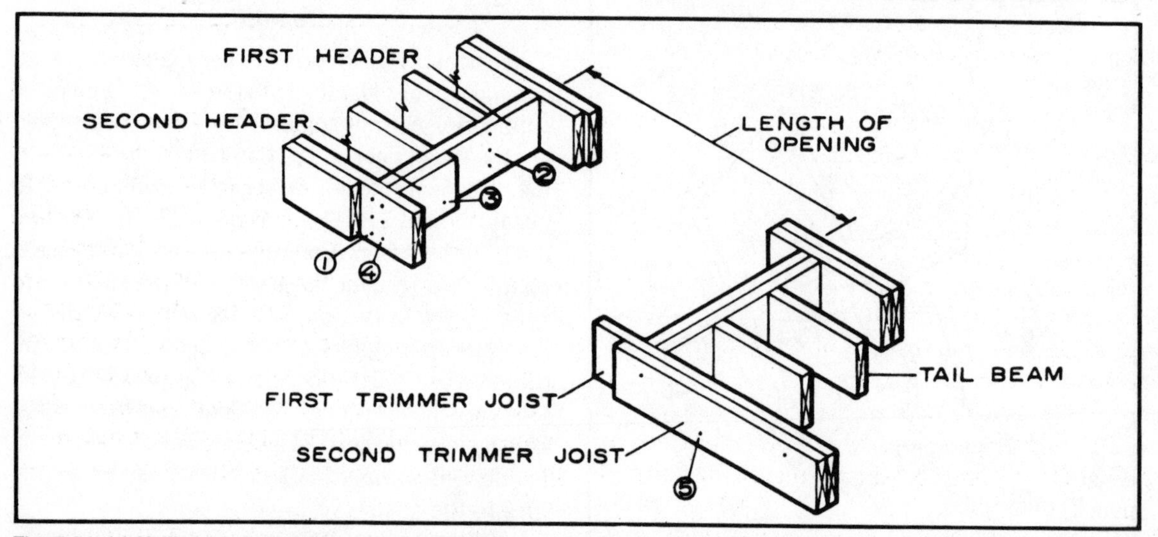

Fig. 7-24. (1) Nailing trimmer to first header; (2) nailing header to tail beams; (3) nailing header together; (4) nailing trimmer to second header; (5) nailing trimmers together.

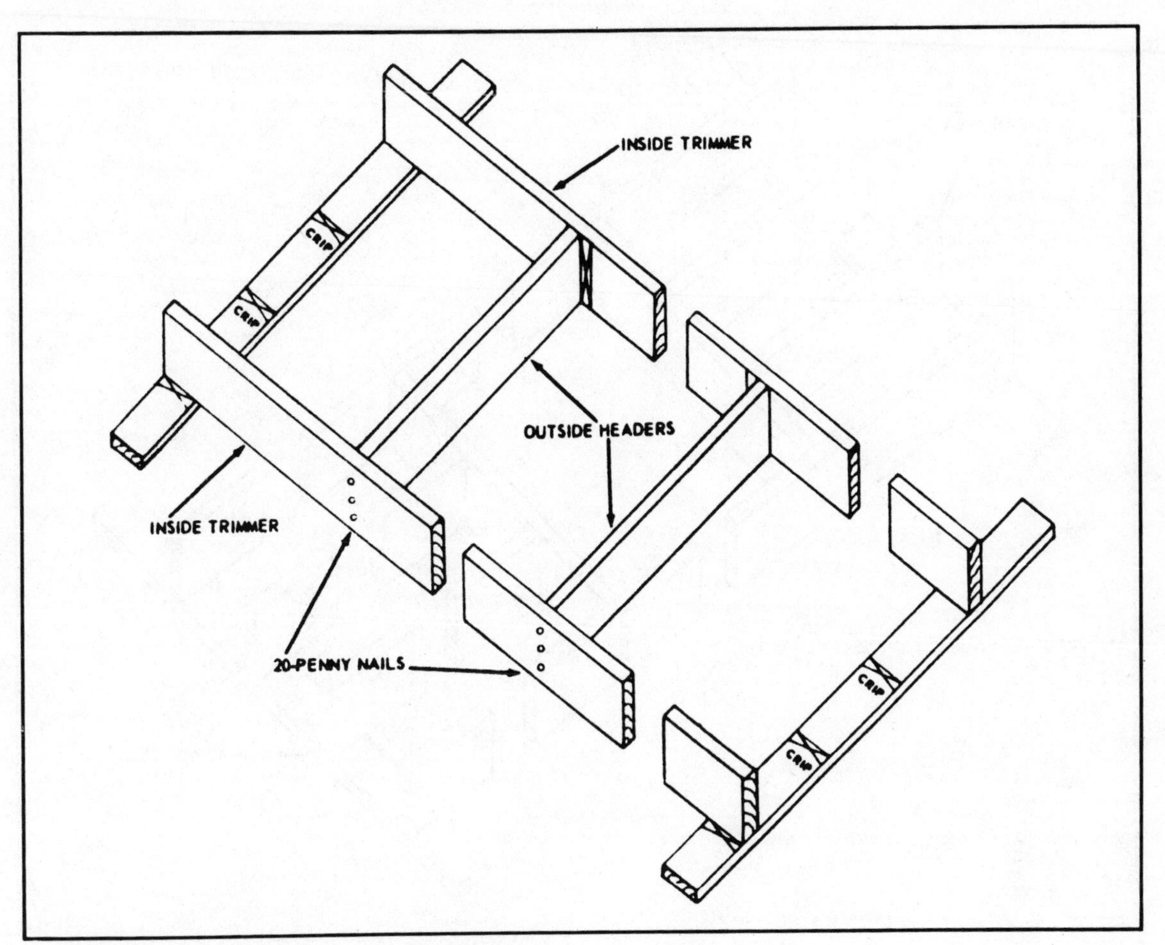

Fig. 7-25. First step in framing around floor opening.

Plywood suitable for subfloor, such as Standard sheathing, Structural I and II, and C-C Exterior grades, has a panel identification index marking on each sheet. These markings indicate the allowable spacing of rafters and floor joists for the various thicknesses when the plywood is used as roof sheathing or subfloor. For example, an index mark of 32/16 indicates that the plywood panel is suitable for a maximum spacing of 32 inches for rafters and 16 inches for floor joists. Thus, no problem of strength differences between species is involved because the current identification is shown for each panel.

Normally, when some type of underlayment is used over the plywood subfloor, the minimum thickness of the subfloor for species such as Douglas fir and southern pine is 1/2 inch when joists are spaced 16 inches on center, and 5/8 inch thick for such plywood as western hemlock, western white pine, ponderosa pine, and similar species. These thicknesses of plywood might be used for 24-inch spacing of joists when a finish 25/32-inch strip flooring is installed at right angles to the joists. It is important to have a solid and safe platform for workmen during construction of the remainder of the house, however. For this reason, some builders prefer a slightly thicker plywood subfloor, especially when joist spacing is greater than 16 inches on center.

Plywood also can serve as combined plywood subfloor and underlayment, thus eliminating separate underlayment because the plywood functions

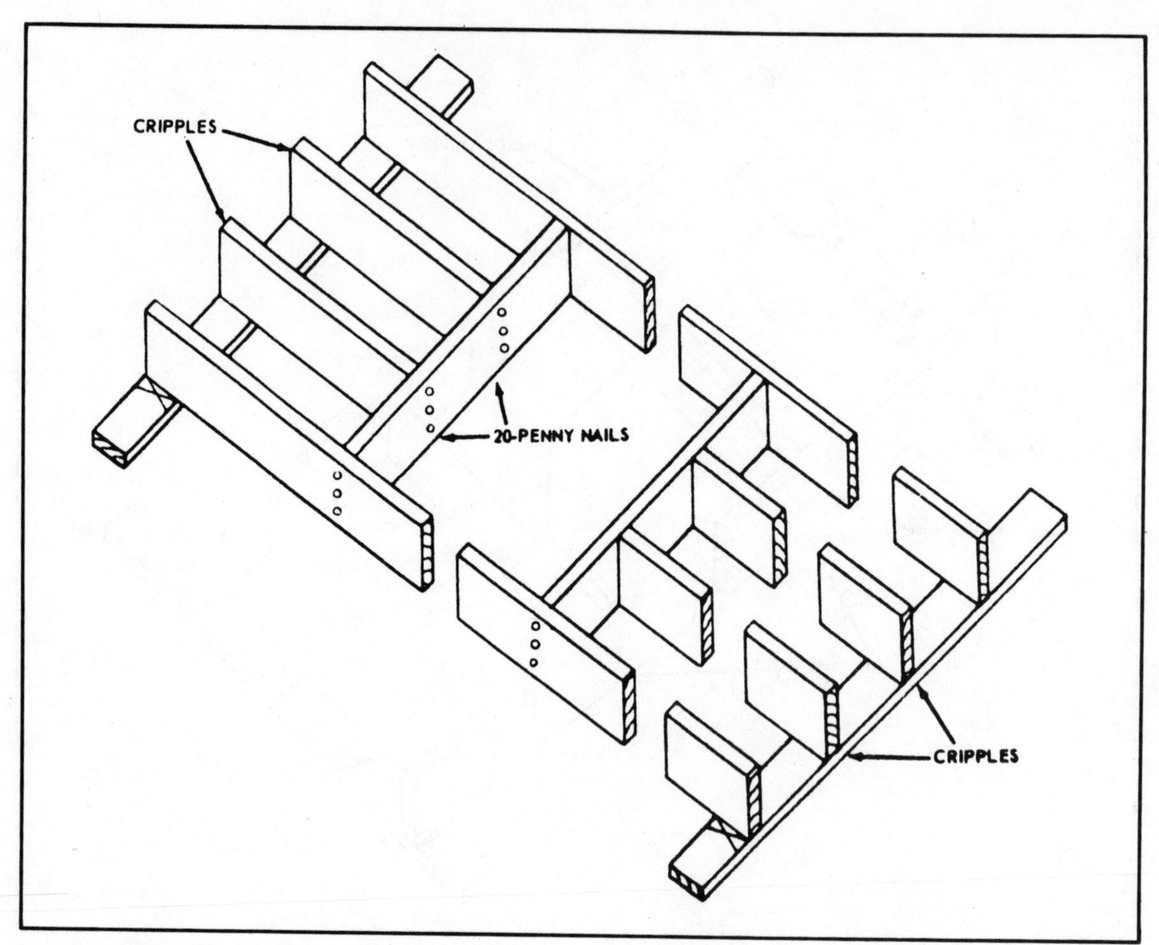

Fig. 7-26. Second step in framing around floor opening.

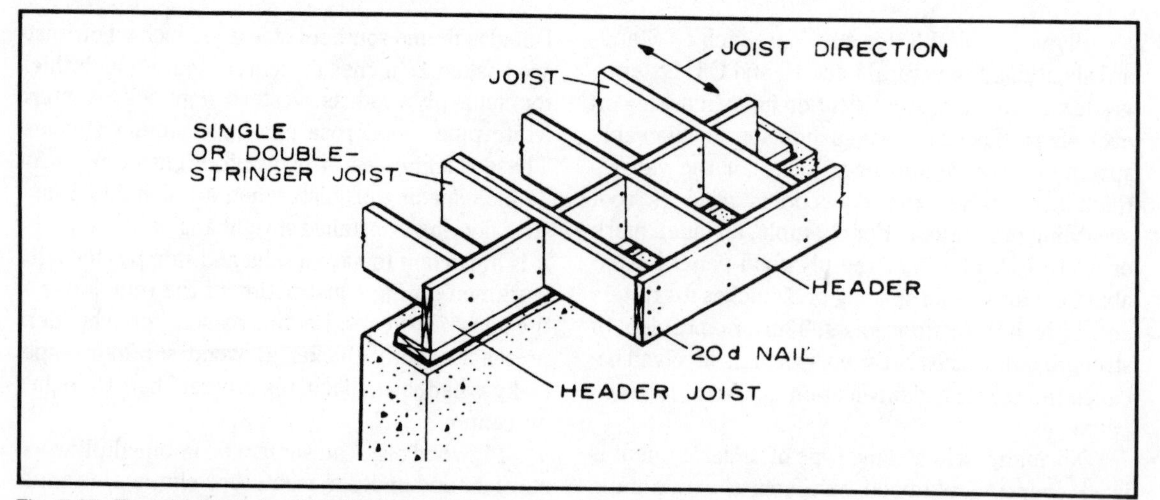

Fig. 7-27. Floor framing at wall projections.

as both structural subfloor and a good substrate. Plywood can be used this way when thin resilient floorings, carpeting, and other nonstructural finish flooring will be used. The plywood used in this manner must be tongued and grooved or blocked with 2-inch lumber along the unsupported edges. Recommendations for its use include:

- ☐ *Grade:* Underlayment, underlayment with exterior glue, C-C plugged
- ☐ *Spacing and thickness:* For species such as Douglas-fir (coast type), and southern pine, 1/2 inch minimum thickness for 16-inch joist spacing, 5/8 inch for 20-inch joist spacing, and 3/4 inch for 24-inch joist spacing. For species such as western hemlock, western white pine, and ponderosa pine, 5/8 inch minimum thickness for 16-inch joist spacing, 3/4 inch for 20-inch joist spacing, and 7/8 inch for 24-inch joist spacing.

Install plywood with the grain direction of the outer plies at right angles to the joists. Stagger the plywood so that end joints in adjacent panels break over different joists. Nail 1/2-inch to 3/4-inch-thick plywood to the joist at each bearing with 8d common or 7d threaded nails. Space nails 6 inches apart along all edges and 10 inches along intermediate members. When plywood serves both subfloor and underlayment, space nails 6 to 7 inches apart at all joists and blocking. Use 8d or 9d common nails, or 7d or 8d threaded nails.

For the best performance, do not lay up plywood with tight joints whether it is used on the interior or exterior. The American Plywood Association recommends 1/32-inch spacing for underlayment or interior wall lining, 1/16-inch spacing for panel sidings and combination subfloor underlayment, and 1/8-inch edge spacing and 1/16-inch end spacing for roof sheathing, subflooring, and wall sheathing. Under wet or humid conditions, double the spacing.

FLOOR FRAMING AT WALL PROJECTIONS

The framing for wall projections, such as bay window or first- or second-floor extensions beyond the lower wall, generally should consist of the floor joists (Fig. 7-27). This extension normally should

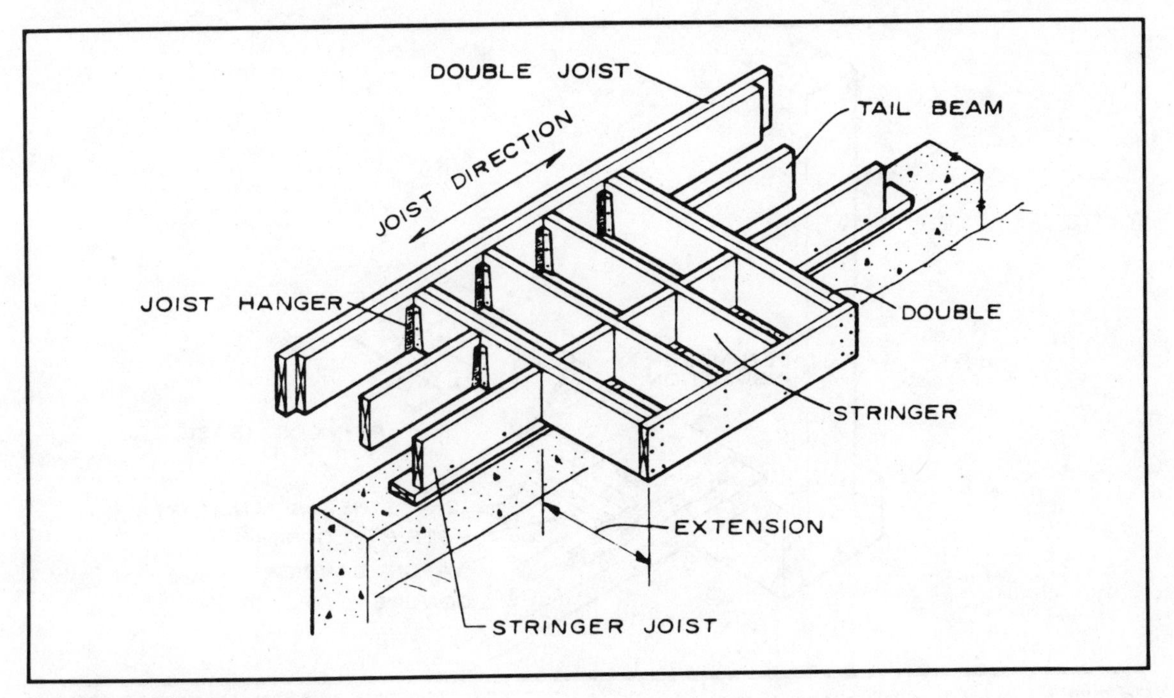

Fig. 7-28. Building an extension.

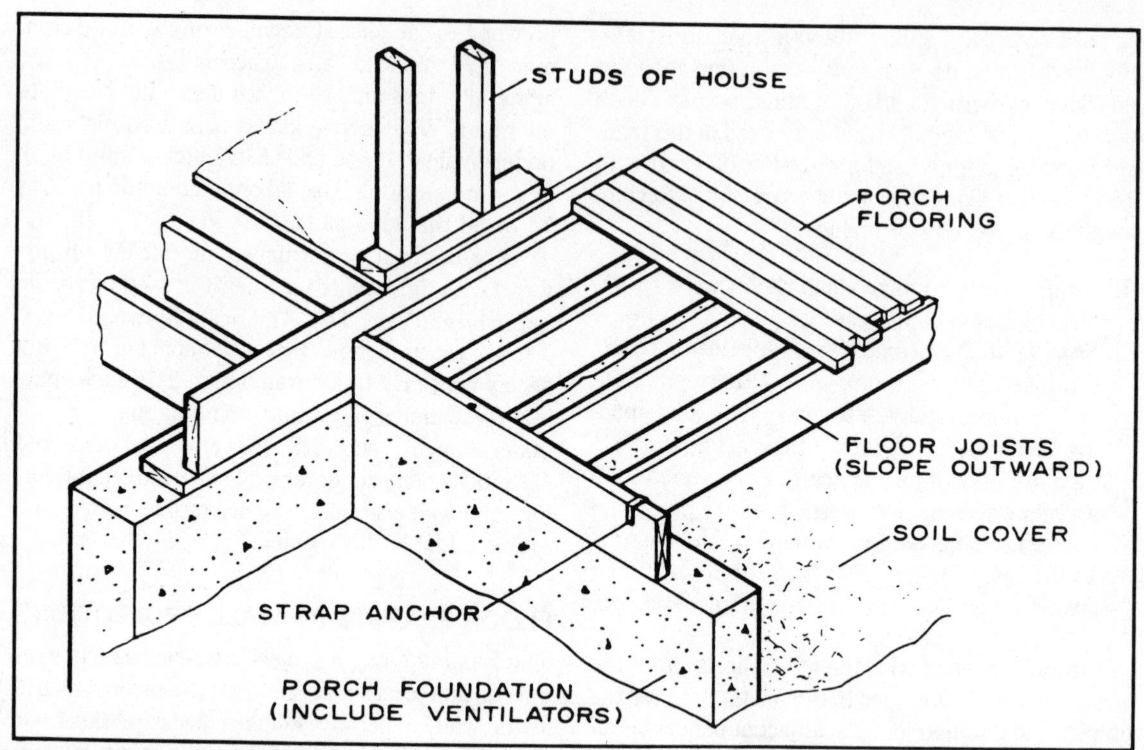

Fig. 7-29. Porch floor with wood framing.

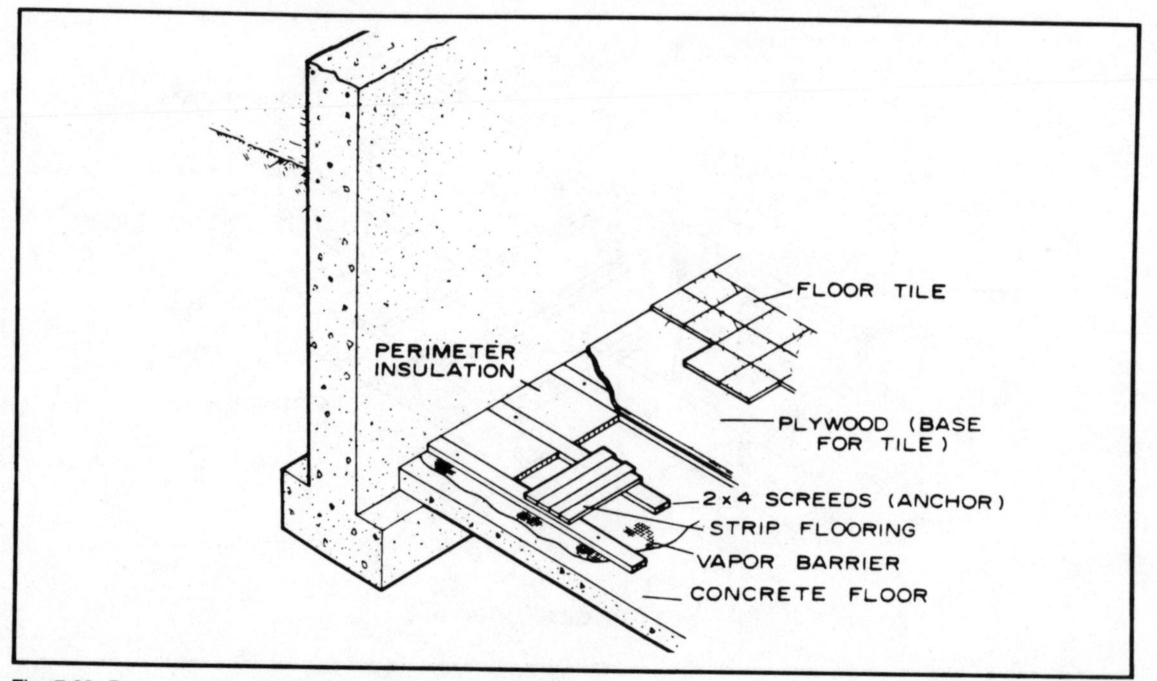

Fig. 7-30. Basement floor details for existing construction.

not exceed 24 inches unless it is designed specifically for greater projections, which might require special anchorage at the opposite ends of the joists. The joists forming each side of the bay should be doubled. Nailing, in general, should conform to that for stair openings. The subflooring is carried to and sawed flush with the outer framing member. Rafters are often carried by a header constructed in the main wall over the bay area, which supports the roof load. Thus, the wall of the bay has less load to support.

Projections at right angles to the length of the floor joists should generally be limited to small areas and extensions of not more than 24 inches. In this construction, the stringer should be carried by double joists (Fig. 7-28). Joist hangers or a ledger will provide good connection for the ends of members.

Figures 7-29 through 7-32 illustrate the installation of porch and basement floors. Finally, Figs. 7-33 and 7-34 show how interior and exterior walls will be connected to your flooring.

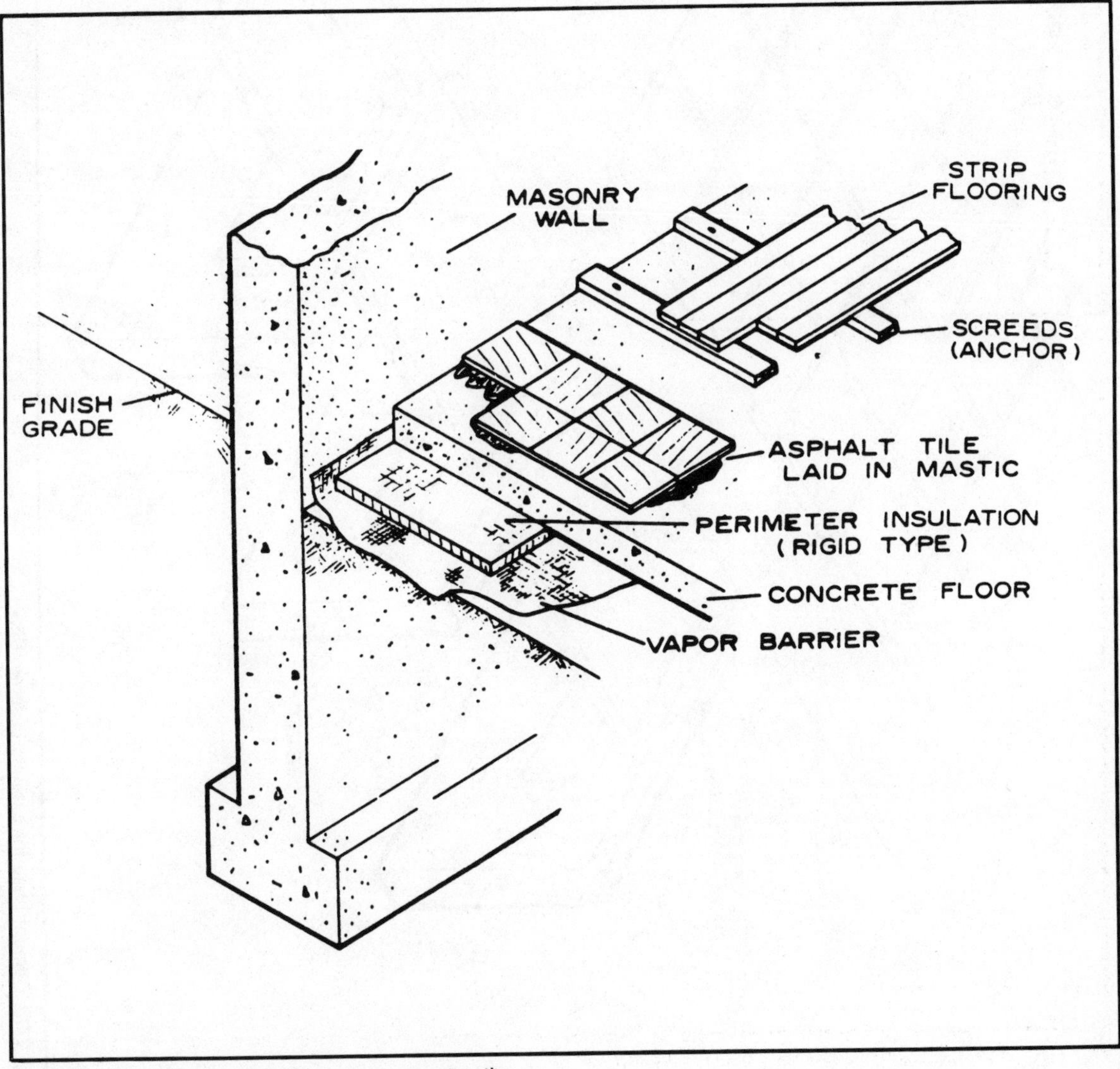

Fig. 7-31. Basement floor details for new construction.

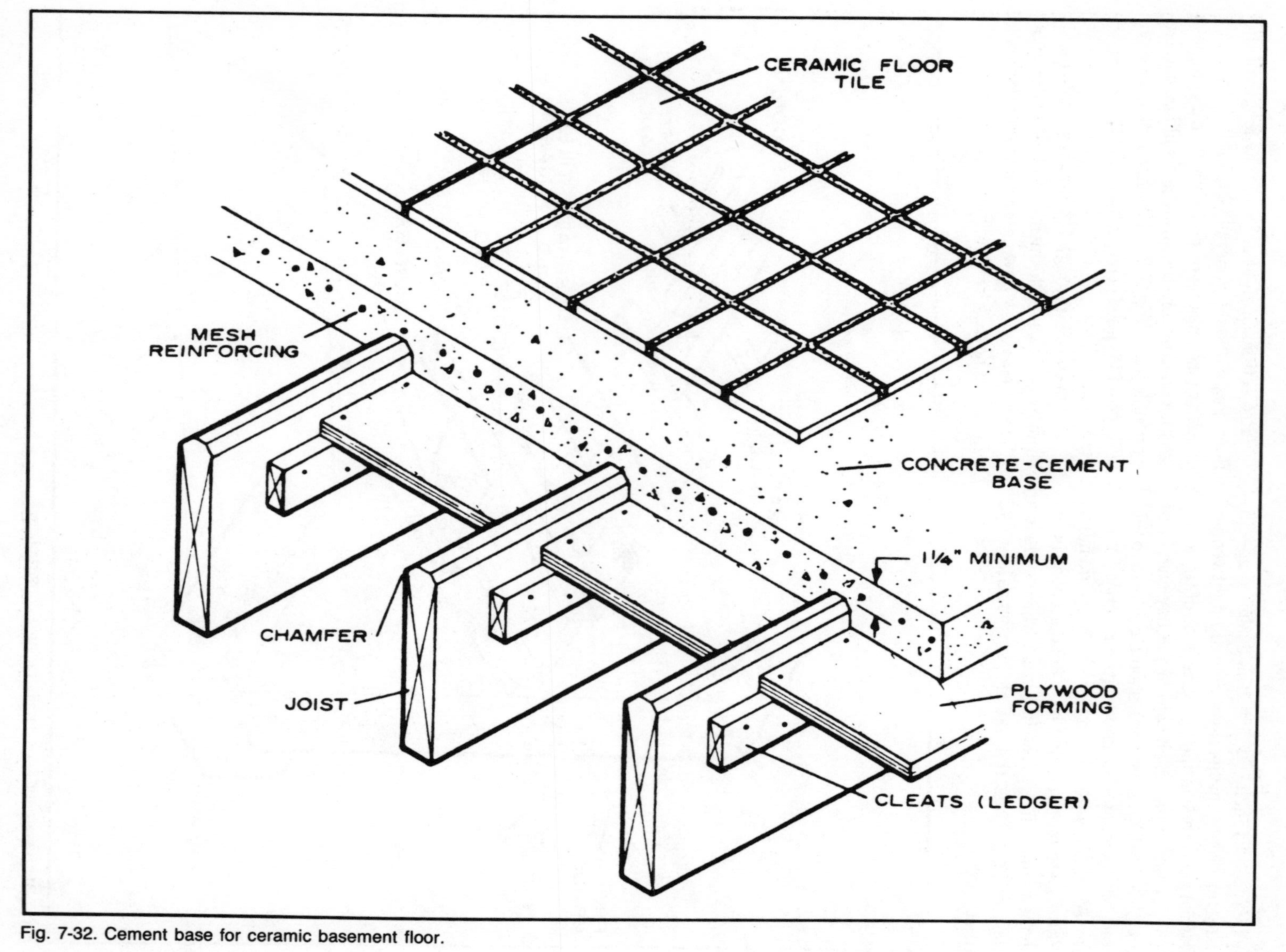

Fig. 7-32. Cement base for ceramic basement floor.

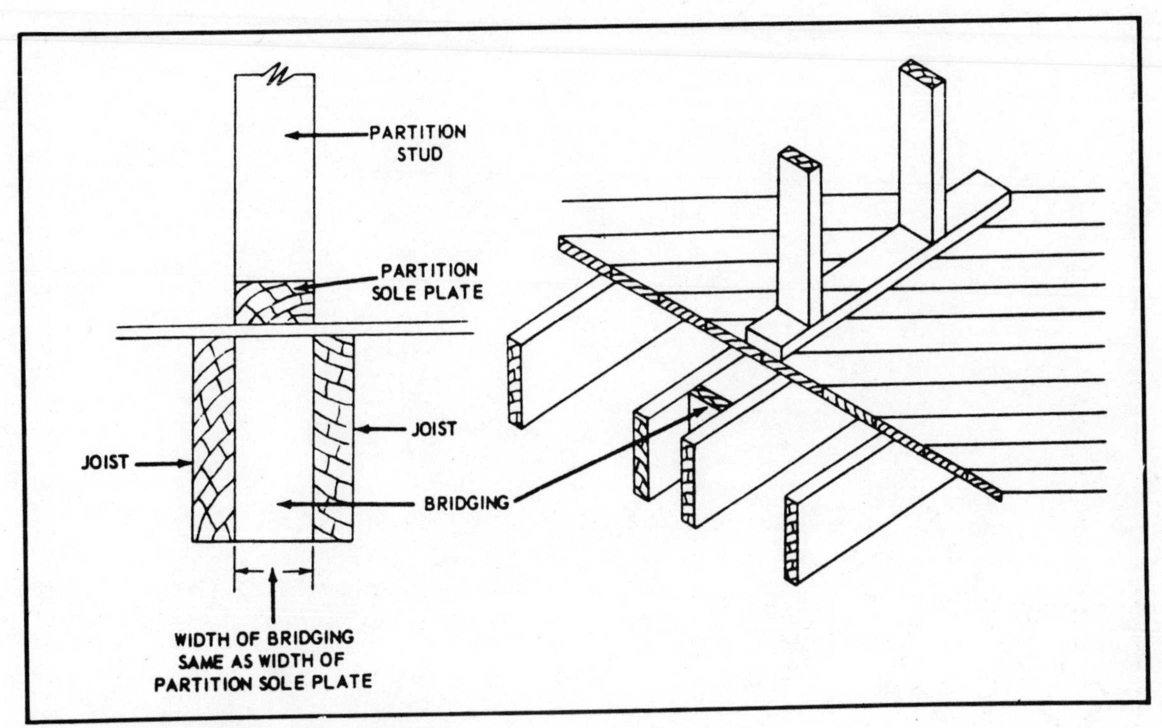

Fig. 7-33. Double joisting under wall partition.

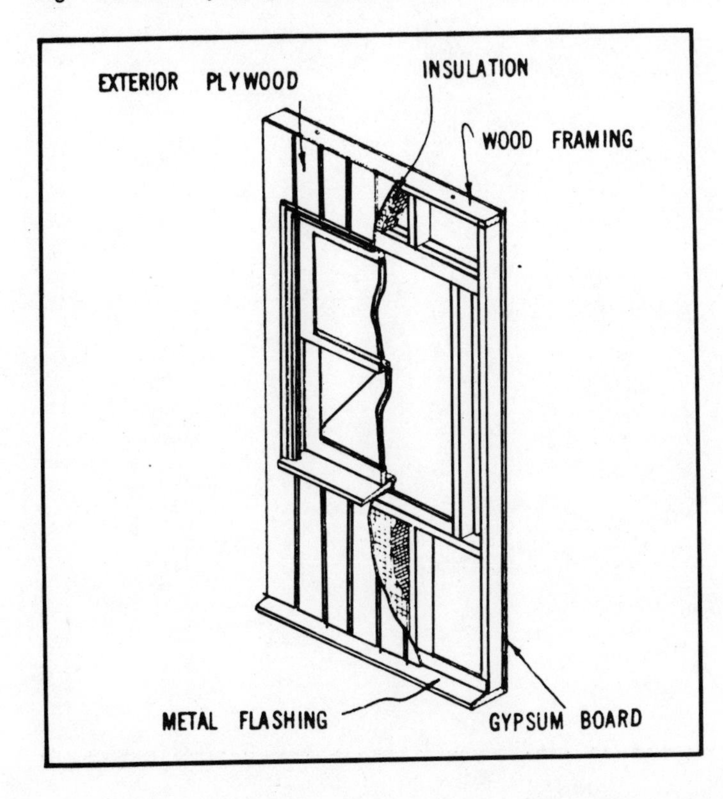

Fig. 7-34. Cross section of wall construction.

Glossary

air-dried lumber—Lumber that has been dried naturally by air and with a minimum moisture content of 15 to 20 percent.

anchor bolt—A bolt used to secure a wooden sill plate to concrete or masonry patio or foundation.

areaway—An open subsurface space adjacent to a building used to admit light or air or as a means of access to a basement.

backfill—To replace excavated earth into a trench around and against a basement foundation.

batten—Narrow strips of wood used to cover joints or as decorative vertical members over wide boards, such as on fences.

batter board—One of a pair of horizontal boards nailed to posts set at the corners of an excavation, used to indicate the desired level; also as a fastening for stretched strings to indicate outlines of foundation walls.

beam—A structural member transversely supporting a load.

bearing wall—A wall that supports any vertical load in addition to its own weight.

blind nailing—Nailing in such a way that the nailheads are not visible on the face of the work.

brace—An inclined piece of lumber applied to a wall or fence section to stiffen the structure. Often used as temporary bracing until framing has been completed.

bridging—Small wood or metal members that are inserted in a diagonal position between the floor joists at midpoint to act both as tension and compression members to brace the joists and spread the action of loads.

butt joint—The junction where the ends of two timbers or other members meet in a square-cut joint.

concrete, plain—Concrete either without reinforcement or reinforced only for shrinkage or temperature change.

corner brace—One of a pair of diagonal brace used at the corners of frame structures to stiffen and strengthen the wall.

cross-bridging—Diagonal bracing between adjacent floor joists, placed near the center of the joist span to prevent joists from twisting.

d—See *penny*.

density—The mass of substance in a unit volume. When expressed in the metric system, it is numerically equal to the specific gravity of the same substance.

direct nailing—Nailing perpendicular to the initial surface or to the junction of the pieces jointed. Also called *face nailing*.

expansion joint—A bituminous fiber strip used to separate blocks or units of concrete to prevent cracking due to expansion as a result of temperature changes. Often used on larger concrete foundation and floor slabs.

footing—A masonry section, usually concrete, in a rectangular form wider than the bottom of the foundation wall or pier it supports.

foundation—The supporting portion of a structure below the first-floor construction or below grade, including the footings.

frost line—The depth of frost penetration in the soil. This depth varies in different parts of the country. Fence footings should be placed below this depth to prevent movement.

girder—A large or principal beam of wood or steel used to support concentrated loads at isolated points along its length.

grain—The direction, size, arrangement, appearance, or quality of fibers in wood.

grain, edge—Lumber that has been sawed parallel to the pith of the log and approximately at right angles to the growth rings.

grain, flat—Lumber that has been sawed parallel to the pith of the log and approximately tangent to the growth rings.

grout—Mortar made of such consistency (by adding water) that it will just flow into the joints and cavities of the masonry work and fill them solid.

heartwood—The wood extending from the pit to the sapwood, the cells of which no longer participate in the life process of the tree.

I beam—A steel beam with a cross section resembling the letter I. It is used for long spans, such as basement beams, or over wide wall openings, such as a double garage door, when wall and roof loads are imposed on the opening.

insulation board—A structural building board made of coarse wood or cane fiber in 1/2- and 25/32-inch thicknesses. It can be obtained in various size sheets, in various densities, and with several treatments.

joint—The space between the adjacent surfaces of two members or components joined and held together by nails, glue, cement, mortar, or other means.

joint cement—A powder that is usually mixed with water and used for joint treatment.

joist—One of a series of parallel beams, usually 2 inches in thickness, used to support floor and ceiling loads and supported in turn by larger beams, girders, or bearing walls.

kiln-dried lumber—Lumber that has been kiln dried often to a moisture content of 6 to 12 percent. Common varieties of softwood lumber, such as framing lumber are dried to a somewhat higher moisture content.

ledger strip—A strip of lumber nailed along the bottom of the side of a girder on which joists rest.

lumber—The product of the sawmill and planing mill not further manufactured other than by sawing, resawing, and passing lengthwise through a standard planing machine, crosscutting to length and matching.

lumber, boards—Yard lumber less than 2 inches thick and 2 or more inches wide. Most commonly used for fencing.

lumber, dimension—Yard lumber from 2 inches to, but not including, 5 inches thick and 2 or more inches wide. Includes joists, rafters, studs, plank, and small timbers.

lumber, matched—Lumber that is dressed and shaped on one edge in a grooved pattern and on the other in a tongued pattern.

lumber, timbers—Yard lumber 5 or more inches in least dimension. Includes beams, stringers, posts, caps, sills, girders, and purlins.

lumber, yard—Lumber of those grades, sizes, and patterns that are generally intended for ordinary construction, such as framework and rough coverage of houses.

masonry—Stone, brick, concrete, hollow tile, concrete block, gypsum block, or other similar building units or materials bonded together with mortar to form a foundation, wall, pier, buttress, or similar mass.

nonbearing wall—A wall supporting no load other than its own weight.

o.c.—Abbreviation for on center; the measurement of spacing for studs, rafters, joists, and posts from the center of one member to the center of the next.

penny—As applied to nails, it originally indicated the price per hundred. The term now serves as a measure of nail length and is abbreviated by the letter *d.*

pier—A column of masonry, usually rectangular, used to support other structural members. Often used as support under decks.

plate, sill—A horizontal member anchored to a masonry wall.

plumb—Exactly perpendicular; vertical.

plywood—A piece of wood made of three or more layers of veneer joined with glue and usually laid with the grain of the adjoining plies at right angles. Usually constructed with an odd number of plies to provide balanced construction.

preservative—Any substance that, for a reasonable length of time, will prevent the action of wood-destroying fungi, borers of various kinds, and similar destructive agents when the wood has been properly coated or impregnated with it.

run—In stairs, the net width of a stem or the horizontal distance covered by a flight of stairs.

sand float finish—Lime mixed with sand, resulting in a textured finish.

sapwood—The outer zone of wood next to the bark. In the living tree it contains some living cells as well as dead and dying cells. In most species, it is lighter colored than the heartwood. In all species, it is lacking in decay resistance.

screed—A small strip of wood, usually the thickness of the plaster coat, used as a guide for plastering.

sealer—A finishing material, either clear or pigmented, that is usually applied directly over uncoated wood for the purpose of sealing the surface.

soil cover—A light covering of plastic film, roll roofing, or similar material used over the soil in crawl spaces of buildings to minimize moisture permeation of the area. Also called *ground cover.*

solid bridging—A solid member placed between adjacent floor joists near the center of the span to prevent joists from twisting.

span—The distance between structural supports, such as walls, columns, piers, beams, girders, and trusses.

splash block—A small masonry block laid with the top close to the ground surface to receive roof drainage from downspouts and to carry it away from the foundation.

square—A unit of measure (100 square feet) usually applied to roofing material.

strip flooring—Wood flooring consisting of narrow, matched strips.

stud—One of a series of slender wood or metal vertical structural members placed as supporting elements in walls and partitions.

subfloor—Boards or plywood laid on joists over which a finish floor is to be laid.

termite shield—A shield, usually of noncorrodible metal, placed in or on a foundation wall or other mass of masonry or around pipes to prevent passage of termites.

toenailing—Driving a nail at a slant to the initial surface in order to permit it to penetrate into a second member.

tongued and grooved—See *lumber, matched.*

truss—A frame or jointed structure designed to act as a beam of long span, while each member is usually subjected to longitudinal stress only, either tension or compression.

vapor barrier—Material used to retard the movement of water vapor into walls and prevent condensation in them. Usually considered as having a perm value of less than 1.0. Applied separately over the warm side of exposed walls or as a part of batt or blanket insulation.

water-repellent preservative—A liquid designed to penetrate into wood and impart water repellency and a moderate preservative protection.

Index

Index

Other Bestsellers From TAB

☐ **KITCHEN REMODELING—A DO-IT-YOURSELFER'S GUIDE—Paul Bianchina**

Create a kitchen that meets the demands of your lifestyle. With this guide you can attractively and economically remodel your kitchen yourself. All the know-how you need is supplied in this complete step-by-step reference, from planning and measuring to installation and finishing. 208 pp., 187 illus.

Paper $14.95 **Hard $23.95**
Book No. 3011

☐ **MASTERING HOUSEHOLD ELECTRICAL WIRING—2nd Edition—James L. Kittle**

Update dangerously old wiring in your house. Add an outdoor dusk-to-dawn light. Repair a malfunctioning thermostat and add an automatic setback. You can do all this and more—easily and safely—for much less than the cost of having a professional do it for you! You can remodel, expand, and modernize existing wiring correctly and safely with this practical guide to household wiring. From testing to troubleshooting, you can do it all yourself. Add dimmer switches and new outlets . . . ground your TV or washer . . . make simple appliances repair . . . set up outside wiring . . . put in new fixtures and more! 304 pp., 273 illus.

Paper $15.95 **Hard $24.95**
Book No. 2987

☐ **PRACTICAL STONEMASONRY MADE EASY—Stephen M. Kennedy**

The current popularity of country-style homes has renewed interest in the use of stone in home construction. Now, with the help of expert stonemason Stephen M. Kennedy, you can learn how to do stonework yourself and actually save money while adding to the value, charm, and enduring quality of your home. This book provides step-by-step guidance in the inexpensive use of stone for the relatively unskilled do-it-yourselfer. 272 pp., 229 illus.

Paper $16.95 **Hard $24.95**
Book No. 2915

☐ **BATHROOM REMODELING—A DO-IT-YOURSELFER'S GUIDE—Paul Bianchina**

This complete step-by-step remodeling reference addresses all aspects of bathroom design. The author's expertise will help you construct the space you need, or improve your use of the space you have. The selection and installation of all traditional bathroom fixtures are covered, and detailed information on contemporary "luxury" bathroom options is presented. 208 pp., 200 illus.

Paper $14.95 **Hard $23.95**
Book No. 3001

☐ **KEEP ITS WORTH: SOLVING THE MOST COMMON BUILDING PROBLEMS—Joseph V. Scaduto and Michael J. Scaduto**

This book outlines how to identify, remedy, and prevent the building problems owners are most often concerned about: wet basements, roof leaks, decay and wood-boring insects, energy maintenance, maintaining mechanical systems, and hazards. 304 pp., 271 illus.

Paper $16.95 **Hard $25.95**
Book No. 2961

☐ **BUILDING A LOG HOME FROM SCRATCH OR KIT—2nd Edition—Dan Ramsey**

This guide to log home building takes you from initial planning and design stages right through the final interior finishing of your new house. There's advice on selecting a construction site, choosing a home that's right for your needs and budget, estimating construction costs, obtaining financing, locating suppliers and contractors, and deciding whether to use a kit or build from scratch. 302 pp., 311 illus.

Paper $14.95 **Book No. 2858**

Other Bestsellers From TAB